Theory of Stability of Continuous Elastic Structures

Library of Engineering Mathematics

Series Editor
Alan Jeffrey, University of Newcastle upon Tyne and University of Delaware

Linear Algebra and Ordinary Differential Equations
A. Jeffrey

Nonlinear Ordinary Differential Equations
R. Grimshaw (University of New South Wales)

Complex Analysis and Applications
A. Jeffrey

Perturbation Methods for Engineers and Scientists
A. W. Bush (University of Teesside)

Theory of Stability of Continuous Elastic Structures
M. Como and A. Grimaldi (University of Rome T.V.)

Qualitative Estimates for Partial Differential Equations
J.N. Flavin (University College Galway) and
S. Rionero (University of Naples)

Asymptotic Methods in Reaction-Diffusion Equations
Z. Peradzýnski (Polish Academy of Sciences, Warsaw)

Continuum Mechanics with Applications
J. Ballmann (RWTH Aachen)

Numerical Methods for Differential Equations
J.R. Dormand (University of Teesside)

Theory of Stability of Continuous Elastic Structures

Mario Como
Antonio Grimaldi

CRC Press
Boca Raton New York London Tokyo

Library of Congress Cataloging-in-Publication Data

Como, Mario.

Theory of stability of continuous elastic structures / by Mario Como and Antonio Grimaldi.

p. cm. — (Engineering mathematics library)

Includes bibliographical references and index.

ISBN 0-8493-8990-9

1. Elastic analysis (Engineering) 2. Structural stability. I. Grimaldi, Antonio. II. Title. III. Series.

TA653.C66 1995

624.1'71—dc20

94-37131

CIP

No claim to original U.S. Government works

International Standard Book Number 0-8493-8990-9

Library of Congress Card Number 94-37131

Printed in the United States of America 1 2 3 4 5 6 7 8 9 0

Printed on acid-free paper

Preface

Euler's pioneering investigations on the "Elastica" were the first studies on the equilibrium bifurcation of elastic structures. The analytic development of the theory of stability may properly be regarded as commencing with Lagrange's work on the stability of equilibrium states of discrete conservative systems. These first fundamental Lagrange's ideas on stability of discrete systems were further analyzed by Dirichlet, who gave the proof of the basic first theorem on stability, also called the energy criterion. New trends in the general theory of stability were then opened by Liapounov, whose studies still inspire further research. In the field of applied mechanics a large number of researchers, particularly Timoshenko, began to study with success the problem of elastic stability of structures. In this context Koiter's nonlinear approach, particularly, has represented fundamental progress.

The extension of the Lagrange-Dirichlet stability criterion to elastic bodies, on the other hand, occurred tacitly and seemed a natural procedure to be taken for granted. The definition of stability for a continuous elastic system, in fact, was not yet fully developed and the same energy criterion has been sometimes used as definition of stability. Even today in the engineering books used most the notion of stability of essential dynamical nature is always tacitly assumed static and is left undefined. Many authors, on the other hand, attempt to connect, for elastic bodies, the energy criterion to the dynamic Liapounov definition of stability. It has been shown that, for a continuous elastic system, the stability concept is dependent on the choice of the distance among configurations. The condition of a minimum of the potential energy has to be replaced by the stronger condition of a "potential well", which requires a potential energy bounded below by a positive strictly increasing function of the norm of the additional displacement from the equilibrium state. However, to date, the general problem of the existence of a potential well at the fundamental equilibrium state of an elastic body is still open, but the problem can be considered solved in the field of the elastic structural models.

The extension to continuous structures of the simpler mechanical ideas on which the theory of stability of elastic discrete systems is founded is the main motivation of this book. A beam, a frame, a plate and a shell are in fact all examples of continuous structures for which the space of the configurations is, as a rule, infinite-dimensional.

After the introductive Chapters 1 and 2 (the latter giving a review of some functional analysis concepts), a formulation of the theory of equilibrium stability of infinite-dimensional elastic systems loaded by conservative forces is given in Chapter 3. A general space and norm of the admissible displacement fields are considered, and in this space the stability definition and the energy criteria are given. In this context, the stability of the unstressed configuration can be considered a crucial starting point. The requirement of stability of the unstressed initial state, via the second differential criterion, implies, in fact, that the stability definition has to be properly given by assuming the "energy norm" as the measure of distances among configurations. In this new framework Chapter 3 formulates the energy criterion of stability, as well as of instability, of the continuous elastic structural systems. Moreover, the undecided case of stability at the critical state is examined, extending, in the functional analysis context, some fundamental results first given by Koiter.

Chapter 4 is devoted to the investigation of the equilibrium stability of several structural elements such as beams, plates and shells, and, as far as possible, of the general elastic continuum: some basic smoothness properties of their potential energy functionals are carefully explored in their energy spaces. First, according to the classical model of the "Elastica", stability and flexural buckling of the axially loaded inextensible elastic rod are investigated. Buckling and stability problems of shafts under constant torque and of deep beams under transversal loads, are then worked out as other particular examples. Then basic stability problems of plates and shells are studied. Finally, the energy space of the tridimensional elastic continuum and the still open problem of the corresponding statement of the energy criterion are examined in detail. A general and useful formulation of the neutral equilibrium of the elastic structures is presented in Chapter 5. The assumption of "small strains with respect to rotations", because of the high rigidity of the structural materials, is considered in this formulation. A short survey of buckling problems, particularly for cylindrical shells, ends the chapter. Chapter 6, strictly connected to Chapter 3, aims to provide an unified analysis of the equilibrium states that occur in the neighborhood of the critical state. The equilibrium evolution under varying loads of a general structural elastic system, with bifurcation or snapping, is analyzed in the first part of the chapter. The second part is dedicated to the study of "linear" or "perfect" elastic structures, represented by those elastic systems that exhibit a linear behavior during loading up to the critical state. The study of "quasi-perfect" elastic structures concludes the analysis. Some examples are then developed involving equilibrium bifurcations of elastic rods, deep beams, frames, and thin cylinders.

Essential references in our work have been the Mikhlin books, the Koiter thesis, the article "Theory of Elastic Stability" by Knops and Wilkes in the *Handbuch der Physik* and several of our papers on the subject, written in these last two decades. The book can be used as a textbook in advanced engineering courses and as a reference for theoretically oriented engineers. Readers of this book should be familiar with calculus and linear algebra, the theory of ordinary and partial differential equations, vectors and matrices, and with the contents of basic courses of functional analysis, mechanics of solids and theory of structures.

We are grateful to Navin Sullivan of CRC Press and to professor Alan Jeffrey for their approval of our project of the book. We are grateful to Paolo Bisegna who has been kind enough to read critically the first three chapters of the book and to the same CRC Press for the linguistic revision of the text. Thanks also to Ugo Janniruberto for his help with the proofreading during the preparation of the book. Whatever errors remain are, of course, the responsibility of the authors. We wish to thank also Marina Como for her attention in preparing the illustrations. We wish to thank Miss Jennifer L. Pate, Project Editor, for her kind and constant assistance also.

Mario Como
Antonio Grimaldi

March 25, 1995
Department of Civil Engineering
University of Rome T.V.

Authors

Mario Como is full professor of Structural Engineering at the University of Rome T.V., (Italy). He became professor in 1971 and has taught mechanics, strength of materials, and structural and earthquake engineering, in various Italian universities. He joined the University of Rome in 1985. He is author and coauthor of more than 150 scientific articles in the field of the theory of elastic stability, plastic buckling, the theory of plasticity, structural and earthquake engineering, the aerodynamics of long-span bridges, and the mechanics of masonry structures and monuments. He authored a book on the theory of stability and is coauthor of two other books on the matrix analysis of structures and earthquake engineering.

Antonio Grimaldi is full professor of Theory of Structures at the University of Rome T.V., (Italy). He became professor in 1976 and has taught continuum mechanics and strength of materials at the University of Calabria. He joined the University of Rome 2 in 1982. He is author and coauthor of about one hundred articles in the fields of plastic behavior of materials, stability and postbuckling of elastic structures, finite elements approximations problems, unilateral contact of elastic structures, and statics and dynamics of bridges. He has also coauthored a book on continuum mechanics.

Contents

To Ida and to Paola

Chapter 1

AN INTRODUCTION TO STABILITY ANALYSIS OF ELASTIC STRUCTURAL SYSTEMS

1.1 INTRODUCTION

A large number of structures can suffer a change in their deformation during loading. This change is usually accompanied by a major or minor reduction of stiffness that does not depend on material yielding. Such a phenomenon occurs because the equilibrium configurations of the structure become unstable during loading and the structure snaps or buckles, seeking new stable equilibrium states. It is also possible that to reach these new states, the structure suffers such high strains as to fail. The various instabilities occurring in structural systems under the action of conservative loads[22] can be subdivided into three different groups[1]:

- Buckling with stable branching
- Buckling with unstable branching
- Snapping-through

In the next pages, we will describe the main features of these forms of instabilities and will begin to examine the current approaches used in their analysis.

1.1.1 BUCKLING WITH STABLE BRANCHING

This form of instability — also called Euler instability — occurs, for instance, in an axially loaded elastic rod or a circular elastic ring under uniform pressure. In these cases the unbuckled and the buckled states are very close to each other. At the same time the buckled structure is still able to sustain loads larger than the critical.

The buckling deformation is completely different from that present in the structure in the prebuckling states. For instance, at buckling, the axially loaded rectilinear rod bends laterally, whereas it remains rectilinear in the prebuckling loaded states. Likewise, the circular ring, which remains circular before the critical state, buckles into circumferential flexural waves.

It is theoretically possible that the structure could remain in equilibrium in the prebuckled deformation under loads larger than the critical. Conversely, these equilibrium states are unstable and cannot be mantained if a disturbance, however small, is applied to the structure.

The existence of various equilibrium branches that depart from the critical state implies that a symmetric and stable equilibrium bifurcation occurs at buckling. So the axially loaded bar of Fig. 1.1, if initially rectilinear, will gradually shorten up until the axial load N reaches the critical value N_c; then, at the critical state, the rod suddenly bends laterally. In the P-v plane, where v is the lateral sagging of the bar, the loading path will first be represented by the set segment OB, the rectilinear equilibrium states of the rod, and then by one of the two symmetric branches BC, corresponding to the buckled bar. The unstable rectilinear equilibrium states are then represented by points on the vertical axis higher in position than the point B; the buckled states, all stable, belong to the ascending branches BC. For the slightly imperfect rod (i.e., for the rod with a small initial curvature, for example) the equilibrium states are represented by the dashed curves of Fig. 1.1.

Other information can be obtained by inspection of the axial load N-axial shortening w diagram of Fig. 1.2. The N-w diagram is first rectilinear, for N increasing from zero up to the critical value N_c, and describes the linearly elastic shortening of the rod. Then, at $N = N_c$, the

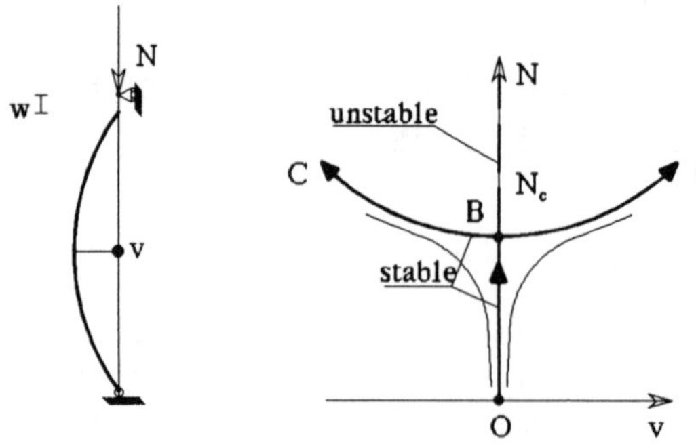

FIGURE 1.1

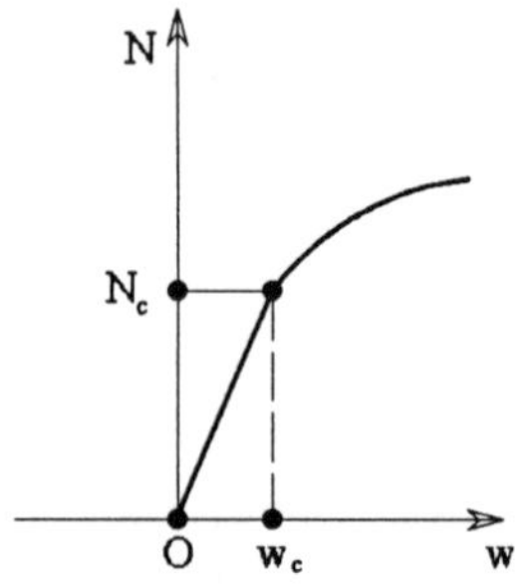

FIGURE 1.2

N-w curve suddenly bends, but N continues to be an increasing function of the shortening w. The same behavior is exhibited by thin plates under inplane compressive or shearing loads.

1.1.2 BUCKLING WITH UNSTABLE BRANCHING

In the previous class of stability problems the imperfect elastic structure never fails under loads lower than the critical one. The curve BC of Fig. 1.1 in the neighborhood of N_c never goes below the point B. In other cases (as for the compressed cylindrical shell) the stiffness reduction occurring at the critical state is so strong that the buckling states can be maintained only by suitably reducing the axial load. Also in this case, there are unbuckled unstable states beyond the critical load.

The various equilibrium branches that depart from the critical state are sketched in the N-v plane of Figs. 1.3 and 1.4. Also in these cases, for perfect structures, the vertical set segment OB represents stable unbuckled equilibrium states.

The equilibrium can bifurcate asymmetrically or symmetrically when the load N reaches N_c as represented in Figs. 1.3 or 1.4. In the first case there are two equilibrium branches — BD stable, BC unstable. In the second case (i.e., when the equilibrium bifurcates symmetrically) there is only one symmetric unstable branch. In both cases the critical state is unstable.

Small imperfections of the structure yield stable or unstable equilibrium paths. In the case of Fig. 1.3 the curves α or β correspond to equilibrium paths of the structure with unstable or stable imperfections. In the first case, the path α reaches in the N-v plane a maximum at a value N^* of the axial load lower than N_c and a snapping-through occurs. Equilibrium does not exist anymore for $N > N^*$ and the structure fails at once if the load N further increases. On the contrary, for stable imperfections, i.e., when the sign of the imperfection is such that the structure is on the curve β, the equilibrium states maintain stable for $N > N_c$. The case of Fig. 1.3 corresponds to an asymmetric unstable bifurcation.

The case of Fig. 1.4 is slightly different. Here a symmetric unstable bifurcation occurs. In this case the whole branch of the buckled states is unstable. Small imperfections always yield

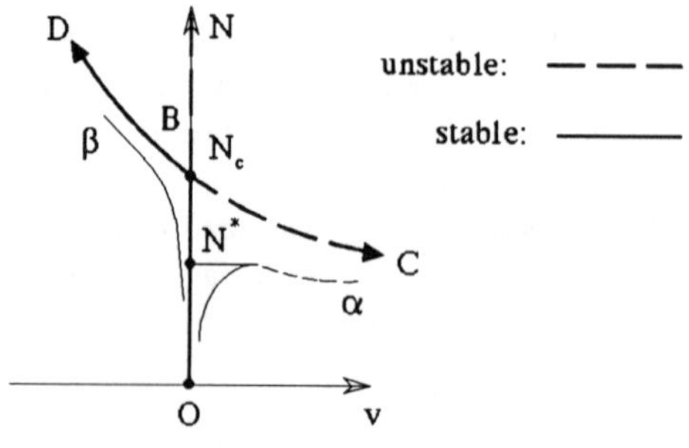

FIGURE 1.3

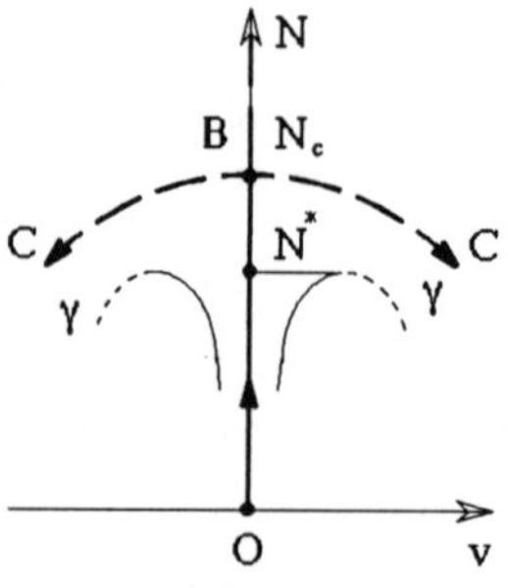

FIGURE 1.4

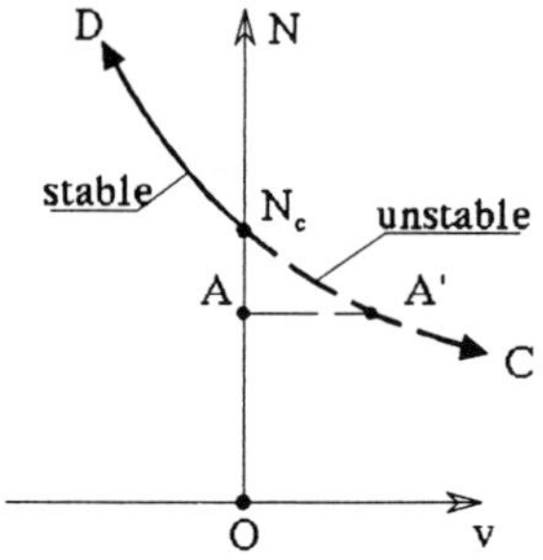

FIGURE 1.5

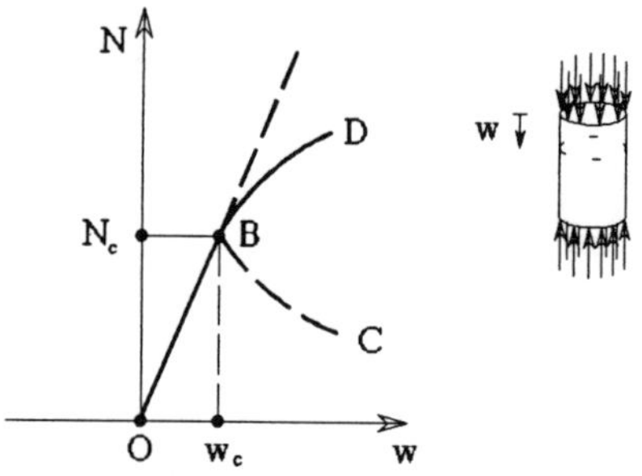

FIGURE 1.6

equilibrium paths, which present a maximum in the N-v plane under a load N* lower than N_c. Also in this case, for perfect structures, unstable equilibrium unbuckled states are possible for $N > N_c$. All the points that belong to the vertical axis and are localized above the point B represent these states. In both cases disturbances with finite amplitude can move the structure from unbuckled to buckled states under axial loads lower than N_c (Fig. 1.5). The diagram of the function N = N(w), represented in Fig. 1.6, is typical of the unstable bifurcation. In this plot, w indicates the "shortening," i.e., the displacement along the direction of the load N. For example, in the case of the compressed cylindrical shell, w represents the axial shortening.[2] In this case the function N = N(w) is linear for $N < N_c$. At $N = N_c$ a discontinuity in the slope occurs, and a new descending curve describes the function N(w) for $w > w_c$. This descending branch can be obtained only by producing a gradual shortening of the cylinder and then measuring the corresponding acting axial load, as in the classical Von Kármán "mechanical press." In the case of the nonsymmetric bifurcation, on the other hand, two different curves BC and BD intersect the rectilinear increasing path at $N = N_c$.

The frame of Fig. 1.7 behaves as an axially compressed cylindrical shell. If the beams of the frame are initially rectilinear and inexstensible, they will remain undeformed up to the critical load N_c. Buckling occurs at $N = N_c$. If ϕ represents the rotation of the joint, positive when counterclockwise, two different branches BC and BD — respectively corresponding to counterclockwise or clockwise rotations ϕ of the joint — describe the buckled equilibrium states of the frame in the N-ϕ plane. Branch BC is unstable. In fact, the counterclockwise rotation of the joint produces bending of the girder that increases the axial load in the strut; the opposite occurs for the buckling mode related to a clockwise rotation of the joint.

Small imperfections, according to their sign, produce the equilibrium paths α or β. With unstable imperfections, collapse of the frame occurs under the load $N^* < N_c$, where the curve α attains its maximum. Symmetric stable or unstable branching frequently also occurs in structures — as, for example, at the antisymmetric buckling of the truss of Fig. 1.8.

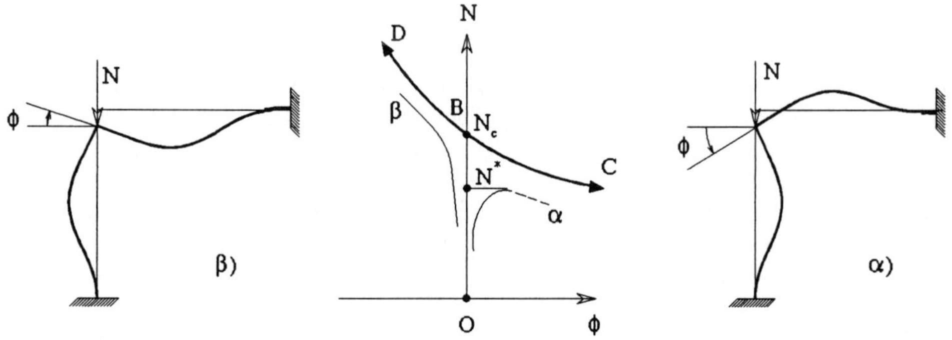

FIGURE 1.7

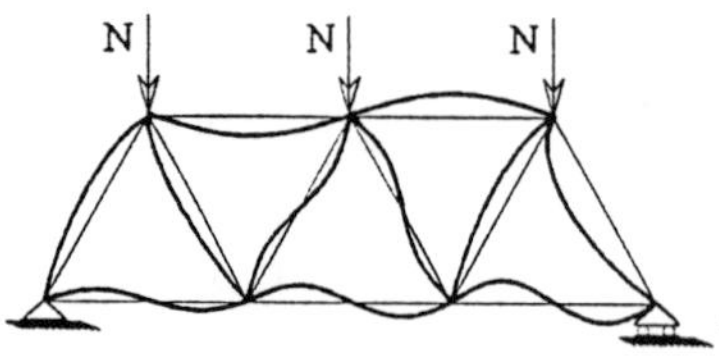

FIGURE 1.8

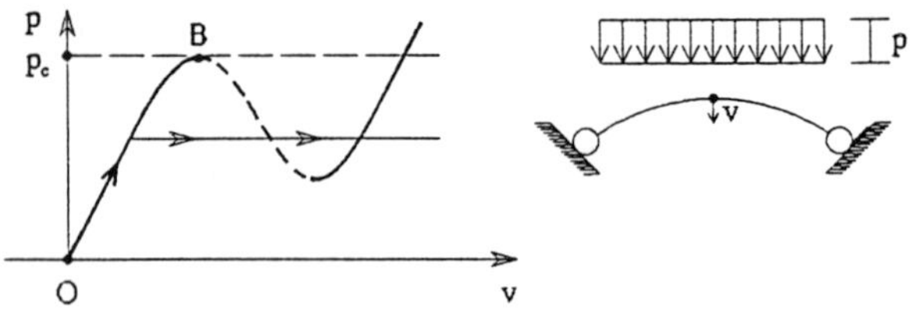

FIGURE 1.9

1.1.3 SNAP-THROUGH INSTABILITY

The third type of unstability is called snap-through.[3] In this case the behavior of the structure is strongly nonlinear from the beginning of the loading. A gradual reduction in stiffness occurs in the structure under increasing loads. At the collapse, the structure behaves as if it has completely lost all its stiffness. Equilibrium is no longer possible under loads larger than the critical one.

Figure 1.9 shows the snap-through of a shallow arch under transversal loads p. The portion OB of the curve load p-vertical displacement v bends with a gradual reduction of the tangent stiffness under increasing p. The snapping of the arch occurs at the point B where the tangent to the p-v curve becomes horizontal. In these cases disturbances of finite amplitude can produce the reversing of the arch. Similar behavior is exhibited by flat cylindrical shells under pressure or by thin tubes in bending.

Euler's pioneering investigations on the "Elastica"[4] were the first studies on the equilibrium bifurcation of elastic structures. On the other hand, the analytic development of the theory of stability may properly be regarded as commencing with Lagrange's work[5] on the stability of equilibrium states of discrete conservative systems. Lagrange showed that if the potential energy is a minimum at a position of equilibrium, then this position is stable, whereas if the potential energy is a maximum, the position is unstable.

These first fundamental ideas of Lagrange on stability of discrete systems were further analyzed by Dirichlet,[6] who gave the following definition of stability reported by Knops and Wilkes[7]: "the equilibrium is stable if, in displacing the points of the system from their equilibrium positions by an infinitesimal amount and giving each one a small initial velocity, the displacements of different points relative to the equilibrium positions of the system, remain, throughout the course of the motion, contained between certain small prescribed limits." The proof of the basic first theorem on stability, also called the energy criterion, is also due to Lagrange and later to Dirichlet. This theorem stated that "for a conservative system with a finite number of degrees of freedom a sufficient condition for stability of an equilibrium position is that the potential energy of the system assumes at the equilibrium position a minimum in the class of virtual displacements satisfying the kinematical constraints." Subsequently Poincaré[8] and then Liapounov[9] developed new ideas in the general theory of stability. These ideas are still current and continue to produce further research.

In the field of the applied mechanics a large number of investigators studied the problem of elastic stability — Bryan,[10] Southwell,[11] Biezeno and Hencky,[12] Trefftz,[13] Marguerre,[14] Kappus,[15] Timoshenko,[16] Krall,[17,18] Koiter,[3] Budiansky,[25] and Hutchinson.[26] The Koiter's nonlinear approach, particularly, has represented fundamental progress in the theory of stability of elastic structures. In this context, the extension of the Lagrange-Dirichlet criterion of stability from discrete elastic systems to elastic bodies occurred tacitly and seemed a natural procedure to be taken for granted. The definition of stability for a continuous elastic system was, as a rule, not yet fully developed, and the energy criterion has been sometimes used as definition of equilibrium stability. So, according to Hadamard,[19] the equilibrium solution is stable if and only if the potential energy attains a minimum value at the equilibrium solution in the class of displacements meeting the boundary conditions. In most engineering books, (see, for instance, Timoshenko and Gere[20] or Vol'mir,[21] where the energy criterion has been successfully applied), the precise notion of stability, always tacitly assumed essentially static in nature, is, in fact, left undefined.

In the next pages we will analyze the major engineering approaches of stability analysis. After a criticism of these procedures, a consistent formulation of the theory of elastic stability of continuous structural systems will be proposed.

1.2 A REVIEW OF THE COMMON APPROACHES TO STABILITY ANALYSIS IN STRUCTURAL ENGINEERING

Let us consider an elastic structure in equilibrium at a certain configuration C_0 under the action of given conservative loads $\lambda\mathbf{q}$, where λ is the loading parameter. A first simple method to analyse stability of the equilibrium configuration C_0 is based on the inspection of the elastic response of the structure subjected to additional infinitesimal deformations.

Equilibrium at C_0 is thus usually defined stable if the structure opposes additional displacements. On the contrary, the equilibrium at C_0 is usually said to be critical when the structure can be infinitesimally moved along a particular deformation, called the buckling mode, without applying any additional action. The equilibrium at the critical state is thus also called "neutral" because the system, under the same loads $\lambda_c\mathbf{q}$, is in equilibrium both at the configuration C_0 and at some infinitesimally close configurations, different from C_0 and all directed along the so-called buckling mode $\mathbf{u}_c$.

Let us consider, for instance, the axially compressed rod where now $\lambda\mathbf{q}= \lambda N$ if N is the acting axial force (Fig. 1.10). At the critical state, i.e., under the Euler critical load,

$$N_c = \frac{EI\pi^2}{L^2} \tag{1.1}$$

the beam is in equilibrium both at the rectilinear position and in the bending positions defined by the transversal displacement component

$$v_c = A \sin \frac{\pi z}{L} \tag{1.2}$$

where the constant A is a very small quantity compared to the beam length L, but still arbitrary.

According to this approach, often called the "method of adjacent equilibrium", the system is displaced into a configuration belonging to an infinitesimal neighborhood of C_0: subsequently,

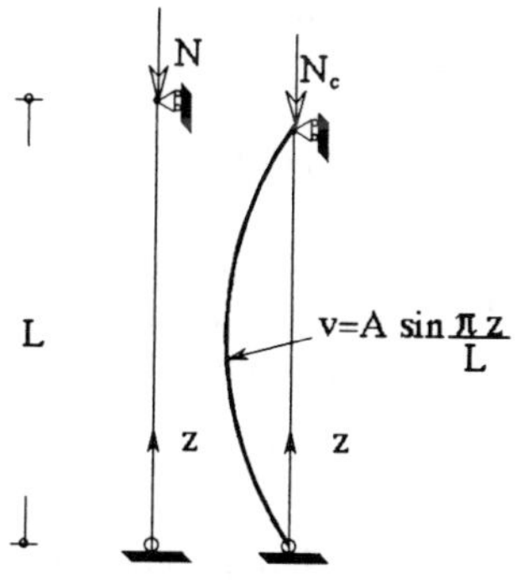

FIGURE 1.10

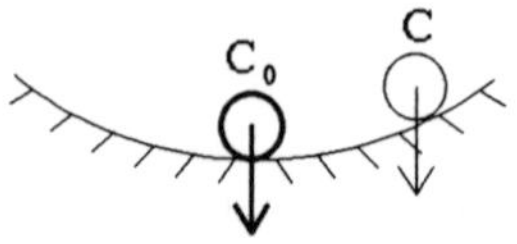

FIGURE 1.11a

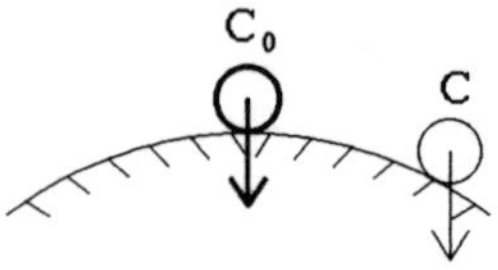

FIGURE 1.11b

the load multiplier λ_c, to which these adjacent equilibrium states correspond, has to be evaluated. The assumption of infinitesimal additional displacements yields to linear differential or integro-differential equations of the adjacent equilibrium.

An equivalent approach is called the energy method. It analyzes the sign of the additional potential energy when, by applying additional loads $\Delta\mathbf{q}$, the elastic structure is displaced from C_0 to another equilibrium configuration C, close to but different from C_0. The system is considered in stable equilibrium under the applied loads $\lambda\mathbf{q}$ when the work $\Delta\mathcal{L}$ of these additional forces $\Delta\mathbf{q}$ is always positive, for any choice of the additional loads $\Delta\mathbf{q}$.

The potential energy of all the forces acting at C_0 will change by the amount $\Delta\mathcal{E}$ due to the imposed additional displacement. The work $\Delta\mathcal{L}$ of the additional forces $\Delta\mathbf{q}$ equals the increment $\Delta\mathcal{E}$, according to the general principle of balance of mechanical work:

$$\Delta\mathcal{L} = \Delta\mathcal{E} \tag{1.3}$$

In the context of systems with a finite number degrees of freedom the simplest example is the heavy small single sphere in equilibrium on a given smooth concave surface (Fig. 1.11a).

If the equilibrium is stable at C_0,whenever the sphere is moved from C_0, $\Delta\mathcal{L} > 0$ and, consequently, $\Delta\mathcal{E} > 0$ and vice versa. The opposite is the case of the heavy sphere placed on the top of a convex smooth surface (Fig. 1.11b).

An elastic structural system subjected to conservative loads, in equilibrium at the configuration C_0, is considered similar to the previous case of the heavy ball. Let $\Delta\mathcal{U}(C)$ and $\Delta\mathcal{V}(C)$ be respectively the increments of the potential energy of the external and of the internal forces in the passage of the structure from C_0 to a close configuration C. Therefore

$$\Delta\mathcal{U}(C) + \Delta\mathcal{V}(C) = \Delta\mathcal{E}(C) \tag{1.4}$$

where

$$\Delta\mathcal{E}(C) = \mathcal{E}(C) - \mathcal{E}(C_0) \tag{1.5}$$

The equilibrium at C_0 is also commonly considered stable if, for any additional displacement of the system in the neighborhood of C_0, the increment $\Delta\mathcal{E}(C)$ of the potential energy of all internal and external forces is positive, i.e, if

$$\Delta\mathcal{E}(C) > 0 \tag{1.6}$$

In fact, if, for any additional disturbance, $\Delta\mathcal{L} > 0$, thus we will have $\Delta\mathcal{E}(C) > 0$, and vice versa.

By definition, $\Delta\mathcal{E}(C_0) = 0$; thus, equivalently, there is stability at C_0 if the additional potential energy $\Delta\mathcal{E}(C)$ achieves its minimum at C_0. On the other hand, if

$$\mathcal{E}(C_0) = 0 \tag{1.7}$$

the principal equilibrium state C_0 is the reference configuration. Consequently

$$\Delta\mathcal{E}(C) = \mathcal{E}(C) \tag{1.5'}$$

If **u** is the additional displacement that moves the structure from C_0 to C, the configuration C_0 is defined by $\mathbf{u} = \mathbf{0}$. The increment (1.3) of the potential energy becomes

$$\Delta\mathcal{E}(C) = \Delta\mathcal{E}(\mathbf{u}) = \mathcal{E}(\mathbf{u}) - \mathcal{E}(\mathbf{0}) = \mathcal{E}(\mathbf{u}) \tag{1.8}$$

where $\mathcal{E}(\mathbf{0}) = 0$.

The common stability condition (1.6) is then equivalent to the assumption that the equilibrium state C_0 is a relative minimum of $\mathcal{E}(\mathbf{u})$. This functional is usually called the additional potential energy of all the external and the internal forces acting on the structure.

The linear theory of elastic stability is formulated on the assumption of infinitesimal displacements **u**. In this case the function **u** displaces the system in an infinitesimal neighborhood of C_0. Hence it is assumed that

$$\mathcal{E}(\mathbf{u}) \approx \mathcal{E}_1(\mathbf{0};\, \mathbf{u}) + \mathcal{E}_2(\mathbf{0};\, \mathbf{u}) \tag{1.9}$$

where $\mathcal{E}_1(\mathbf{0};\, \mathbf{u})$ and $\mathcal{E}_2(\mathbf{0};\, \mathbf{u})$, also indicated as $D\mathcal{E}(\mathbf{0};\, \mathbf{u})$ and $D_2\mathcal{E}(\mathbf{0};\, \mathbf{u})$, are the first and second differential of the total potential energy $\mathcal{E}(\mathbf{u})$, evaluated at C_0 and along the additional displacement **u**. Thus, for example,

$$\mathcal{E}_1(\mathbf{0};\mathbf{u}) = D\mathcal{E}(\mathbf{0};\mathbf{u}) = \lim_{\alpha \to 0} \frac{\mathcal{E}(\mathbf{0}+\alpha\mathbf{u}) - \mathcal{E}(\mathbf{0}+\alpha\mathbf{u})}{\alpha} = \left[\frac{\partial\mathcal{E}(\alpha\mathbf{u})}{\partial\alpha}\right]_{\alpha=0} \tag{1.10}$$

With different notation the first differential of the energy is also called, according to the variational calculus, the variation of the functional $\mathcal{E}$ in the direction **u** and is written as

$$\mathcal{E}_1\,(\mathbf{0};\, \mathbf{u}) = \delta\mathcal{E}\,(\mathbf{0};\, \mathbf{u}) \tag{1.10$'$}$$

On the other hand, taking into account the equilibrium at C_0, the principle of virtual works gives

$$\mathcal{E}_1(\mathbf{0};\, \mathbf{u}) = 0 \tag{1.11}$$

for any direction **u**. Hence, according to the linear theory of elastic stability, from (1. 9) we get

$$\mathcal{E}(\mathbf{u}) \approx \mathcal{E}_2(\mathbf{0};\, \mathbf{u}) \tag{1.12}$$

For the sake of simplicity in the expression of $\mathcal{E}_2(\mathbf{0};\, \mathbf{u})$ the zero can be removed and, at the same time, the functional $\mathcal{E}_2(\mathbf{0};\, \mathbf{u})$ depends on the load multiplier λ, representing the loads acting at the equilibrium configuration C_0. Therefore

$$\mathcal{E}(\mathbf{u}) \approx \mathcal{E}_2(\lambda;\, \mathbf{u}) \tag{1.12$'$}$$

So, for example, in the case of the axially compressed rod of length L, with transversal displacements v(z),

$$\mathcal{E}_2(\lambda;v) = \frac{EI}{2}\int_0^L \left(\frac{d^2v}{dz^2}\right)^2 dz - \frac{\lambda N}{2}\int_0^L \left(\frac{dv}{dz}\right)^2 dz \tag{1.13}$$

The stability condition at C_0 is thus given in the following usual form

$$\mathcal{E}_2(\lambda;\, \mathbf{u}) > 0 \tag{1.14}$$

which has to be satisfied along any nonzero additional displacement **u**. In order to evaluate the buckling mode $\mathbf{u}_c$, the critical state is characterized by

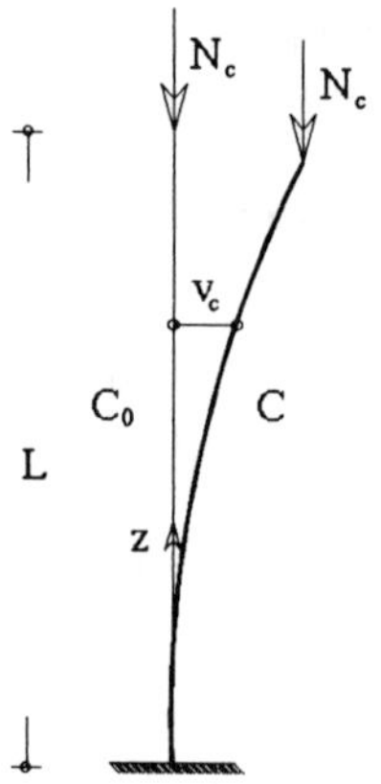

FIGURE 1.12

$$\mathcal{E}_2(\lambda_c\,;\,\mathbf{u}_c\,) = 0 \tag{1.15}$$

$$\mathcal{E}_2(\lambda_c\,;\,\mathbf{u}\,) > 0 \tag{1.16}$$

This last condition has to be satisfied for any displacement **u** not zero and not aligned with $\mathbf{u}_c$. The critical state is thus attained when the loads $\lambda\mathbf{p}$ reach the value $\lambda_c\mathbf{p}$ at which the structure is incapable of exhibiting any positive reaction along the infinitesimal additional displacement $\mathbf{u}_c$. The functional $\mathcal{E}_2(\lambda;\,\mathbf{u})$ becomes semi-definite positive when the critical state is attained. The quadratic functional $\mathcal{E}_2(\lambda;\,\mathbf{u})$ also attains its minimum at $\lambda = \lambda_c$ and at the buckling mode $\mathbf{u}_c$. Hence

$$D\mathcal{E}_2(\lambda_c\,;\,\mathbf{u}_c,\mathbf{v}) = \delta\mathcal{E}_2(\lambda_c;\,\mathbf{u}_c,\,\mathbf{v}) = 0 \tag{1.17}$$

or, according to a different symbology,

$$\mathcal{E}_{11}(\lambda_c;\,\mathbf{u}_c,\mathbf{v}) = 0 \tag{1.17$'$}$$

along any displacement **v**. The functional $\mathcal{E}_{11}(\lambda_c;\,\mathbf{u}_c,\mathbf{v})$ is the bilinear form in **u** and **v** associated to the quadratic functional $\mathcal{E}_2(\lambda;\,\mathbf{u})$. Condition (1.17) implies (1.15). Actually, from (1.17′) and taking $\mathbf{v} = \mathbf{u}_c$, we get

$$\mathcal{E}_{11}(\lambda_c\,;\,\mathbf{u}_c,\,\mathbf{u}_c) = 2\,\mathcal{E}_2(\lambda_c;\,\mathbf{u}_c) = 0 \tag{1.18}$$

Taking into account assumption (1.12), condition (1.17) shows that at the critical state the neutral equilibrium is attained. Condition (1.17), in fact, describes the occurrence of the equilibrium at every configuration C in an infinitesimal neighborhood of C_0 and having the direction of the critical displacement $\mathbf{u}_c$ (Fig. 1.12).

Assume therefore that an equilibrium position of an elastic structure is initially stable and let the load parameter be progressively increased. Under the increasing load, equilibrium will remain stable until the load parameter λ reaches the value λ_c. At $\lambda = \lambda_c$, $\mathcal{E}_2(\lambda;\,\mathbf{u})$ becomes positive semi-definite. The value λ_c is said to be the critical multiplier and corresponds to the buckling load. This value is also said to give the limit of stability. In applying this stability test, it is supposed that for values of λ larger than λ_c the primary equilibrium state is unstable. This statement is usually justified because λ_c represents the first value of the load parameter at which the potential energy ceases to have a minimum at the primary state.

Condition (1.17), which yields to linear equations of the additional equilibrium, is the basis of the classical linear theory of elastic stability of structures. A large number of problems have been solved according to this theory. In this framework, a classical example is offered by the axially compressed rod. In this case from (1.13) we have

$$\mathcal{E}_{11}(v,u) = EI\int_0^L \frac{d^2u}{dz^2}\frac{d^2v}{dz^2}dz - \lambda N\int_0^L \frac{du}{dz}\frac{dv}{dz}dz \tag{1.19}$$

Thus, from the variational condition $\mathcal{E}_{11}(u;\ v) = 0$, for any admissible displacement v(z) integration yields

$$|EIv''u'|_0^L - |(\lambda Nv' + EIv''')u|_0^L + \int_0^L \left(EIv^{Iv} + \lambda Nv''\right)u\,dz = 0 \tag{1.20}$$

from which the well-known equations follow

$$EIv^{Iu} + \lambda Nv'' = 0 \quad |EIv''u'|_0^L = 0 \quad |(\lambda Nv' + EIv''')u|_0^L = 0 \tag{1.20$'$}$$

They admit the solution $v = Av_c$, where v_c is the buckling mode and A is an arbitrary constant; A is not identically zero only for particular values of λ, the eigenvalues of (1.20′). The smallest critical multiplier λ_c of λ corresponds to the smallest eigenvalue, i.e., to the smallest value of λ at which we have $\mathcal{E}_2(\lambda;\ \mathbf{u}_c) = 0$.

In many cases the principal equilibrium configuration C_0 is near to the undeformed configuration C_i. Thus the second differential $\mathcal{E}_2(\mathbf{u})$ can be considered linear in λ and assumes the usual form

$$\mathcal{E}_2(\lambda;\ \mathbf{u}) = -\lambda\mathcal{L}_2(\mathbf{u}) + \mathcal{V}_2(\mathbf{u}) \tag{1.21}$$

where $\mathcal{V}_2(\mathbf{u})$, positive definite, is the elastic energy corresponding to the additional displacement **u** and evaluated as if C_0 were the stress-free configuration of the structure, and $\mathcal{L}_2(\mathbf{u})$, is a generally positive quadratic functional, representative of the destabilizing effects of the loads. Thus

$$\mathcal{E}_2(\lambda;\mathbf{u}) = \mathcal{L}_2(\mathbf{u})\left[\frac{\mathcal{V}_2(\mathbf{u})}{\mathcal{L}_2(\mathbf{u})} - \lambda\right] \tag{1.22}$$

Hence the critical multiplier λ_c is given by the smallest value of the ratio[20]

$$\frac{\mathcal{V}_2(\mathbf{u})}{\mathcal{L}_2(\mathbf{u})} \tag{1.23}$$

for any admissible additional displacement **u**. Thus it is usually assumed

$$\lambda_c = \min\left[\frac{\mathcal{V}_2(\mathbf{u})}{\mathcal{L}_2(\mathbf{u})}\right] \tag{1.24}$$

among all the admissible additional displacement fields. The definition of λ_c according to (1.24) is equivalent to the definition of the critical state given by (1.15) and (1.16). Furthermore, the research of the minimum λ_c yields a variational condition that is equivalent to the condition (1.17), corresponding to the neutral equilibrium. We have in fact

$$\left[\frac{\mathcal{V}_2(\mathbf{u}_c + t\mathbf{v})}{\mathcal{L}_2(\mathbf{u}_c + t\mathbf{v})} - \lambda_c\right] \geq 0 \tag{1.25}$$

This condition has to be satisfied for any real number *t* and for any admissible additional displacement **v**. But

$$\mathcal{V}_2(\mathbf{u}_c + t\mathbf{v}) = \mathcal{V}_2(\mathbf{u}_c) + t\mathcal{V}_{11}(\mathbf{u}_c, \mathbf{v}) = t^2\mathcal{V}_2(\mathbf{v}) \tag{1.26}$$

$$\mathcal{L}_2(\mathbf{u}_c + t\mathbf{v}) = \mathcal{L}_2(\mathbf{u}_c) + t\ \mathcal{L}_{11}(\mathbf{u}_c, \mathbf{v}) = t^2\mathcal{L}_2(\mathbf{v}) \tag{1.26′}$$

where

$$\mathcal{V}_{11}(\mathbf{u},\mathbf{u}) = 2\ \mathcal{V}_2(\mathbf{u})\ \ \mathcal{L}_{11}(\mathbf{u},\mathbf{u}) = 2\ \mathcal{L}_2(\mathbf{u}) \tag{1.27}$$

Thus

$$\frac{t\left[\mathcal{V}_{11}(\mathbf{u}_c, \mathbf{v}) - \lambda_c\mathcal{L}_{11}(\mathbf{u}_c, \mathbf{v})\right] + t^2\left[\mathcal{V}_2(\mathbf{v}) - \lambda_c\mathcal{L}_2(\mathbf{v})\right]}{\mathcal{L}_2(\mathbf{u}_c + t\mathbf{v})} \geq 0 \tag{1.28}$$

According to (1.24) the coefficient of t^2 is ≥ 0. Thus, to satisfy condition (1.28) it is necessary that

$$\mathcal{V}_{11}(\mathbf{u}_c, \mathbf{v}) - \lambda_c\mathcal{L}_{11}(\mathbf{u}_c + \mathbf{v}) = 0 \tag{1.29}$$

for any admissible additional displacement **v**. Thus condition (1.17) also derives from condition (1.24).

A third approach, very common in technical literature, (see, for example, Bolotin[24] or Krall[18]) is based on the inspection of the small motions of the structure about C_0. These motions can be analyzed assuming (1.12) by using the Hamilton principle[18], i.e., by setting

$$D\mathcal{H}(\mathbf{m}, \delta\mathbf{m}) = D\int_{t_0}^{t_1} \left[\mathcal{T}(\dot{\mathbf{u}}) - \mathcal{E}_2(\mathbf{u})\right]dt = 0 \tag{1.30}$$

with

$$\mathbf{m} = \begin{bmatrix} \mathbf{u} \\ \dot{\mathbf{u}} \end{bmatrix} \qquad \delta\mathbf{m} = \begin{bmatrix} \delta\mathbf{u} \\ \delta\dot{\mathbf{u}} \end{bmatrix} \tag{1.31}$$

i.e., the vanishing of the variation $D\mathcal{H}$ of the Hamilton integral for every varied motion $\delta\mathbf{m}(P,t)$, zero at instants t_0 and t_1. But, according to (1.10)

$$\mathcal{T}(\dot{\mathbf{u}}) = \frac{1}{2}\int_\Omega \mu\dot{\mathbf{u}}^2 d\Omega \cdot D\mathcal{T}(\dot{\mathbf{u}}, \delta\dot{\mathbf{u}}) = \int_\Omega \mu\ddot{\mathbf{u}} \cdot \delta\mathbf{u}\ d\Omega = -\int_\Omega \mu\ddot{\mathbf{u}} \cdot \delta\mathbf{u}\ d\Omega \tag{1.32}$$

and condition (1.30) becomes

$$D\mathcal{H}(\mathbf{m}, \delta\mathbf{m}) = \int_{t_0}^{t_1}\left[-\int_\Omega \mu\ddot{\mathbf{u}} \cdot \delta\mathbf{u}\, d\Omega - \mathcal{E}_{11}(\mathbf{u}, \delta\mathbf{u})\right] dt = 0 \tag{1.31′}$$

Now let us analyze the stationary motions of the system in the neighborhood of the configuration C_0. These small oscillations are defined by

$$\mathbf{u}(P,t) = \mathbf{v}(P)\ e^{(a + i\omega)t} \tag{1.33}$$

Taking $\delta\mathbf{u}(P,t) = \mathbf{v}(P) \sin \omega t$, with $t_0 = 0$, $t_1 = \frac{2\pi}{\omega}$, putting (1.33) into (1.31′) we get

$$\omega^2 - 2i\alpha\omega - \alpha^2 = \frac{\mathcal{E}_2(\mathbf{v})}{\frac{1}{2}\int_\Omega \mu \mathbf{v}^2 d\Omega} \tag{1.34}$$

According to this last approach, the elastic structural system is said to be in stable equilibrium if, when it is perturbed in its principal equilibrium state C_0, it indefinitely oscillates about C_0 with amplitude that we can make as small as we like by suitably reducing the magnitude of the disturbance. Thus to have stability of the small perturbed motions (1.33) it is required that $\omega^2 > 0$ and $\alpha = 0$, i.e., $\mathcal{E}_2(\mathbf{u}) > 0$. On the contrary, when $\alpha^2 > 0$ and $\omega^2 = 0$, there is instability and we have $\mathcal{E}_2(\mathbf{u}) < 0$. Thus, by inspection of (1.34), we recognize the equivalence between this last approach and that of the additional potential energy. The critical state, as sketched in Fig. 1.13, then corresponds to the passage from states of motion where $\omega^2 > 0$, $\alpha = 0$ to states where, on the contrary, $\alpha^2 > 0$, $\omega^2 = 0$. The critical state is then attained when

$$\omega^2 = 0 \quad \alpha = 0 \tag{1.35}$$

i.e., when $\mathcal{E}_2(\mathbf{u}) = 0$.

The three different approaches more or less manifestly contain a "static" definition of stability. Even the third procedure, which at first sight may be considered to be based on a dynamic definition of stability, taking into account only the small stationary motions in the neighborhood of the critical state, is equivalent to the other ones. On the contrary, stability has a deep dynamic meaning, and it has to be correctly defined, according to Liapounov[9], as we will recall in the next pages. In any case, the three different approaches yield to the linear theory of elastic stability and, as a rule, are able to characterize the critical state, with sufficient accuracy. In fact, all these approaches use the approximate form of the additional total potential energy expressed by the second differential $\mathcal{E}_2(\mathbf{u})$. This approximation is usually justified by the fact that the additional displacements $\mathbf{u}$ have infinitesimal amplitude. In this framework the critical state is identified by the occurrence of the condition of neutral equilibrium for infinitesimal additional displacements analyzed above. Consequently, the equilibrium bifurcation at the critical state will be defined by the occurrence of an horizontal branch, as shown in Fig. 1.14.

In Fig. 1.14a the parameter v indicates the magnitude of the critical displacement. The curved equilibrium paths that branch out at the critical state, shown in Figs. 1.14b, 1.14c, and 1.14d, and that correspond to the various cases discussed at Section 1.1, therefore cannot be

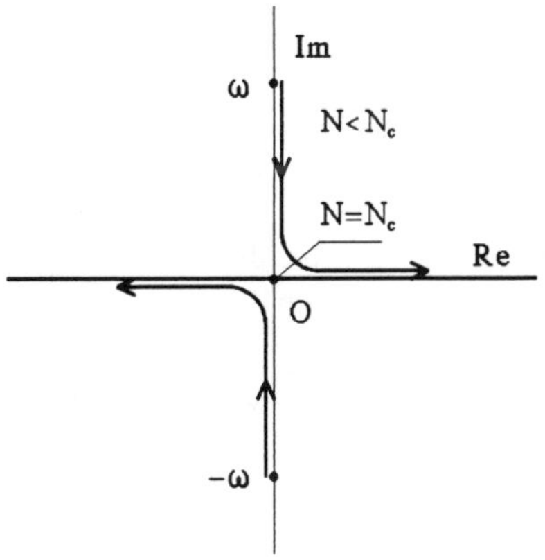

FIGURE 1.13

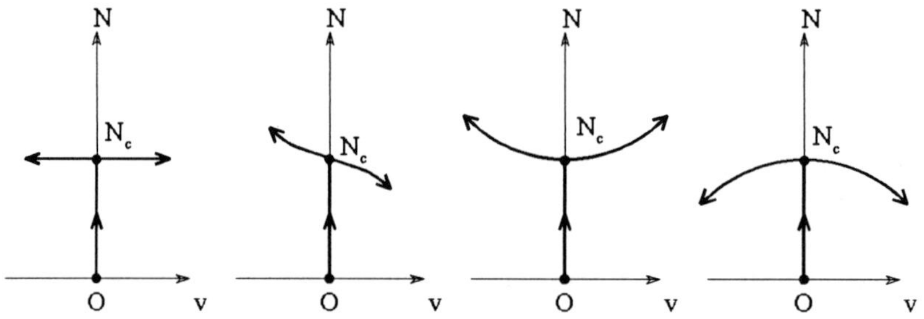

FIGURE 1.14.a **FIGURE 1.14.b** **FIGURE 1.14.c** **FIGURE 1.14.d**

correctly described by the linear theory of stability. The question of whether the critical state is stable or unstable cannot receive any answer from the linear theory of stability.

The most serious failure of the linear theory of stability lies in its inability to define the slope of the equilibrium branches in the neighborhood of the critical state. In fact in many cases, the tangent to these branches cannot be horizontal near $\lambda = \lambda_c$. This discrepancy is due to the approximation (1.12) in the evaluation of the first differential of $\Delta\mathcal{E}(\lambda_c,\mathbf{u}_c)$ at the critical state and along the direction $\mathbf{u}_c$.

In brief, even though the functionals $\mathcal{E}_2(\lambda,\mathbf{u})$ and $\mathcal{E}(\lambda,\mathbf{u})$, evaluated for small additional displacements $\mathbf{u}$, more or less take the same values, their first differentials $D\mathcal{E}_2(\lambda,\mathbf{u})$ and $D\mathcal{E}(\lambda,\mathbf{u})$ could be very different the one from the other. Koiter's nonlinear approach[3], in this context, has represented a fundamental progress in the theory of stability of elastic structures.

Deeper criticism can also be raised to the assumed static definition of stability if we point out that the structural systems, being "*continuous*", have an infinite number of degrees of freedom. Here it follows a very significant example. Let us consider the unloaded rod shown in Fig. 1.15. Let us assume that the set of all the admissible configurations of the rod are represented by all the functions, continuous with their derivatives up to the fourth order in the interval [0,L]. We will then assume that the functions $v(z) \in C^4[0,L]$ and, at the same time, that they satisfy the boundary conditions $v = v'' = 0$ at $z = 0$, $z = L$.

The rectilinear configuration C_0 of the unloaded rod (i.e., the rectilinear configuration $v(z) \equiv 0 \;\forall\, z \in [0,L]$ should certainly be stable. Such a conjecture can also be supported by the fact that, if we apply one of the three previous stability tests, for instance the energy criterion, the stability condition is verified. In fact in this case, considering additional flexural displacements $v(z)$ of the bar, we have

$$\mathcal{E}_2(v) = \frac{1}{2} EI \int_0^L \left(\frac{d^2v}{dz^2}\right)^2 dz \tag{1.36}$$

Consequently, the stability condition $\mathcal{E}_2(v) > 0$ is satisfied. All the disturbed motions of the rod starting from the rectilinear configuration should thus remain bounded inside a neighborhood of C_0 that can be made arbitrarily small by suitably reducing the magnitude of the disturbance.

It is natural to assume, as a measure of the distance among two different states of motion $\mathbf{m}_1(t)$, $\mathbf{m}_2(t)$ of the rod at time t, the function

$$d[\mathbf{m}_1(t),\mathbf{m}_2(t)] = \max_{[0,L]}\left\{|v_1(z,t) - v_2(z,t)| + \max_{[0,L]}|\dot{v}_1(z,t) - \dot{v}_2(z,t)|\right\} \tag{1.37}$$

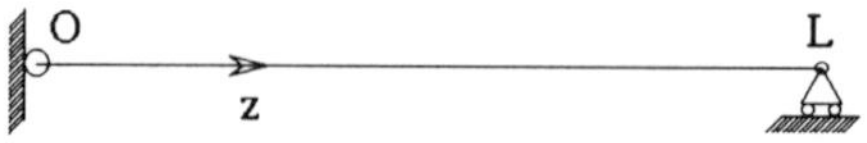

FIGURE 1.15

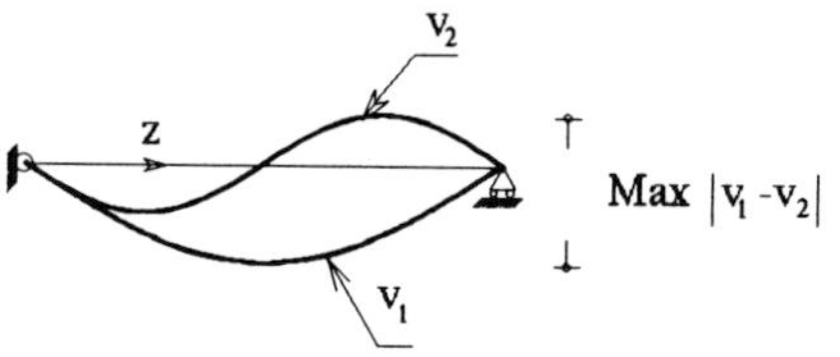

FIGURE 1.16

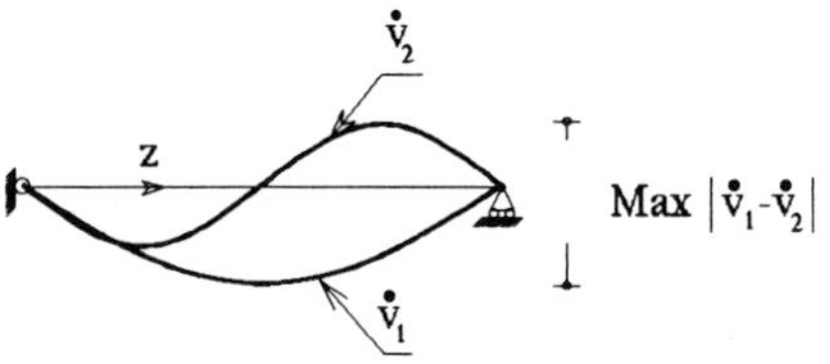

FIGURE 1.17

as shown in Figs. 1.16 and 1.17.Thus the distance between a generic perturbed state of the motion **m**(t) and the initial rectilinear configuration of the bar is

$$d[\mathbf{m}(t),0] = \max_{[0,L]}\left\{|v(z,t)| + \max_{[0,L]}|\dot{v}(z,t)|\right\} \tag{1.38}$$

Let us examine now the motion of the rod subsequent to the following initial disturbance

$$v(z,0) = C\sin\frac{n\pi z}{L} \quad \dot{v}(z,0) = 0,\ C > 0, n \in N \tag{1.39}$$

representing an initial deformation of the bar, with the assumption that $C << 1$, i.e., that v(z,0) is small in the sense of the metric (1.38).

To analyze the disturbed motion of the rod, let us evaluate the small oscillations of the rod described by the linear differential equation

$$EIv^{IV} + \mu\ddot{v} = 0 \tag{1.40}$$

with the usual significance of the symbols. A solution of this equation satisfying the initial condition $[v(z,t)]_{z=0} = C\sin(n\pi z/L)$, $[\dot{v}(z,t)]_{t=0} = 0$, is

$$v(z,t) = C\sin\frac{n\pi z}{L}\cos\omega_n t \tag{1.41}$$

with

$$\omega_n^2 = \frac{EI}{\mu}\frac{n^4\pi^4}{L^4} \tag{1.42}$$

According to the metric (1.38), the distance between the initial unstressed state of the rod and the impressed initial disturbance (i.e., in brief, the magnitude of the initial deformation) is

$$d(0) = C \tag{1.43}$$

The distance between the initial state of the unloaded bar and the generic motion at time t, on the other hand, is given by

$$d(t) = C\{|\cos\omega_n t| + \omega_n|\sin\omega_n t|\} \tag{1.44}$$

Consequently

$$\sup_{t>0} d(t) \geq C\omega_n \tag{1.45}$$

and

$$\lim_{n\to\infty}\ \sup d(t) \geq \lim_{n\to\infty} C\omega_n = \infty \tag{1.45'}$$

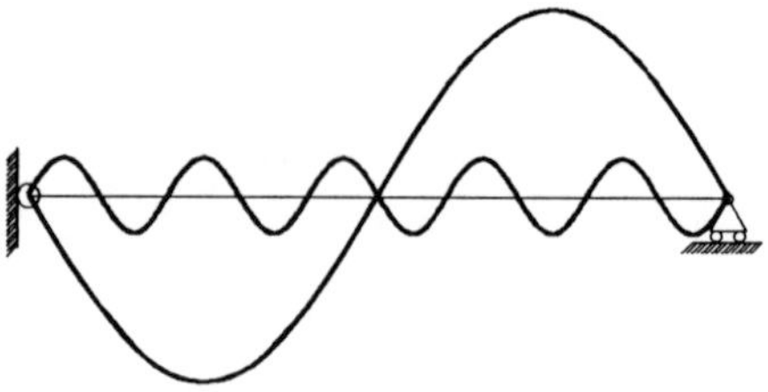

FIGURE 1.18

Thus, as small as we take the constant C, the distance d among the initial state of the rod and all the disturbed motions, with n sufficiently large, will be larger than any positive number arbitrarily chosen. Thus we have to conclude, to our surprise, that the unloaded state of the rod is unstable.

For a better understanding of this result it is useful to analyze some aspects of the initial deformation imparted to the rod in more detail. Let us examine the disturbances having magnitude $d(0) \leq C$. In this set we include all the waves as

$$C' \sin \frac{n\pi z}{L}$$

with $C' \leq C$ and $\forall\, n \in N$, i.e., all the waves with bounded magnitude having an unbounded number of nodes. The consequent strain energy of the deformed rod is

$$\mathcal{V}_2(v) = \mathcal{E}_2(v) = C'^2 \frac{EI}{4} \frac{n^4\pi^4}{L^3} \tag{1.46}$$

Hence, as shown in Fig. 1.18, although we can take the constant C' as small as we like, if we don't limit the magnitude of n, i.e., the number of nodes, unbounded quantities of energy can be initially imparted to the rod.

With the chosen measure, the amount of the energy initially imparted to the rod does not come out at all. On the contrary, if we measure the distance at time t, between the motion and the rectilinear configuration of the bar, this amount comes out. In fact, during the motion high velocities are present at the various sections of the rod if the energy initially imparted is large.

The strange result of instability of the unloaded rod so obtained is thus due to the assumed measure of the disturbance: this measure is not physically acceptable because "small" disturbances ("small" in the sense of the assumed measure) don't correspond to small amounts of the imparted energy.

If, on the contrary, we assume as a measure of the motions the total energy, we will be immediately able to recognize that at its initial unloaded state the rod turns out to be stable. In fact, with

$$d(t) = \frac{EI}{2} \int_0^L v''^2 dz + \frac{1}{2} \int_0^L \mu \dot{v}^2 dz \tag{1.47}$$

we have

$$d(0) = d(t)\ \forall\, t > 0 \tag{1.48}$$

With this new assumed measure at any time t, to small initial disturbances correspond small distances among the disturbed motions and the initial state of the rod. We remark that for systems with a finite number of degrees of freedom (Fig. 1.19), the above discrepancies obtained by assuming different measures of the distances among the motions, don't occur any more. In this case, in fact, if we limit the magnitude of the initial displacements, the imparted

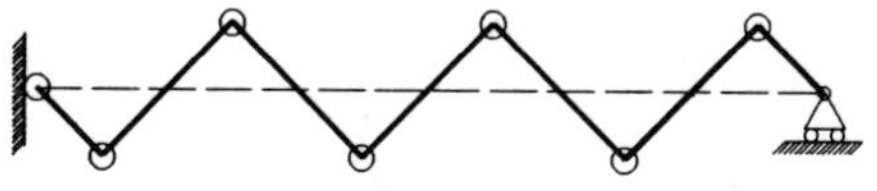

FIGURE 1.19

strain energy is consequently bounded too. The above example points out the deep difference existing between systems with finite and infinite degrees of freedom, as far as stability analysis is concerned.

We have to answer the question why the energy criterion fails as a test for stability in the previous example if we assume the distance (1.38). According to (1.36), in fact, $\mathcal{E}_2(v) > 0$ for any $v \neq 0$. Thus the energy criterion (1.6), taken independently on the chosen measure to evaluate distances among motions, yields the erroneous conclusion that the rectilinear state of the rod is stable. We will examine this last question in the following pages.

In order to have a clearer idea of the significance of the energy criteria as a stability test, however, it is first necessary to recall the correct statements of these criteria for systems with finite degrees of freedom.

1.3 A REVIEW OF THE ENERGY CRITERIA OF STABILITY FOR FINITE DEGREES OF FREEDOM SYSTEMS

For systems with a finite number of degrees of freedom the space R^n is the space of the admissible configurations; likewise, the space R^{2n} is the space of the states of motion. The configurations of an elastic system with n degrees of freedom are defined by the n lagrangian coordinates $q_1, q_2, ..., q_n$. Likewise a generic state **m** of motion of the system is defined by the vector[23]

$$\mathbf{m} = \{q_1, q_2, \dots q_n\ ; \dot{q}_1, \dot{q}_2, \dots \dot{q}_n\} \tag{1.49}$$

This vector can be also written as

$$\mathbf{m} = \mathbf{P} = \dot{\mathbf{P}} \tag{1.49'}$$

where

$$\mathbf{P} = \{q_1, q_2, \dots, q_n, 0, 0, \dots 0\},\ \ \dot{\mathbf{P}} = \{0, 0, \dots 0, \dot{q}_1, \dot{q}_2, \dots \dot{q}_n\} \tag{1.49''}$$

considering the space R^{2n} of the states of motion as direct sum of the spaces of the displacements and of the space of velocities. In the space R^{2n} the measure of the distance between two different states of motion **m**', **m**'' of the system, for instance, can be taken as

$$d(\mathbf{m}', \mathbf{m}'') = \left\{ \Sigma\left(q_i^{'} - q_i^{''}\right)^2 + \Sigma\left(\dot{q}_i^{'} - \dot{q}_i^{''}\right)^2 \right\}^{1/2} \tag{1.50}$$

Under the action of given conservative forces the system is in equilibrium at the configuration $\mathbf{P}_0$. We assume that $\mathbf{P}_0$ is idefined by the coordinates $q_1 = 0$, $q_2 = 0$, ..., $q_n = 0$. Likewise, let

$$\mathbf{0} = \{0, 0, \dots, 0; 0, 0, \dots, 0\} \tag{1.51}$$

be that particular state of motion representing the state of rest of the system, taken as the origin of the coordinates axes of the space R^{2n}. The position vector **P**, whose coordinates are q_1, q_2, ..., q_n, thus represents the additional displacement of the system starting from the fundamental

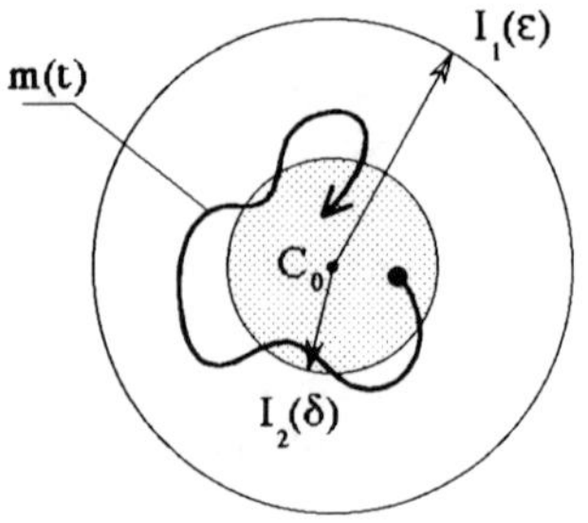

FIGURE 1.20

configuration $\mathbf{P}_0$. Let $\mathbf{m}(t)$ indicate the generic state of motion at time t, consequent to the imparted initial disturbance. We have

$$\mathbf{m}(t) = \{q_1(t), q_2(t), \ldots q_n ; \dot{q}_1(t), \dot{q}_2(t), \ldots \dot{q}_n(t)\} \tag{1.52}$$

According to the classical Liapounov definition of stability, we say that the equilibrium configuration $\mathbf{P}_0$ is stable if, as small as we take the quantity $\varepsilon > 0$, it is possible to pick out a $\delta > 0$ so that, for any initial disturbance $\mathbf{m}(0)$ such that $|\mathbf{m}(0) - \mathbf{0}| < \delta$, at any instant t yields $|\mathbf{m}(t) - \mathbf{0}| < \varepsilon$. In brief the equilibrium configuration C_0 is Liapounov stable if, given any spherical neighborhood $I_1(\varepsilon)$ of P_0 in R^{2n}, it is possible to find a corresponding spherical neighborhood $I_2(\delta)$ in R^{2n} such that any initial disturbance starting within $I_2(\delta)$ yields a disturbed motion at any time contained in $I_1(\varepsilon)$. (Fig. 1.20).

As will be discussed in the next chapter, in finite-dimensional spaces all the distances are equivalent; i.e., if $d_I(\mathbf{m}', \mathbf{m}'')$ and $d_{II}(\mathbf{m}', \mathbf{m}'')$ are two different distances, we can always determine positive numbers α and β such that

$$d_I(\mathbf{m}', \mathbf{m}'') \leq \alpha\, d_{II}(\mathbf{m}', \mathbf{m}'') \quad d_{II}(\mathbf{m}', \mathbf{m}'') \leq \beta\, d_I(\mathbf{m}', \mathbf{m}'') \tag{1.53}$$

We infer therefore that for systems with finite degrees of freedom the definition of stability is norm independent. Thus if a system, at a given configuration C_0, is in stable equilibrium with respect to an assumed distance, it will also be stable with respect to any other chosen norm. For instance in place of the distance (1.50) we can take, as a measure of the distance between two different states of motion $\mathbf{m}'$ and $\mathbf{m}''$, the quantity

$$|\mathbf{m}' - \mathbf{m}''| = \left\{\max\left|q_i' - q_i''\right| + \max\left|\dot{q}_i'(t) - \dot{q}_i''(t)\right|\right\} \; (i = 1, 2, \ldots, n) \tag{1.54}$$

With the assumption of conservative loads let $\mathcal{E}(q_1,q_2,\ldots,q_n)$ be the function representing the increment of the potential energy of all the external and internal loads in the passage of the system from $\mathbf{P}_0$ to a close configuration $\mathbf{P}$. An equilibrium state $\mathbf{P}_0$ corresponds to a stationary point of $\mathcal{E}(q_1,q_2,\ldots,q_n)$, where

$$\left(\frac{\partial \mathcal{E}}{\partial q_i}\right)_{P_0} = 0 \;\; (i = 1, 2, \ldots, n) \tag{1.55}$$

According to the Lagrange-Dirichlet theorem, the equilibrium state $\mathbf{P}_0$ of the system is thus Liapounov stable if $\mathbf{P}_0$ is a relative minimum for the function $\mathcal{E}(q_1,q_2,\ldots,q_n)$, assumed *continuous* in a region containing $\mathbf{P}_0$.

For simplicity, let us also assume that the configuration $\mathbf{P}_0$ coincides with the zero $\mathbf{0}$ of the space R^n. The existence of a relative minimum of $\mathcal{E}(q_1,q_2,\ldots,q_n)$ at $\mathbf{0}$ can thus be expressed as:

$\exists\, k > 0: 0 < |\mathbf{P} - \mathbf{0}| < k \Rightarrow \mathcal{E}(\mathbf{P}) - \mathcal{E}(\mathbf{0}) > 0$. The function $\mathcal{E}(\mathbf{P})$ is defined within an arbitrary constant: thus we can put $\mathcal{E}(\mathbf{0}) = 0$. The theorem is based on two assumptions:

1. The existence of a spherical neighborhood of radius k of the origin **0** within which the additional potential energy $\mathcal{E}(\mathbf{P})$, but at **0**, is positive.

2. The continuity of the function $\mathcal{E}(\mathbf{P})$ in a spherical neighborhood S_k of **0**. Thus the statement of the Lagrange-Dirichlet theorem is: The conditions

$$1.\quad \exists k > 0: 0 < |\mathbf{P} - \mathbf{0}| < k \Rightarrow \mathcal{E}(\mathbf{P}) > 0 \tag{1.56}$$

$$2.\quad \lim_{|\mathbf{P}-\mathbf{Q}|\to 0} \mathcal{E}(\mathbf{P}) = \mathcal{E}(\mathbf{Q}),\ \ \mathbf{Q} \in S_k = \{\mathbf{Q} : |\mathbf{Q} - \mathbf{0}| < k,\ k > 0\} \tag{1.57}$$

are sufficient to stability of the state **0**, i.e.,

$$\forall\, \varepsilon > 0, \exists\, \delta(\varepsilon) > 0: \forall\, \mathbf{m}(0) : |\mathbf{m}(0) - \mathbf{0}| < \delta \Rightarrow |\mathbf{m}(t) - \mathbf{0}| < \varepsilon, \forall\, t > 0 \tag{1.58}$$

Let us consider the space R^{2n} of the states of motion $\mathbf{m} = \{q_1, q_2, \ldots, q_n;\ \dot{q}_1, \dot{q}_2, \ldots, \dot{q}_n\}$ A spherical neighborhood S_k in R^{2n} centered at **0** and with radius k is the set of the points **m** such that $|\mathbf{m} - \mathbf{0}| < k$. Thus we get

$$|\mathbf{m} - \mathbf{0}| = \left(\mathbf{P}^2 + \dot{\mathbf{P}}^2\right)^{1/2} \tag{1.59}$$

In this neighborhood let us analyse the behavior of the total energy

$$E_T(\mathbf{m}) = \mathcal{T}(\dot{\mathbf{P}}) + \mathcal{E}(\mathbf{P}) \tag{1.60}$$

where

$$\mathcal{T}(\dot{\mathbf{P}}) = \Sigma\, m_i\, \dot{q}_i^2 \tag{161}$$

obtained by summing the kinetic energy $\mathcal{T}(\dot{\mathbf{P}})$ and the potential energy $\mathcal{E}(\mathbf{P})$ of the system. At $\mathbf{m} = \mathbf{0}$ we have $E_T(\mathbf{0}) = 0$; also for the function $E_T(\mathbf{m})$, as well as for $\mathcal{E}(\mathbf{P})$, the origin **0** is thus a point of relative minimum.

Like the function $\mathcal{E}(\mathbf{P})$, $E_T(\mathbf{m})$ will be continuous at the origin, taking into account the form of the kinetic energy $\mathcal{T}(\dot{\mathbf{P}})$.Thus we have

$$\lim_{|\mathbf{m}|\to 0} E_T(\mathbf{m}) = 0 \tag{1.62}$$

Hence, however we take a number $\theta > 0$, it will always be possible to pick out a $\delta > 0$ such that $|\mathbf{m} - \mathbf{0}| < \delta$ yields $E_T(\mathbf{m}) < \theta$.

Let us consider now a spherical surface ∂S_ε of radius $\varepsilon < k$, i.e., the set of the states of motion **m** such that $|\mathbf{m} - \mathbf{0}| = \varepsilon < k$. This set is closed and bounded, i.e., compact in R^{2n}. According to the Weierstrass theorem, the continuous function $E_T(\mathbf{m})$ will have a minimum over ∂S_ε, i.e.,

$$\exists\, \tilde{\mathbf{m}} \in \partial S_\varepsilon : E_T(\tilde{\mathbf{m}}) = \operatorname*{Ifn}_{\partial S_\varepsilon} E_T(\mathbf{m}) = \operatorname*{Min}_{\partial S_\varepsilon} E_T(\mathbf{m}) \tag{1.63}$$

Let us take a $\theta > 0$, such that $\theta < E_T(\tilde{\mathbf{m}})$. Thus, according to assumption (1.62), there will certainly exist a spherical neighborhood S_δ of the origin such that for any state of motion **m** contained in S_δ, $E_T(\mathbf{m}) < \theta < E_T(\tilde{\mathbf{m}})$.

Thus, if the disturbance $\mathbf{m}(0)$ is such that $|\mathbf{m}(0) - \mathbf{0}| < \delta$, the total potential energy of the disturbance will be such that $E_T(\mathbf{m}(0)) < E_T(\widetilde{\mathbf{m}})$. On the other hand, the total energy is constant during the motion. At any time $t > 0$, thus, $E_T(\mathbf{m}(t)) = E_T(\mathbf{m}(0))$. Consequently, at any time $t > 0$ we will also have $E_T(\mathbf{m}(t)) < E_T(\widetilde{\mathbf{m}})$. The point representing the state of the motion $\mathbf{m}(t)$ will never be able to reach the spherical surface of radius ε where $E_T(\mathbf{m}) \geq E_T(\widetilde{\mathbf{m}})$ because at any time $E_T(\mathbf{m}(t)) < E_T(\widetilde{\mathbf{m}})$. By a suitable control of the initial disturbance magnitude we can restrict the disturbed motion of the system within an arbitrarily small neighborhood S_ε of $\mathbf{P}_0$. In conclusion, if the potential energy $\mathcal{E}(\mathbf{P})$ is continuous in a neighborhood of $\mathbf{P}_0$ and has there a relative minimum, the equilibrium of the system is Liapounov stable at $\mathbf{P}_0$.

Let us admit now that the additional potential energy is also at least twice differentiable at $\mathbf{P}_0$. By using the Taylor theorem, thus we can write

$$\mathcal{E}(\mathbf{P}) = \mathcal{E}_2(\mathbf{P}) + o(|\mathbf{P}|^2) \tag{1.64}$$

where $\mathbf{P} = \{q_1, q_2, ..., q_n\}$ is the additional displacement starting from $C_0 = \mathbf{0}$ and $\mathcal{E}_2(\mathbf{P})$ the second differential of the function $\mathcal{E}(\mathbf{P})$, i.e.,

$$\mathcal{E}_2(\mathbf{P}) = 1/2\left(\frac{\partial^2 \mathcal{E}}{\partial q_i \partial q_j}\right)_0 q_i q_j \tag{1.65}$$

and where,

$$\lim_{|\mathbf{P}|\to 0} \frac{o\left(|\mathbf{P}|^2\right)}{|\mathbf{P}|^2} = 0 \tag{1.66}$$

The existence of a neighborhood S_ρ of the origin $\mathbf{0} = \mathbf{P}_0$ such that $\forall\, \mathbf{P} \neq \mathbf{0} \in S_\rho$ we have $\mathcal{E}_2(\mathbf{P}) > 0$, yields the existence of a neighborhood S_k of $\mathbf{0}$ where $\forall\, \mathbf{P} \neq \mathbf{0} \in S_k$, $\mathcal{E}(\mathbf{P}) > 0$. The condition

$$\mathcal{E}_2(\mathbf{P}) > 0 \tag{1.67}$$

to be satisfied for any additional displacement $\mathbf{P} \in S_\rho$, together with the assumption of the second differentiability of $\mathcal{E}(\mathbf{P})$, is thus sufficient for the stability at $\mathbf{P}_0$.

From these last results we can state that all the three approaches previously examined are sufficient conditions for stability, when applied to conservative systems with finite degrees of freedom and with a potential energy at least two times differentiable at $\mathbf{P}_0$. Energy criteria were originally established by Lagrange[5] and Dirichlet[6] for discrete systems. Only later were these criteria tacitly extended to infinite-dimensional systems. For these systems the same concept of stability was not yet completely clear and the static definition of stability, corresponding to the existence of the weak minimum (1.6) of the additional potential energy — no longer coupled to the continuity of the additional potential energy — was frequently taken for granted.

1.4 FIRST PROBLEMS TO SOLVE IN EXTENDING STABILITY CRITERIA FROM DISCRETE TO CONTINUOUS ELASTIC SYSTEMS

You can now better understand which difficulties you can meet analyzing the stability of continuous systems if you tacitly extend to them the criteria of stability of discrete systems. Indeed, many assumptions or results of the Lagrange-Dirichlet theorem are not verified any more for systems with infinite degrees of freedom. For instance, condition (1.63) does not hold any more. Also if we have $E_T(\mathbf{m}) > 0$ over ∂S_ε, we can have $\inf E_T(\mathbf{m}) = 0$. The perturbed system can therefore escape from the neighborhood S_ε.

At the same time we remark that, if we apply the criterion of stability involving the second differential $\mathcal{E}_2(\mathbf{P})$ of the potential energy, the simple fulfillment of the condition $\mathcal{E}_2(\mathbf{P}) > 0$ doesn't make sense. As well as for finite degrees of freedom systems, the potential energy $\mathcal{E}(\mathbf{P})$ has to be at least two times differentiable at $\mathbf{0}$ for the condition $\mathcal{E}_2(\mathbf{P}) > 0$ entails stability. Thus, if we write

$$\mathcal{E}(\mathbf{u}) = \mathcal{E}_1(\mathbf{u}) + \mathcal{E}_2(\mathbf{u}) + o(|\mathbf{u}|^2) \tag{1.68}$$

or, taking into account the equilibrium at $\mathbf{P}_0$,

$$\mathcal{E}(\mathbf{u}) = \mathcal{E}_2(\mathbf{u}) + o(|\mathbf{u}|^2) \tag{1.68'}$$

$\mathcal{E}_2(\mathbf{u})$ has to be quadratic and continuous at $\mathbf{0}$ and

$$\lim_{|\mathbf{u}|\to 0} \frac{o\left(|\mathbf{u}|^2\right)}{|\mathbf{u}|^2} = 0 \tag{1.69}$$

In the previous example the functional $\mathcal{E}_2(\mathbf{u})$, given by (1.36), is not continuous with respect to the metric (1.37). In fact, if we verify the continuity of $\mathcal{E}(\mathbf{u})$ at $\mathbf{u} = \mathbf{0}$, $\forall z \in [0,L]$, i.e., at the rectilinear configuration of the bar, being $\mathcal{E}_2(\mathbf{0}) = 0$, the following condition has to hold

$$\lim_{\mathbf{u}\to\mathbf{0}} \mathcal{E}_2(\mathbf{u}) = 0 \tag{1.70}$$

where $\mathbf{u} \to \mathbf{0}$ in the sense of the metric (1.37). Thus if we take a sequence of displacements $\{v_n\}_{n\in N}$ converging to the rectilinear configuration of the bar in the sense of the chosen metric, i.e., such that

$$\lim_{n\to\infty} d\left(v_n, 0\right) = \lim_{n\to\infty} \max_{z\in[0,L]} |v_n(z)| = 0 \tag{1.71}$$

the limit of the sequence $\{\mathcal{E}_2(v_n)\}_{n\in N}$ has to be equal to zero. Then, if we choose the sequence (Fig. 1.21)

$$v_n = \frac{1}{n}\sin\frac{n\pi z}{L} \quad n \in N \tag{1.72}$$

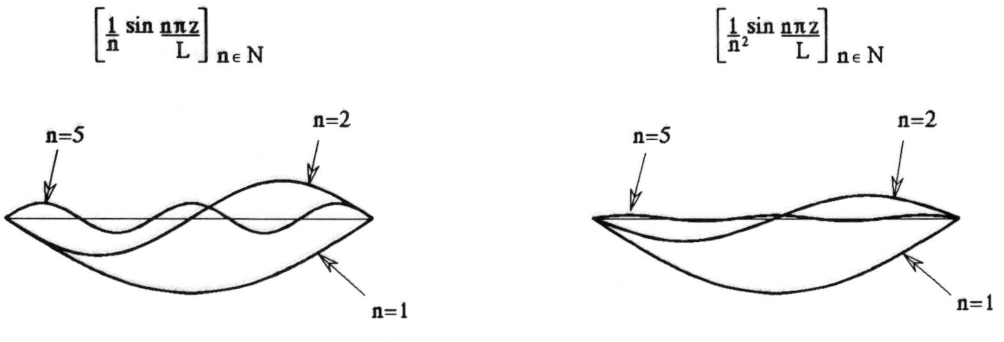

FIGURE 1.21

it will certainly converge to zero for any z in the interval [0,L] and in the sense of the chosen metric because

$$\lim_{n\to\infty} d(v_n,0) = \lim_{n\to\infty} \max_{z\in[0,L]} |v_n(z)| = 0 \tag{1.73}$$

But

$$\mathcal{E}_2(v_n) = \frac{EI}{4}\frac{n^2\pi^4}{L^4} \tag{1.74}$$

and we get

$$\lim_{v_n\to 0} \mathcal{E}_2(v_n) = +\infty \tag{1.75}$$

We can take other sequences of displacements converging to the rectilinear configuration of the bar, i.e., to the function $v(z) = 0$, $\forall\ z \in [0,L]$. If we take in fact (Fig. 1.21)

$$v_n = \frac{1}{n^2}\sin\frac{n\pi z}{L} \tag{1.76}$$

we have

$$\lim_{n\to\infty} d(v_n, 0) = 0 \tag{1.77}$$

But now

$$\mathcal{E}_2(v_n) = \frac{1}{n^4}\frac{EI}{4}\frac{n^4\pi^4}{L^4} \tag{1.78}$$

and

$$\lim_{v_n\to 0} \mathcal{E}_2(v_n) = \frac{EI}{4}\frac{\pi^4}{L^3} \tag{1.79}$$

Comparing (1.75) and (1.79) yields that $\mathcal{E}_2(v)$ is discontinuous at $v = 0$.

You can now begin to understand how many problems we have to solve in order to establish a consistent theory of stability of elastic continuous structural systems. First of all we have to define the most appropriate set of admissible displacements of the system that can be represented by a linear normed space. Here we have to define the most appropriate norm to measure distances among configurations of the system. In this space we also have to define the potential energy functional $\mathcal{E}(\mathbf{u})$. In this space, this functional should be sufficiently smooth to formulate the various criteria of stability.

The problem of the choice of the metric is thus fundamental. For systems with an infinite number of degrees of freedom, the property of equivalence of all the norms is in fact not verified. For continuous systems the stability definition is norm dependent. You immediately recognize the necessity to use some fundamental concepts of functional analysis.

REFERENCES

1. Hutchinson, J. W. and Koiter, W. T., Postbuckling theory. Appl. Mech. Rev. 23, 1970.
2. Karman, V. T. and Tsien, H. S., The buckling of thin cylindrical shells under axial compression. J. Appl. Mech. 17(1), 1950.
3. Koiter, W. T., Elastic stability and postbuckling behavior. Proc. Symp. Non-linear Problems, ed. by R. E. Langer. University of Wisconsin Press, Madison, 1963.

4. Euler, L., Methodus inveniendi lineas curvas maximi minimive proprietate gaudentes, App. De curvis elasticis. Lausanne and Geneva, 1744. English transl. Isis 20(58), 1, 1933.
5. Lagrange, J. L., Mecanique analitique. 1st Ed. Veuve Desaint, Paris, 1788.
6. Dirichlet, G., Lejeune, Über die Stabilität des Gleichgewichts. J. Reine Angew. Math. 32, 1846.
7. Knops, R. J. and Wilkes, E. W., Theory of Elastic Stability. Handbuch der Physik, Band VIa/3, Springer-Verlag, Berlin, 1973.
8. Poincaré, H., Mémoire sur les courbes définies par les equations differentielles, J. Math. Pures et Appl.(3)8, 1882.
9. Liapounov, A. M., Problème general de la stabilité du mouvement. Societé Mathématique de Kharkow, Transl. by Davaux, E., Ann. Fac. Sci. Toulouse, 2nd Ser. 9, Reprinted at Princeton: Princeton University Press (Ann. of Math.), 1949.
10. Bryan, G. H., On the stability of elastic systems. Proc. Cambridge Philos. Soc. 6, 1888.
11. Southwell, R. V., On the general theory of elastic stability. Philos. Trans. R. Soc. London, Ser. A 213, 1914.
12. Biezeno, C. B., and Hencky, H., On the general theory of elastic stability. Proc. K. Ned. Akad. Wet. 31, 1928.
13. Trefftz, E., Über die Ableitung der Stabilitätskriterien des elastischen Gleichgewichts aus der Elastizitätstheorie endlicher Deformationen. Proc. Third Int. Congr. Appl. Mech., Stockholm, 1930.
14. Marguerre, K., Über die Behandlung von Stabilitätsproblemen mit Hilfe der energetischen Methode. Z. Angew. Math. Mech. 18, 1938.
15. Kappus, R., Zur Elastizitätstheorie endlicher Verschiebungen. I und II. Z. Angew. Math. Mech. 19, 1939.
16. Timoshenko, S. P., Sur la stabilité des systemes elastiques. Ann. Ponts Chaussées, 9(15–17), 1913.
17. Krall, G., Osservazioni sui principi variazionali per la stabilità in elastostatica e loro applicazioni. Atti Accad. Naz. Lincei Mem. Cl. Sci. Fis. Mat. Nat. 8, 1966–1967.
18. Krall, G., Meccanica Tecnica delle Vibrazioni. Zanichelli Bologna 1940
19. Hadamard, J., Leçons sur la Propagation des Ondes et les Equations de l'Hydrodynamique. Paris, 1903. Reprinted in New York, Chelsea, 1949.
20. Timoshenko, S. P. and Gere, J. M., Theory of elastic stability. McGraw-Hill, New York, 1961.
21. Vol'mir, A. S., Stability of Elastic Systems. NTIS U.S. Department of Commerce, 1965.
22. Ziegler, H., Principles of Structural Stability. Blaisdell Publishing Company, Waltham, MA, 1968.
23. Gantmacher, F., Lectures in Analytical Mechanics. MIR Publishers, Moscow, 1970.
24. Bolotin V. V. Nonconservative Problems of the Theory of Elastic Stability. Pergamon Press, New York, 1963.
25. Budiansky, B., Buckling behavior of elastic structures, Adv. Appl. Mech., 14, 1974.
26. Budiansky, B., and Hutchinson J. W., Dynamic Buckling of Imperfection Sensitive Structures, Proc. XI Int. Congr. Appl. Mech., Munich, Springer-Verlag, Berlin, 1964.

Chapter 2

A REVIEW OF SOME FUNCTIONAL ANALYSIS CONCEPTS

2.1 INTRODUCTION

The theory of stability presented in this book is derived from the few simple, intuitive, mechanical considerations discussed in the previous chapter. The extension of these relations to infinite-dimensional spaces is the motivation for the mathemathics of functional analysis, which, in a sense, enables us to extend the simpler insights into the elastic stability of systems with finite degrees of freedom to continuous ones. A beam, a frame, a plate, a shell are, in fact, all examples of continuous structures for which the space of the configurations is, as a rule, infinite-dimensional.

Thus we will study the functional spaces in order to find in our stability analysis answers to questions like: "Which is the most appropriate space of the configurations to define equilibrium stability of continuous structural systems? What is the stability definition? What is the correct statement of the energy criterion of stability and of its variational formulation?"

Step by step, in order to define the space of the configurations of the continuous elastic structural systems and to analyse their energy functionals, we necessarily have to recall those essential concepts of functional analysis that have direct use in the next chapters of the book.

2.2 METRIC SPACES, NORMED SPACES, BANACH SPACES

The definition of the space of the configurations of continuous structural systems is a crucial point to formulate a consistent theory of stability. We have, in fact, first to define, in the most appropriate way, distances and angles between displacement and velocity fields of the system.

In calculus we study functions defined on the real line R. A little reflection shows that in the limit process, and so on, we use the fact that on R we have available a distance function, call it d, which associates a distance $d(x,y) = | x - y |$ with every pair of points $x,y \in R$. In functional analysis we study more general spaces and functions defined on them. We replace, the set of real numbers underlying R by an abstract set X of elements x and introduce on X a distance function, the metric, which has only a few of the most fundamental properties of the distance function on R.

A metric space is a set (X,d) with a metric d on it. The metric d associates with any pair of elements of X a distance d that is a function defined on $X \times X$ such that for all $u,v,w \in X$ we have

M_1. d is real valued and nonnegative
M_2. $d(u,v) = 0$ if and only if $u = v$
M_3. $d(u,v) = d(v,u)$
M_4. $d(u,v) \leq d(u,w) + d(w,v)$

Particularly important metric spaces are obtained if we define a metric on a vector space X. In fact, if there is no relation between the algebraic structure and the metric, we cannot obtain an useful theory that combines algebraic and metric concepts. To obtain such a relation between algebraic and geometric properties of X, we define a metric on X by means of a norm. This is an auxiliary concept that uses the algebraic operations of the vector spaces X. The resulting space is called a normed space. Here a norm on the vector space X is a real-valued function on X whose value at a $\mathbf{u} \in X$ is denoted by

$$\|u\| \tag{2.1}$$

and has the following properties, for every $\mathbf{u},\mathbf{v} \in X$, $\alpha \in R$

N_1. $\|\mathbf{u}\| \geq 0$
N_2. $\|\mathbf{u}\| = 0$ if and only if $\mathbf{u} = 0$
N_3. $\|\alpha\mathbf{u}\| = |\alpha| \, \|\mathbf{u}\|$
N_4. $\|\mathbf{u} + \mathbf{v}\| \leq \|\mathbf{u}\| + \|\mathbf{v}\|$

A norm on X defines a metric d on X, which is given by

$$d(\mathbf{u},\mathbf{v}) = \|\mathbf{u} - \mathbf{v}\| \tag{2.2}$$

and is called the metric induced by the norm. The latter generalizes the length of a vector in the plane or in the tridimensional space. Notice that inequality N_4 implies

$$\left| \, \|\mathbf{u}\| - \|\mathbf{v}\| \, \right| \leq \|\mathbf{u} - \mathbf{v}\| \tag{2.3}$$

In fact

$$\left| \, \|\mathbf{u}\| - \|\mathbf{v}\| \, \right| = \left| \, \| \mathbf{u} - \mathbf{v} + \mathbf{v} \| - \|\mathbf{v}\| \, \right| \leq \left| \, \|\mathbf{u} - \mathbf{v}\| + \|\mathbf{v}\| - \|\mathbf{v}\| \, \right| = \|\mathbf{u} - \mathbf{v}\|$$

According to inequality (2.3) we can immediately recognize that the norm is continuous, i.e., the mapping that associates $\|\mathbf{u}\|$ to u is a continuous mapping of $(X, \| \, \|)$ into R.

A sequence $\{\mathbf{u}_n\}$ in a normed space $(X, \| \, \|)$ is said to be Cauchy sequence if for every $\varepsilon > 0$ there is an $N(\varepsilon)$ such that

$$\|\mathbf{u}_n - \mathbf{u}_m\| < \varepsilon \text{ for every m, n} > N$$

The space X is said to be complete if every Cauchy sequence in X converges, that is, has a limit that is an element, say **u**, of X. In this case $\|\mathbf{u}_n - \mathbf{u}\| \to 0$ and we will write $\mathbf{u} = \lim \mathbf{u}_n$. A complete normed linear vector space is called a Banach space.

Prototypes of normed spaces are the familiar spaces of all vectors in the plane and in the tridimensional space. Particular important for the structural systems with a finite number of degrees of freedom is the Euclidean space R^n (Fig. 2.1) where the vectors have components $\xi_1, \xi_2, \dots \xi_n$. It is a Banach space with norm defined by

$$\|\mathbf{u}\| = \left(\sum_{i=1}^{n} |\xi_i|^2 \right)^{1/2} \tag{2.4}$$

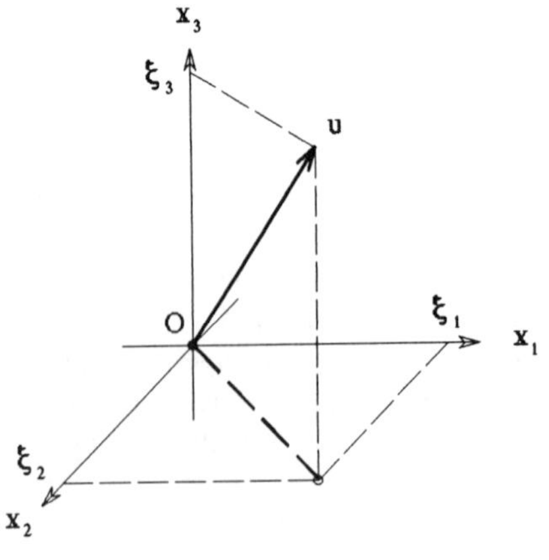

FIGURE 2.1

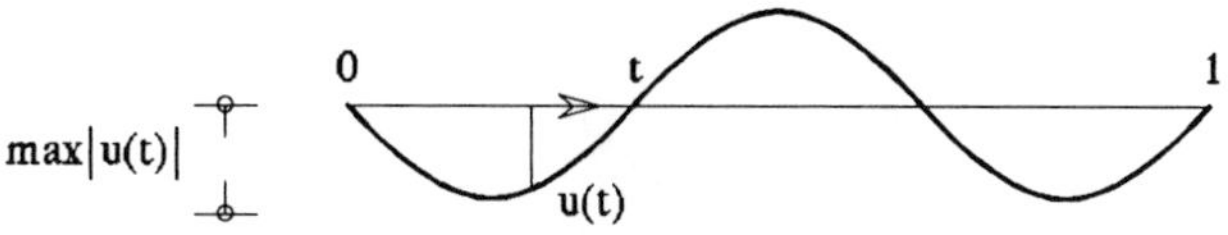

FIGURE 2.2

The space R^n is in fact complete and yields the metric

$$d(\mathbf{u},\mathbf{v}) = \|\mathbf{u}-\mathbf{v}\| = \left(\sum_{i=1}^{n} |\xi_i - \eta_i|^2\right)^{1/2} \tag{2.5}$$

for any pair of points **u**, **v** whose components in R^n are $\xi_1, \xi_2, \ldots, \xi_n; \eta_1, \eta_2, \ldots, \eta_n$

2.2.1 OTHER EXAMPLES OF NORMED SPACES

We will analyze some examples of infinite-dimensional spaces that could be useful to define the configurations of continuous structural systems, for example, beams, as depicted in Fig. 2.2.

Example 1. The normed linear space C[a,b] consists of continuous functions u(t) on the real interval [a,b] together with the norm

$$\|u\| = \max_{a \le t \le b} |u(t)| \tag{2.6}$$

called the uniform norm. The null vector **0** is the function identically zero on [a,b]. We can verify that norm (2.6) satisfies the four required axioms. Obviously, $\|u\| \ge 0$ and is zero only for the function that is identically zero. The triangle inequality follows from the relation

$$\max |u(t) + v(t)| \le \max [|u(t)| + |v(t)|] \le \max |u(t)| + \max |v(t)|$$

Finally, the last axiom follows from the relation

$$\max |\alpha\, u(t)| = \max |\alpha|\, |u(t)| = |\alpha| \max |u(t)|$$

This space with the uniform norm is complete and therefore is a Banach space.

Example 2. The normed linear space D[a,b] consists of all functions u(t) on the interval [a,b] that are continuous and have continuous derivatives on [a,b]. The norm on the space D[a,b] is defined as

$$\|u\| = \max_{a \le t \le b} |u(t)| + \max_{a \le t \le b} |\dot{u}(t)|$$

Also this space is complete.

Example 3. The space of continuous functions u(t) on the interval [a,b] becomes a normed space with the integral norm defined as

$$\|u\| = \int_a^b |u(t)|\, dt \tag{2.7}$$

called the L_1 norm. This space is, of course, different from the normed space C[a,b] and is incomplete. Define a sequence of elements in X by the equation ($n \ge 2$):

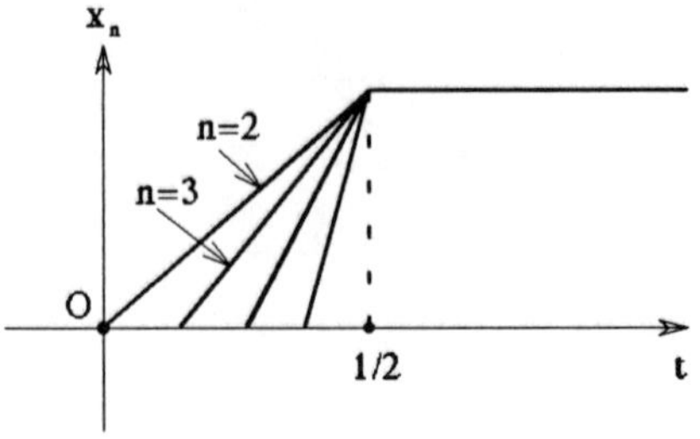

FIGURE 2.3

$$x_n(t) = 0 \qquad \text{for } 0 \le t \le \frac{1}{2} - \frac{1}{n}$$

$$x_n(t) = nt - \frac{n}{2} + 1 \qquad \text{for } \frac{1}{2} - \frac{1}{n} \le t \le \frac{1}{2}$$

$$x_n(t) = 1 \qquad \text{for } t \ge \frac{1}{2}$$

This sequence of functions is illustrated in Fig. 2.3. Each member of the sequence is a continuous function and thus a member of the space X. The sequence is Cauchy, since, as is easily verified, $\|x_n - x_m\| = \frac{1}{2}\left[\frac{1}{n} - \frac{1}{m}\right] \to 0$.

It is obvious, however, that there is no continuous function to which the sequence converges. It is interesting to reconcile this result with the completeness of C[a,b]: the sequence considered here is not Cauchy with respect to the uniform norm of C[a,b].

Example 4. The space $L_2[a,b]$ of the square integrable functions u(t) on [a,b], in the Lebesgue sense, is a normed space with norm

$$\|u\|_{L_2} = \left(\int_a^b u^2(t)\,dt\right)^{1/2} \tag{2.7'}$$

2.3 FINITE AND INFINITE-DIMENSIONAL NORMED SPACES: EQUIVALENCE AND NONEQUIVALENCE OF NORMS

The fact that the finite-dimensional normed spaces are simpler than the infinite-dimensional ones has many implications in the theory of stability. In this context we want first to recall the following general lemma, which is a source for very important results. It substantially states that in finite-dimensional spaces we cannot find a combination of linearly independent vectors that involves large scalars but represents a small vector.

Let $\{\mathbf{u}_1, \mathbf{u}_2, ..., \mathbf{u}_n\}$ e a linearly independent set of vectors in a normed space X of dimension n. Then there is a number c > 0 such that for every choice of scalars $\alpha_1, \alpha_2, \ldots, \alpha_n$ we have

$$\|\alpha_1 \mathbf{u}_1 + \alpha_2 \mathbf{u}_2 + \ldots + \alpha_n \mathbf{u}_n\| \ge c(|\alpha_1| + |\alpha_2| + \ldots + |\alpha_n|) \tag{2.8}$$

We can write $s = |\alpha_1| + |\alpha_2| + \cdots + |\alpha_n|$; if s = 0, all α_j are zero, so that the previous inequality holds for any c. Then (2.8) is equivalent to the inequality that we obtain by dividing (2.8) by s and writing $\beta_j = \alpha_j/s$, that is,

$$\|\beta_1 \mathbf{u}_1 + \beta_2 \mathbf{u}_2 + \cdots + \beta_n \mathbf{u}_n\| \ge c \quad (\Sigma\, |\beta_j| = 1) \tag{2.8'}$$

Hence it suffices to prove the existence of a c > 0 such that (2.8′) holds for every n-tuple of scalars $\beta_1, \beta_2, ..., \beta_n$ with $\Sigma|\beta_j| = 1$.

Suppose that this is false. Thus there exists a sequence $\{\mathbf{y}_m\}$ of vectors $\mathbf{y}_m = \beta_1^{(m)}\mathbf{u}_1 + \beta_2^{(m)}\mathbf{u}_2 + \cdots + \beta_n^{(m)}\mathbf{u}_n$ with $\Sigma|\beta_j^{(m)}| = 1$ such that $\|\mathbf{y}_m\| \to 0$ as $m \to \infty$.

We reason as follows: since $\Sigma|\beta_j^{(m)}| = 1$, we have $|\beta_j^{(m)}| \le 1$. Hence for each fixed j the sequence $\{\beta_j^{(m)}\} = \{\beta_j^{(1)}, \beta_j^{(2)},\ldots\}$ is bounded. Consequently, by the Bolzano-Weierstrass theorem, $\{\beta_1^{(m)}\}$ has a convergent subsequence. Let β_1^* denote the limit of that subsequence, and let us discard the $\mathbf{y}_m$ not in this subsequence. Let $\{\mathbf{y}_{1,m}\}$ be the new subsequence. By the same argument $\{\mathbf{y}_{1,m}\}$ has a subsequence $\{\mathbf{y}_{2,m}\}$ for which the corresponding sequence of scalars $\{\beta_2^{(m)}\}$ converges; let β_2^* denote the limit. Continuing in this way, after n steps we obtain a subsequence $\{\mathbf{y}_m^0\}$ of $\{\mathbf{y}_m\}$ whose terms are of the form

$$\mathbf{y}_m^0 = \sum_{j=1}^{n} \gamma_j^{(m)}\,\mathbf{u}_j \qquad \text{with}\left(\sum_{j=1}^{n}\left|\gamma_j^{(m)}\right| = 1\right)$$

with scalars $\gamma_j^{(m)}$ satisfying $\gamma_j^{(m)} \to \beta_j^*$ as $m \to \infty$. Hence, as $m \to \infty$, $\mathbf{y}_m^0 \to \mathbf{y} = \Sigma\,\beta_j^*\mathbf{u}_j$ where $\Sigma\,|\beta_j^*| = 1$, so that not all β_j^* are zero. Since $\{\mathbf{u}_1, \mathbf{u}_2,\ldots,\mathbf{u}_n\}$ is a linearly independent set, we thus have $\mathbf{y} \ne \mathbf{0}$. On the other hand, $\mathbf{y}_m^0 \to \mathbf{y}$ implies $\|\mathbf{y}_m^0\| \to \|\mathbf{y}\|$, by the continuity of the norm. Since $\|\mathbf{y}_m\| \to 0$ by assumption and $\{\mathbf{y}_m^0\}$ is a subsequence of $\{\mathbf{y}_m\}$, we must have $\|\mathbf{y}_m^0\| \to 0$. Hence $\|\mathbf{y}\| = 0$, so $\mathbf{y} = \mathbf{0}$ by N_2. This contradicts $\mathbf{y} \ne \mathbf{0}$ and the lemma is proved.

As consequence of this lemma it is possible to prove that every finite-dimensional normed space is complete and that all the norms on a finite dimensional vector space X are equivalent.

We recall that a norm $\|.\|_1$ on a vector space X is said to be equivalent to a norm $\|.\|_2$ on X if there are positive numbers α and β such that for all $\mathbf{u} \in X$ we have

$$\alpha\,\|\mathbf{u}\|_2 \le \|\mathbf{u}\|_1 \le \beta\,\|\mathbf{u}\|_2 \tag{2.9}$$

To prove this statement let dim X = n and $\{\mathbf{e}_1,\mathbf{e}_2,\ldots,\mathbf{e}_n\}$ be any basis for X. Then every $\mathbf{u} \in X$ has an unique representation

$$\mathbf{u} = \alpha_1\mathbf{e}_1 + \alpha_2\mathbf{e}_2 + \cdots + \alpha_n\mathbf{e}_n$$

Let us evaluate the norm $\|\mathbf{u}\|_1$ of the vector $\mathbf{u}$. By the previous lemma there is a positive constant c such that

$$\|\mathbf{u}\|_1 \ge c(|\alpha_1| + |\alpha_2| + \cdots + |\alpha_n|)$$

On the other hand the triangle inequality gives

$$\|\mathbf{u}\|_2 \le \sum_{j=1}^{n} |\alpha_j|\,\|e_j\|_2 \le k\sum_{j=1}^{n}|\alpha_j| \qquad k = \max\|e_j\|_2$$

Hence, a $\|\mathbf{u}\|_2 \le \|\mathbf{u}\|_1$, where $a = c/k > 0$. The other inequality in (2.9) is now obtained by an interchange of the roles of $\|\mathbf{u}\|_1$ and $\|\mathbf{u}\|_2$ in the preceding argument.

This theorem is no more valid in infinite-dimensional space. We can recognize the consequence of this result in the stability definition passing from systems with finite degrees of freedom to continuous ones.

2.4 DENSE SUBSETS: TOTAL BOUNDED AND COMPACT SPACES

An important theorem due to Weierstrass states that if f(t) is a continuous function on the closed interval [0,1], then for every number $\varepsilon > 0$ there exists a polynomial function p(t) such that $|f(t) - p(t)| < \varepsilon$ for all $t \in [0,1]$. The theorem means that we can approximate a continuous function in [0,1] arbitrarily closely in modulus by a polynomial.

The polynomials p(t) are a subspace of the metric space of continuous functions with uniform metric. We can see that there is a subset of functions — the polynomials — such that the distance between an arbitrary element of the space and an element of the subset can be made as small as we please. This idea can be extended to a general metric space by the following definition:

A subset M of the metric space (S,d) is said to be *dense* in (S,d) if, for every $x \in (S,d)$ and $\varepsilon > 0$, there is a $y \in M$ such that $d(x,y) < \varepsilon$. Let us take an arbitrary element x of the metric space (S,d). A sequence $\{u_n\}$ of elements of the dense set M exists such that $\lim d(u_n,x) = 0$ or $\lim u_n = x$. With this definition we can assert that the set of all polynomials on [0,1] is dense in the space of continuous functions with the uniform metric. We can also say that the rational numbers are a dense subset in the real line R with metric (2.5), with n = 1.

The rational numbers are an incomplete metric space, but if we add the limits of all Cauchy sequences of rationals we obtain the complete metric space of the reals. This suggests that any incomplete metric space can be considered to be a dense subset of a complete metric space. In fact any incomplete metric space (S,d) can be embedded, as a dense subset, in a complete metric space. The latter is called a completion of (S,d).

The rational numbers are countable, that is, they can be put in one-to-one correspondence with the positive integers; the real numbers, on the contrary, are uncountable. Any irrational can be approximated by a rational number; likewise, any element of an uncountable set can be approximated by an element of a countable set. A generalization of this idea brings us to the definition of separability of a space: a metric space is said to be *separable* if it contains a dense countable set.

A separable space may contain more than one set that is dense and countable. A separable space is not too large; i.e., we can reach every element of the space through a sequence of elements of a countable set. We can go further and we can ask whether there is a finite set of points in a metric space such that every point of the space is "near" a point of the finite set. This is possible if the metric space is totally bounded.

Let A be a subset of a metric space X and let $\varepsilon > 0$. A finite set of points $N = \{e_1, e_2, \ldots, e_m\}$ is called an ε-net for A if for every point $p \in A$ there exists an $e_i \in N$ with $d(p,e_i) < \varepsilon$. Let, for example, $A\{(x,y) : x^2 + y^2 < 4\}$, i.e., A is the open disk centered at the origin and of radius 2. If $\varepsilon = 3/2$, then the set

$$N = \{(1,-1),(1,0),(1,1),(0,-1),(0,0),(0,1),(-1,-1),(-1,0),(-1,1)\}$$

is an ε-net for A. On the other hand, if $\varepsilon = 1/2$, then N is not an ε-net for A. For example, p = (1/2,1/2) belongs to A but the distance between p and any point in N is greater than 1/2 (Fig. 2.4). Another example: the set $x_n = a + n(b - a) / N$, $n = 1,2,\ldots,N-1$, is an ε-net for the

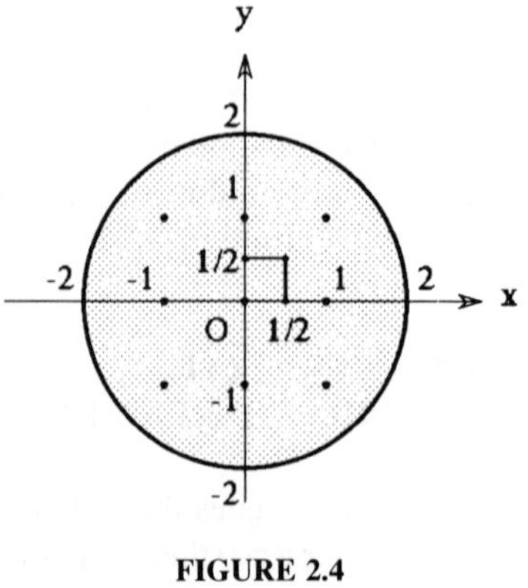

FIGURE 2.4

following definition: a subset A of a metric space (S,d) is said to be totally bounded if, for every $\varepsilon > 0$, A contains a finite set called A_ε-net such that for each $x \in (S,d)$ there is a $y \in A_\varepsilon$ such that $d(x,y) < \varepsilon$.We notice that the definition specifies "every $\varepsilon > 0$": making ε smaller will enlarge the finite ε-net but it still will exist. It is possible to show that a totally bounded metric space is separable. We recall few other basic properties that are related to the concept of compactness. Compactness can be defined in several ways. We recall the usual definition: a metric space is said to be *compact* if every sequence in X has a convergent subsequence.

A compact set is totally bounded. Also interesting in this context are the connections between compactness, closedness, and boundedness. A compact set M of a metric space is closed and bounded. The converse is in general false, but not in finite-dimensional spaces: in these spaces, in fact, the compact sets are precisely closed and bounded.

2.5 OPERATORS

A mapping from a normed space X into a normed space Y is called an *operator*. A mapping from X into the scalar field R is called a *functional*. Of particular importance are the bounded linear operators and bounded linear functionals, since they are continuous and take advantage of the vector space structure. In fact, a fundamental theorem states that a linear operator is continuous if and only if it is bounded. In analysis, infinite-dimensional normed spaces are more important than finite-dimensional ones. The latter are simpler, and linear operators or functionals on them are always bounded and can be represented by matrices.

2.5.1 LINEAR BOUNDED TRANSFORMATIONS ON NORMED VECTOR SPACES

An operator from a vector space X into a vector space Y is said to be linear if for every $\mathbf{x}_1,\mathbf{x}_2 \in X$ and all scalars α_1,α_2 we have

$$T(\alpha_1\mathbf{x}_1 + \alpha_2\mathbf{x}_2) = \alpha_1 \; T(\mathbf{x}_1) + \alpha_2(\mathbf{x}_2) \;\; \forall\, \alpha_1, \alpha_2 \in R \;\; \forall\, \mathbf{x}_1, \mathbf{x}_2 \in X \tag{2.10}$$

The linear operator T is said bounded if there is a real number c such that for all $\mathbf{x} \in D(T) = X$:

$$\|T\mathbf{x}\| \le c\, \|\mathbf{x}\| \tag{2.11}$$

where the norm on the right is that on the space X and the norm on the left is that on the space Y. Formula (2.11) shows that a bounded linear operator maps bounded sets in D(T) onto bounded sets in Y. This motivates the term "bounded operator".

We can thus recognize the present use of the word "bounded" that is different from that used in calculus, where a bounded function is one whose range is a bounded set. Which is the smallest possible c such that (2.11) still holds for all nonzero $\mathbf{x} \in D(T)$? By division,

$$\frac{\|T\mathbf{x}\|}{\|\mathbf{k}\|} \le c \tag{2.11$'$}$$

and this shows that c must be at least as large as the supremum of the expression on the left taken over $D(T) - \{\mathbf{0}\}$. This quantity is denoted by $\|T\|$; Thus

$$\|T\| = \sup_{D(T)-(\mathbf{0})} \frac{\|T\mathbf{x}\|}{\|\mathbf{x}\|} \tag{2.12}$$

$\|T\|$ is called the norm of the operator T. Finally we note that with $c = \|T\|$ from (2.10) we get

$$\|T\mathbf{x}\| \le \|T\|\ \|\mathbf{x}\| \tag{2.13}$$

An equivalent definition of the norm of the linear operator T is

$$\|T\| = \sup_{\mathbf{x}\in D(T),\|\mathbf{x}\|=1} \|T\mathbf{x}\| \tag{2.13'}$$

Let T: X → Y be a linear transformation that is continuous at the origin; then we can show that T is continuous everywhere in X. In fact we have, by hypothesis, that for any $\varepsilon > 0$ there does exist a $\delta > 0$ such that if $\|\mathbf{x}\| < \delta$, $\|T\mathbf{x}\| < \varepsilon$. But assuming $\mathbf{x} = \mathbf{x}' - \mathbf{x}_0$ gives $\|T(\mathbf{x}' - \mathbf{x}_0)\| = \|T\mathbf{x}' - T\mathbf{x}_0\| < \varepsilon$, if $\|\mathbf{x}' - \mathbf{x}_0\| < \delta$, which means that T is continuous at $\mathbf{x}_0$, an arbitrary point of X. We now connect continuity with boundedness. A necessary and sufficient condition that a linear operator T:X → Y be continuous is that it is bounded.

Assume that T is bounded; then

$$\|T\mathbf{x} - T\mathbf{x}_0\| = \|T(\mathbf{x} - \mathbf{x}_0)\| \le c\ \|\mathbf{x} - \mathbf{x}_0\|$$

hence $\|\mathbf{x} - \mathbf{x}_0\| < \delta$ implies $\|T\mathbf{x} - T\mathbf{x}_0\| \le c\delta$ and T is continuous. Conversely, assume that T is continuous at an arbitrary $\mathbf{x}_0 \in D(T)$. Then, given any $\varepsilon > 0$, there is a $\delta > 0$ such that

$$\|T\mathbf{x} - T\mathbf{x}_0\| \le \varepsilon \tag{2.14}$$

for all $\mathbf{x} \in D(T)$ satisfying $\|\mathbf{x} - \mathbf{x}_0\| \le \delta$. Let us take any $\mathbf{y} \neq \mathbf{0}$ in D(T) and set $x = \mathbf{x}_0 + (\delta/\|\mathbf{y}\|)\ \mathbf{y}$. Then $\mathbf{x} - \mathbf{x}_0 = (\delta/\|\mathbf{y}\|)\ \mathbf{y}$. Hence $\|\mathbf{x} - \mathbf{x}_0\| = \delta$, so we may use (2.14). Since T is linear we have

$$\|T\mathbf{x} - T\mathbf{x}_0\| = \|T(\mathbf{x} - \mathbf{x}_0)\| = \left\|T\left(\frac{\delta}{\|\mathbf{y}\|}\mathbf{y}\right)\right\| = \frac{\delta}{\|\mathbf{y}\|}\|T\mathbf{y}\| \tag{2.15}$$

and (2.14) implies $(\delta/\|\mathbf{y}\|)\ \|T\mathbf{y}\| \le \varepsilon$. Thus $\|T\mathbf{y}\| \le (\varepsilon/\delta)\,\|\mathbf{y}\|$. This can be written $\|T\mathbf{y}\| \le c\|\mathbf{y}\|$, where $c = \varepsilon/\delta$, and shows that T is bounded.

2.5.2 INVERSE TRANSFORMATIONS

If a linear transformation is one-to-one, D(T) →R(T), then it possesses a linear inverse T^{-1}: R(T) →D(T). Of course the requirement that T be one-to-one is equivalent to the requirement that the null space of T, N(T) = {**0**}. We can now determine whether T^{-1} is continuous.

Let T: X → Y be a linear transformation. If there exists a constant $m > 0$ such that

$$\|T\mathbf{x}\| \ge m\ \|\mathbf{x}\|\ \forall\, \mathbf{x} \in X \tag{2.16}$$

then T has a continuous inverse T^{-1} : R(T) →X and

$$\|T^{-1}\mathbf{y}\| \le 1/m\ \|\mathbf{y}\| \quad \forall\ \mathbf{y} \in R(T) \tag{2.17}$$

Furthermore,

$$\|T^{-1}\| \le 1/m \tag{2.18}$$

Vice-versa, if T has a continuous inverse T^{-1}, condition (2.16) is satisfied.

Suppose in fact m exists. $T\mathbf{x} = \mathbf{0}$ uniquely implies $\mathbf{x} = \mathbf{0}$, since otherwise the inequality (2.16) would not be satisfied: hence N(T) = {**0**}, and T^{-1} exists. Now if $T\mathbf{x} = \mathbf{y}$, $\mathbf{x} = T^{-1}\mathbf{y}$ and $\|\mathbf{x}\| = \|T^{-1}\mathbf{y}\| \le 1/m\ \|T\mathbf{x}\| = 1/m\ \|\mathbf{y}\|$, and T^{-1} is bounded and therefore continuous, then clearly

$$\left\|T^{-1}\right\| = \sup_{\|\mathbf{y}\| = 1} \left\|T^{-1}\mathbf{y}\right\| \le \frac{1}{m} \tag{2.19}$$

Conversely, if T^{-1} exists and is continuous

$$\|\mathbf{x}\| = \|T^{-1}\,\mathbf{y}\| \le \|T^{-1}\|\,\|\mathbf{y}\| = \|T^{-1}\|\,\|T\mathbf{x}\| \tag{2.20}$$

and we can take $m = \dfrac{1}{\left\|T^{-1}\right\|}$.

2.5.3 COMPACT OPERATORS

We remember that for a compact space it is possible to find a finite subset of points such that every point of the space is near a point of the finite subset. In this sense a compact space is not much larger than a finite-dimensional space. We say that a linear transformation $T: X \to Y$ over the Banach spaces X,Y is said to be compact (or completely continuous) if it maps every bounded set in X into a compact set in Y. The compact operators have the effect of "smoothing" sequences. Let $\{\mathbf{x}^{(n)}\}$ be a bounded sequence in X, then $\{T\mathbf{x}^{(n)}\}$ is a bounded sequence in a compact set of Y and contains a convergent subsequence. Hence, a compact transformation takes a nonconvergent sequence and smoothes it sufficiently to make it converge. In a certain sense compact operators have properties analogous to those of matrix operators.

2.5.4 LINEAR FUNCTIONALS

A functional is an operator whose range lies on the real line R. We recall that a functional $\mathcal{F}$ on a vector space X is linear if for any two vectors $\mathbf{u}$ and $\mathbf{v} \in X$ and any two scalars α, β there holds

$$\mathcal{F}(\alpha\mathbf{u} + \beta\mathbf{v}) = \alpha\mathcal{F}(\mathbf{u}) + \beta\mathcal{F}(\mathbf{v}) \tag{2.21}$$

We are primarily concerned with continuous linear functionals. There is the peculiar property that if a functional on a normed space X is continuous at a single point, it is continuous throughout X. We need the following two definitions because most of the functionals to be considered will be linear and bounded.

For a bounded linear functional $\mathcal{F}$ there exists a real number c such that for all $\mathbf{u}$ in the domain $D(\mathcal{F}) = X$

$$|\mathcal{F}(\mathbf{u})| \le c\,\|\mathbf{u}\| \tag{2.22}$$

According to the previous definition of the norm of an operator we have

$$\|\mathcal{F}\| = \sup_{\mathbf{u} \in D(\mathcal{F}), \mathbf{u} \ne \mathbf{0}} \frac{|\mathcal{F}(\mathbf{u})|}{\|\mathbf{u}\|} \tag{2.23}$$

$\|\mathcal{F}\|$ is called the norm of the functional $\mathcal{F}$. An alternative formula for the norm of $\mathcal{F}$ is

$$\|\mathcal{F}\| = \sup_{\mathbf{u} \in D(\mathcal{F}), \|\mathbf{u}\| = 1} |\mathcal{F}(\mathbf{u})| \tag{2.24}$$

The boundedness is equivalent to the continuity. In fact a linear functional $\mathcal{F}$ with domain $D(\mathcal{F})$ in a normed space is continuous if and only if is bounded.

2.6 INNER PRODUCT SPACES, HILBERT SPACES

The concept of orthogonality between displacement fields is very useful to define the space of the configurations of a continuous structural system. It is therefore necessary to introduce inner product spaces and then Hilbert spaces.

An inner product space or a pre-Hilbert space is a vector space X with an inner product defined on X. A Hilbert space is a complete inner product space (complete in the metric defined by the inner product). Here an inner product on X is a mapping of X × X into R; that is, with every pair of vectors **u** and **v** there is an associated scalar, which is written

$$< \mathbf{u}, \mathbf{v} > \tag{2.25}$$

and is called the inner product of **u** and **v**, such that for all vectors **u**, **v**, and **w** and scalar α we have

H_1. $< \mathbf{u} + \mathbf{v}, \mathbf{w} > = < \mathbf{u}, \mathbf{w} > + < \mathbf{v}, \mathbf{w} >$
H_2. $< \alpha\mathbf{u}, \mathbf{w} > = \alpha < \mathbf{u}, \mathbf{w} >$
H_3. $< \mathbf{u}, \mathbf{v} > = < \mathbf{v}, \mathbf{u} >$
H_4. $< \mathbf{u}, \mathbf{u} > \geq 0$
H_5. $< \mathbf{u}, \mathbf{u} > = 0 \Leftrightarrow \mathbf{u} = 0$

An inner product on X defines a norm on X, given by

$$\|\mathbf{u}\| = \sqrt{\mathbf{u}, \mathbf{u}} \tag{2.26}$$

and a metric on X given by

$$d(\mathbf{u}, \mathbf{v}) = \|\mathbf{u} - \mathbf{v}\| = \sqrt{(\mathbf{u} - \mathbf{v}, \mathbf{u} - \mathbf{v})} \tag{2.27}$$

A norm on an inner product space satisfies the important parallelogram equality

$$\|\mathbf{u} + \mathbf{v}\|^2 + \|\mathbf{u} - \mathbf{v}\|^2 = 2\,(\|\mathbf{u}\|^2 + \|\mathbf{v}\|^2) \tag{2.28}$$

which represents a generalization of a result for parallelograms in two-dimensional geometry: the sum of the squares of the lengths of the diagonals of a parallelogram is equal to twice the sum of the squares of two adjacent sides (Fig. 2.5). If a norm does not satisfy (2.28) it cannot be obtained from an inner product by the use of (2.26). Hence, Hilbert spaces are Banach spaces, but not all normed spaces are inner product spaces.

Hilbert spaces, equipped with their inner products, possess the key property of orthogonality, which is not available in normed spaces. An element **u** of an inner product space X is said to be orthogonal to an element **v** ∈ X if

$$< \mathbf{u} , \mathbf{v} > = 0 \tag{2.29}$$

We say that **u** and **v** are orthogonal and we write $\mathbf{u} \perp \mathbf{v}$. Similarly, for subsets A and B ⊂ X we write $\mathbf{u} \perp A$ if $\mathbf{u} \perp \mathbf{a}$ for all $\mathbf{a} \in A$, and $A \perp B$ if $\mathbf{a} \perp \mathbf{b}$ for all $\mathbf{a} \in A$ and all $\mathbf{b} \in B$.

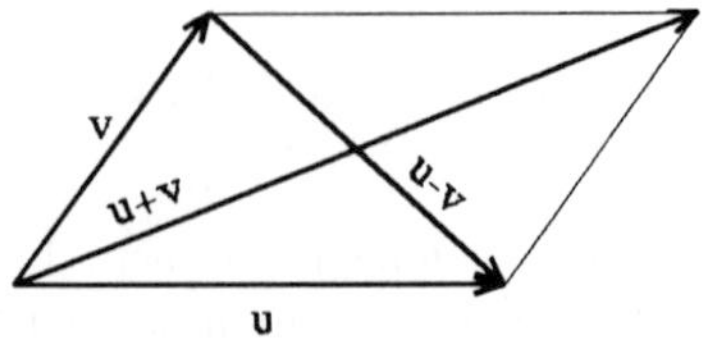

FIGURE 2.5

2.6.1 EXAMPLES OF INNER PRODUCT AND HILBERT SPACES

Example 1. The euclidean space R^n.

The space R^n is a Hilbert space with inner product defined by

$$< \mathbf{u} , \mathbf{v} > = \xi_1 \eta_1 + \xi_2\eta_2 + \cdots + \xi_n\eta_n \tag{2.30}$$

where $\mathbf{u} = (\xi_i) = (\xi_1, \xi_2,\dots,\xi_n)$, $\mathbf{v} = (\eta_i) = (\eta_1,\eta_2,\dots,\eta_n)$. In fact from (2.30) we obtain

$$\|\mathbf{u}\| = \sqrt{\mathbf{u}, \mathbf{u}} = \left(\sum_{i=1}^{n} |\xi_i|^2 \right)^{1/2} \tag{2.31}$$

The euclidean space R^n is also complete.

Example 2. The space $L_2[a,b]$ of the square integrable functions u(t) on [a,b] is a pre-Hilbert space with inner product

$$< u, v > = \int_a^b u(t)\, dt \tag{2.32}$$

where the integral is taken in the Lebesgue sense. The corresponding norm is given by (2.9).

Example 3. The space C[a,b] is not an inner product space, hence is not a Hilbert space. We show that the norm defined by

$$\|u\| = \max_{t\in I} |u(t)| \quad I = [a, b]$$

cannot be obtained from an inner product because this norm does not satisfy the parallelogram equality (2.13). Indeed, if we take

$$u(t) = 1 \quad v(t) = \frac{t-a}{b-a}$$

we have $||u|| = 1$, $||v|| = 1$ and

$$u(t) + v(t) = 1 + \frac{t-a}{b-a} \quad u(t) - v(t) = 1 - \frac{t-a}{b-a}$$

Hence $||u + v|| = 2$, $||u - v|| = 1$, and $||u + v||^2 + ||u - v||^2 = 5$, but $2(||u||^2 + ||v||^2) = 4$.

Example 4. Let Γ be an open region in R^2 and let $\mathbf{C}^{(1)}(\Gamma)$ be the space of real-valued functions having continuous first partial derivatives in Γ. Take

$$< u, v > = \int_\Gamma \left(uv + \frac{\partial u}{\partial t_1}\frac{\partial v}{\partial t_1} + \frac{\partial u}{\partial t_2}\frac{\partial v}{\partial t_2} \right) d\Gamma \tag{2.33}$$

$$\|u\|^2 = \int_\Gamma \left[u^2 + \left(\frac{\partial u}{\partial t_1} \right)^2 + \left(\frac{\partial u}{\partial t_2} \right)^2 \right] d\Gamma \tag{2.34}$$

This space is an inner product space.

Example 5. Let Γ be an open region in R^n and let $C^{(n)}(\Gamma)$ be the space of real-valued functions having in Γ continuous partial derivatives in all variables up to and including order n. Let k = ($k_1,k_2,...,k_n$) be a vector with integer entries, and let

$$|k| = \sum_{i=1}^{n} k_i \tag{2.35}$$

We define the differential operator

$$D^k u = \frac{\partial^{|k|} u}{\partial_{t_1}^{k_1} \partial_{t_2}^{k_2} \cdots \partial_{t_n}^{k_n}} \tag{2.36}$$

and take

$$< u, v >^{(m)} = \int_\Gamma \sum_{|k| \le m} D^k u \; D^k v \; d\Gamma \tag{2.37}$$

$$\|u\|^{(m)} = \left\{ \int_\Gamma \sum_{|k| \le m} \left(D^k u\right)^2 \; d\Gamma \right\}^{1/2} \tag{2.38}$$

This inner product space will be considered in the definition of Sobolev spaces.

2.6.2 FURTHER PROPERTIES OF INNER PRODUCT SPACES

We recall the Schwarz inequality, the triangle inequality, and some other important properties of the inner product spaces. An inner product and the corrresponding norm satisfy the Schwarz inequality and the triangle inequality as follows:

$$|<\mathbf{u},\mathbf{v}>| \le \|\mathbf{u}\| \, \|\mathbf{v}\| \tag{2.39}$$

where the equality sign holds if and only if $\{\mathbf{u},\mathbf{v}\}$ is a linearly dependent set. The norm also satisfies the triangle inequality

$$\|\mathbf{u} + \mathbf{v}\| \le \|\mathbf{u}\| + \|\mathbf{v}\| \tag{2.40}$$

where the equality sign holds if and only if $\mathbf{v} = 0$ or $\mathbf{u} = c\mathbf{v}$ $(c > 0)$, a linearly dependent set.

2.6.3 CONTINUITY OF INNER PRODUCT

If in an inner product space $\mathbf{u}_n \to \mathbf{u}$ and $\mathbf{v}_n \to \mathbf{v}$, then $<\mathbf{u}_n,\mathbf{v}_n> \to <\mathbf{u},\mathbf{v}>$. We obtain in fact

$$|<\mathbf{u}_n,\mathbf{v}_n> - <\mathbf{u},\mathbf{v}>| = |<\mathbf{u}_n,\mathbf{v}_n> - <\mathbf{u}_n,\mathbf{v}> + <\mathbf{u}_n,\mathbf{v}> - <\mathbf{u},\mathbf{v}>|$$

$$\le |<\mathbf{u}_n,\mathbf{v}_n - \mathbf{v}>| + |<\mathbf{u}_n - \mathbf{u}, \mathbf{v}>| \le \|\mathbf{u}_n\| \, \|\mathbf{v}_n - \mathbf{v}\| + \|\mathbf{u}_n - \mathbf{u}\| \, \|\mathbf{v}\| \to 0 \tag{2.41}$$

since $\mathbf{v}_n - \mathbf{v} \to 0$ and $\mathbf{v}_n - \mathbf{v} \to 0$ as $n \to \infty$. As a consequence of this result, it is easy to prove that every inner product space can be completed.

2.6.4 TOTAL ORTHONORMAL SETS AND SEQUENCES

First formulate the fundamental lemma of variational calculus: an element orthogonal to all the elements of a dense set is equal to zero. Let M be a dense set in a Hilbert space H and let

$$<\boldsymbol{\phi},\mathbf{u}> = 0 \qquad \forall\ \mathbf{u} \in \mathrm{M},\ \boldsymbol{\phi} \in \mathrm{H} \tag{2.42}$$

We take an arbitrary element $\mathbf{v} \in$ H. By the definition of a dense set, there exists a sequence $\{\mathbf{u}_n\}$ of elements of the set M such that lim $\mathbf{u}_n = v$ or $\|\mathbf{u}_n - \mathbf{v}\| \to 0$. But

$$<\boldsymbol{\phi},\mathbf{u}_n> = 0 \qquad \forall\ \mathbf{u}_n \in \{\mathbf{u}_n\} \tag{2.43}$$

On the other hand, because of the continuity of the scalar product

$$\lim_{n\to\infty} <\boldsymbol{\phi}, \mathbf{u}_n> = <\boldsymbol{\phi},\mathbf{v}> = 0 \tag{2.44}$$

Thus $\boldsymbol{\phi}$ is orthogonal to any element of the Hilbert space H and, in particular, to itself. This gives $<\boldsymbol{\phi},\boldsymbol{\phi}> = 0$ and, by axiom H_5, $\boldsymbol{\phi} = 0$.

An orthonormal set in an inner product space is a subset with elements of unitary norm and which is mutually orthonormal. The truly interesting orthonormal sets in inner product spaces and Hilbert spaces are those that consist of sufficiently many elements that every element in space can be represented or sufficiently accurately approximated by the use of these orthonormal sets.

In finite-dimensional spaces the situation is simple: all we need is an orthonormal set of n elements. The question is, what can be done to take care of infinite-dimensional spaces. Relevant concepts follow.

A total orthonormal set in a inner product space X is a subset $M \subset X$ whose span is dense in X. In any Hilbert space a total orthonormal set exists.

The Hilbert spaces are much like the Euclidean spaces. The separable Hilbert spaces are even closer to Euclidean spaces. The separability ensures that the infinite-dimensional Hilbert space is not too "large". A Hilbert space H is in fact said to be separable if there exists a countable set of elements $\{\boldsymbol{\phi}_n\}$ in H whose finite linear combinations are dense in H. In other words, given an element $\mathbf{u}$ in H and $\forall\ \varepsilon > 0$, there exists an index N and scalars $\alpha_1,\alpha_2,\ldots,\alpha_N$ such that

$$\| \mathbf{u} - \sum_{i=1}^{N} \alpha_i\ \boldsymbol{\phi}_i \| < \varepsilon$$

Separable Hilbert spaces are therefore Hilbert spaces with countable orthonormal sets or orthonormal bases.

The set L(S) of all finite linear combinations of vectors from the set S is a linear subspace, which in general is not closed. If L(S) contains only a finite number of linearly independent vectors, then L(S) is a finite-dimensional linear subspace and hence is closed. If L(S) is infinite-dimensional, L(S) and its closure $\overline{LS}$ differ. For example, the set $S = \{1,x,x^2,\ldots\}$ generates the space P of all polynomials, and the closure of P is the whole of $L_2[a,b]$.

2.6.5 THE PROJECTION THEOREM

The concept of orthogonality has many of the consequences in Hilbert spaces that it has in plane geometry. For example, the Pythagorean theorem is true in Hilbert spaces. In fact

$$\|\mathbf{u} + \mathbf{v}\|^2 = \|\mathbf{u}\|^2 + \|\mathbf{v}\|^2 \Leftrightarrow \mathbf{u} \perp \mathbf{v} \tag{2.45}$$

Another important result, in this context, is given by the projection theorem. In plane geometry the shortest distance from a point to a line is given by the perpendicular from the point to the line (Fig. 2.6).

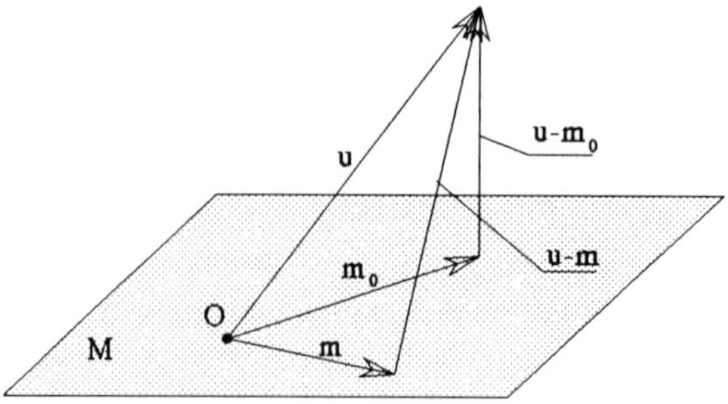

FIGURE 2.6

This result is generalized in Hilbert spaces by the projection theorem: let H be a Hilbert space and M a closed subspace of H. Corresponding to any vector $\mathbf{u} \in$ H, there is a unique vector $\mathbf{m}_0 \in$ M such that $\|\mathbf{u} - \mathbf{m}_0\| \leq \|\mathbf{u} - \mathbf{m}\|$ for all $\mathbf{m} \in$ M. Furthermore, a necessary and sufficient condition that $\mathbf{m}_0 \in \mathbf{M}$ be the unique minimizing vector is that $\mathbf{u} - \mathbf{m}_0$ be orthogonal to M.

2.6.6 REPRESENTATION OF FUNCTIONALS ON HILBERT SPACES

Of practical importance is the general form of bounded linear functionals on various spaces. For a Hilbert space every bounded linear functional $\mathcal{F}$ can be represented in terms of the inner product, namely

$$\mathcal{F}(\mathbf{u}) = < \mathbf{u},\mathbf{w} > \tag{2.46}$$

where $\mathbf{w}$ depends on $\mathcal{F}$, is uniquely determined by $\mathcal{F}$ and has norm $\|\mathbf{w}\| = \|\mathcal{F}\|$. This result is known as the Riesz theorem.

2.7 SYMMETRIC POSITIVE, POSITIVE-DEFINITE OPERATORS AND THEIR ENERGY SPACES

Let A: D $\subset$ H $\rightarrow$ H be a linear operator defined in the domain D dense in a Hilbert space H. The operator A is called *symmetric* in D $\subset$ H if for every pair of elements $\mathbf{u}$, $\mathbf{v}$ from D we have

$$<Au, v> = <u, Av> \tag{2.47}$$

A symmetric operator is said to be positive in its domain D $\subset$ H if for all $\mathbf{u}$ in D the following relations hold

$$\begin{gathered} <A\mathbf{u}, \mathbf{u}> \geq 0 \\ <A\mathbf{u}, \mathbf{u}> = 0 \Rightarrow \mathbf{u} = 0 \end{gathered} \tag{2.48}$$

Further, if we can find a constant $\gamma > 0$ such that for all $\mathbf{u}$ in D the relation

$$<A\mathbf{u}, \mathbf{u}> \geq \gamma^2 \|\mathbf{u}\|^2 \tag{2.49}$$

is satisfied, the operator A is called positive-definite in D. If A is a positive operator, the scalar product $<A\mathbf{u},\mathbf{u}>$ is called the *energy* of the element $\mathbf{u}$ relative to the operator A.

This terminology is used because in all cases in which the element $\mathbf{u}$ can be treated as a deformation of some elastic system, the quantity $^1/_2 <A\mathbf{u},\mathbf{u}>$ coincides with the potential energy of deformation of this system if the operator A is the "stiffness operator" of the system. Positivity of the operator A thus means that the system cannot deform without consumption of energy. If the magnitude of deformations is estimated in terms of a norm, a positive-definite operator means that the system can support a large deformation only with the expense of a sufficiently large quantity of energy. If, however, the operator is positive and not positive-definite, the system may admit a deformation that is arbitrarily large in norm, but produced by an arbitrarily small expense of energy.

2.7.1 THE MINIMUM OF A QUADRATIC FUNCTIONAL

2.7.1.1 The Basic Variational Problem

Definite positive operators play a fundamental role in the theory of elastic structures. In this framework a fundamental problem is the research of solutions of the equation

$$A\mathbf{u} = \mathbf{f} \tag{2.50}$$

where A is a symmetric operator whose domain is a subspace D that is dense in a Hilbert space H, and **f** is an element of this space. Usually D is a space of functions continuous with all their derivatives in a smooth region Ω of the tridimensional space and satisfying linear and homogeneous boundary conditions. The space H in which the subspace D is embedded is, as a rule, the space $L_2(\Omega)$.

First of all we recall that when the operator A is positive, if a solution **s** of the equation (2.50) exists in D, it is unique. In fact, let $\mathbf{u}_1$ and $\mathbf{u}_2$ be two solutions of (2.50), i.e., such that $A\mathbf{u}_1 = \mathbf{f}$ and $A\mathbf{u}_2 = \mathbf{f}$. Thus we get $A(\mathbf{u}_1 - \mathbf{u}_2) = 0$. But A is positive; hence $\mathbf{u}_1 - \mathbf{u}_2 = 0$, and $\mathbf{u}_1 = \mathbf{u}_2$.

Furthermore, if A is positive, it is possible to show that the solution of equation (2.50), if it exists in D, minimizes the functional

$$\mathcal{F}(\mathbf{u}) = \frac{1}{2} <A\mathbf{u},\mathbf{u}> - <\mathbf{u},\mathbf{f}> \tag{2.51}$$

Conversely, if in the subspace D a vector **u** exists that minimizes $\mathcal{F}(\mathbf{u})$, **u** is the solution of (2.50). For any $\mathbf{u} \in$ D, in fact, we have

$$\mathcal{F}(\mathbf{s} + \mathbf{u}) - \mathcal{F}(\mathbf{s}) = <A\mathbf{s} - \mathbf{f},\mathbf{u}> + \frac{1}{2} <A\mathbf{u},\mathbf{u}> \geq <A\mathbf{s} - \mathbf{f}, \mathbf{u}> \tag{2.52}$$

If **s** satisfies the equation $A\mathbf{s} = \mathbf{f}$, $\mathcal{F}(\mathbf{u})$ achieves its global minimum at **s**. Conversely, if $\mathcal{F}(\mathbf{u})$ achieves its minimum at **s** it is required that

$$<As - f,u> = 0 \ \forall \ u \in D \tag{2.53}$$

But D is dense in H, hence, according to the lemma of variational calculus recalled in Section 2.6.4, $A\mathbf{s} - \mathbf{f} = \mathbf{0}$ and **s** the solution of (2.50).

The problem of minimizing a functional of type (2.51) is usual in linear elasticity. It appears that we should search for the minimum of $\mathcal{F}(\mathbf{u})$ in the subspace D. Thus restriction is not necessary, and with the assumption that A is definite positive we can use a class of functions, the energy space H_A, which is larger than D. We will show now that in this space H_A the problem of finding the minimum of $\mathcal{F}(\mathbf{u})$ always has a solution.

We first introduce the energy product <Au,v> associated with the definite positive operator A into the domain D. The subspace D is an inner product space with the new scalar product

$$[\mathbf{u},\mathbf{v}] = <A\mathbf{u},\mathbf{v}> \tag{2.54}$$

and norm, called the "energy" norm of **u**,

$$|||\mathbf{u}||| = (<A\mathbf{u},\mathbf{u}>)^{1/2} \tag{2.55}$$

We will denote this new space by D_A. Like (2.3), (2.39), and (2.40), which hold in H, we have the following inequalities $\forall \ \mathbf{u},\mathbf{v} \in D_A$

$$| \, |||\mathbf{u}||| - |||\mathbf{v}||| \, | \leq |||\mathbf{u} - \mathbf{v}|||; \; |[\mathbf{u}, \mathbf{v}]| \leq |||\mathbf{u}||| \; |||\mathbf{v}|||; \; |||\mathbf{u} + \mathbf{v}||| \leq |||\mathbf{u}||| + |||\mathbf{v}||| \tag{2.56}$$

The subspace D_A, will in general, be incomplete, but we may complete it in the usual way by adjoining elements that are the limits of Cauchy sequences in D_A. Let $\{\mathbf{u}_n\}$ be one of these Cauchy sequence in D_A. Then

$$\lim_{m,n \to \infty} \langle A(\mathbf{u}_n - \mathbf{u}_m), \mathbf{u}_n - \mathbf{u}_m \rangle = \lim_{m,n \to \infty} |||\mathbf{u}_n - \mathbf{u}_m||| = 0 \tag{2.57}$$

On the other hand, according to (2.49)

$$\langle A(\mathbf{u}_n - \mathbf{u}_m), \mathbf{u}_n - \mathbf{u}_m \rangle = |||u_n - u_m|||^2 \geq \gamma^2 ||u_n - u_m||^2 \tag{2.58}$$

and so $\{\mathbf{u}_n\}$ is also Cauchy in H. But H is complete, so there exists an element $\mathbf{u} \in$ H such that $||\mathbf{u}_n - \mathbf{u}|| \to 0$ or $\lim_H \mathbf{u}_n = \mathbf{u}$. This element $\mathbf{u}$ is adjoined to the subspace D_A.

We can evaluate the energy norm of this element $\mathbf{u}$. In fact, according to the first part of (2.56), $| \, |||\mathbf{u}_n||| - |||\mathbf{u}_m||| \, | \leq |||\mathbf{u}_n - \mathbf{u}_m|||$ and

$$\lim_{m,n \to \infty} \left| |||\mathbf{u}_n||| - |||\mathbf{u}_m||| \right| = 0 \tag{2.59}$$

Thus $\lim |||\mathbf{u}_n|||$ exists and we assume

$$|||\mathbf{u}||| = \lim_{n \to \infty} |||\mathbf{u}_n||| \tag{2.60}$$

With this procedure, by using elements from the space H, the space D_A is made complete and becomes a new Hilbert space, which we indicate by H_A. It is clear that D is dense in H and D_A is dense in H_A. We have $D_A \subset H_A \subset H$.

Since H_A contains points that are not in D_A, the inner product $\langle A\mathbf{u}, \mathbf{v} \rangle$ is not defined in the whole H_A. We define the inner product $\langle \mathbf{u}, \mathbf{v} \rangle_A$ on the Hilbert space H_A by

$$[\mathbf{u}, \mathbf{v}] = \langle \mathbf{u}, \mathbf{v} \rangle_A = \lim_{m,n \to \infty} \langle A\mathbf{u}_n, \mathbf{v}_m \rangle \tag{2.61}$$

where $\{\mathbf{u}_n\}$ and $\{\mathbf{v}_n\}$ are Cauchy sequences in D_A. Also, the limit (2.61) exists. In fact

$$|\langle A\mathbf{u}_n, \mathbf{v}_m \rangle - \langle A\mathbf{u}_r, \mathbf{v}_s \rangle| = |[\mathbf{u}_n, \mathbf{v}_m] - [\mathbf{u}_r, \mathbf{v}_s]|$$

$$= |[\mathbf{u}_n - \mathbf{u}_r, \mathbf{v}_m - \mathbf{v}_s] + [\mathbf{v}_s, \mathbf{u}_n - \mathbf{u}_r] + [\mathbf{u}_r, \mathbf{v}_m - \mathbf{v}_s]|$$

$$\leq |||\mathbf{u}_n - \mathbf{u}_r||| \; |||\mathbf{v}_m - \mathbf{v}_s||| + |||\mathbf{u}_n - \mathbf{u}_r||| \; |||\mathbf{v}_s||| + |||\mathbf{v}_m - \mathbf{v}_s||| \; |||\mathbf{u}_r||| \tag{2.62}$$

and

$$\lim_{m,n,r,s \to \infty} |[\mathbf{u}_n, \mathbf{v}_m] - [\mathbf{u}_r, \mathbf{v}_s]| \tag{2.63}$$

Thus

$$\lim_{m,n \to \infty} |[\mathbf{u}_n, \mathbf{v}_m]| = 0$$

exists and we have (2.61). In brief, the three inequalities (2.56) are extended from D_A to the whole space H_A. Since the operator A is positive definite, we have, for any Cauchy sequence $\{\mathbf{u}_n\} \in D_A$,

$$//\mathbf{u}_n//^2 \leq (1/\gamma^2) \langle A\mathbf{u}_n, \mathbf{u}_n \rangle \tag{2.64}$$

and since every point in H_A is also in H, we have, in the limit

$$\|\mathbf{u}\| \leq (1/\gamma)\, |||\mathbf{u}||| \tag{2.65}$$

Hence convergence in energy in H_A implies convergence in H.

We can show now that if we seek for a minimum of $\mathcal{F}(\mathbf{u})$ in the enlarged space H_A, this minimum always exists but cannot be attained in D_A. With this goal, first we extend the functional $\mathcal{F}(\mathbf{u})$ defined on D_A to the whole space H_A.

According to (2.51) the first term on the right side of $\mathcal{F}(\mathbf{u})$ is replaced by the energy product $[\mathbf{u},\mathbf{u}]$, $u \in H_A$. The second term, for a fixed f and any $\mathbf{u} \in H_A$ is a bounded linear functional $f(\mathbf{u})$ on H_A, since, using the Schwarz inequality, we get

$$|f(\mathbf{u})| = |<\mathbf{f}, \mathbf{u}>| \leq \|\mathbf{f}\|\, \|\mathbf{u}\| \leq (1/\gamma)\, \|\mathbf{f}\|\, |||\mathbf{u}||| \tag{2.66}$$

Now, as we have recalled in Section 2.6.1.4, to every bounded linear functional $f(\mathbf{u})$ on a Hilbert space H there corresponds a unique element $\mathbf{w} \in H$ such that

$$f(\mathbf{u}) = <\mathbf{w}, \mathbf{u}> \tag{2.67}$$

Hence, there exists an element $\mathbf{w}_f \in H_A$ such that

$$<\mathbf{f}, \mathbf{u}> = [\mathbf{w}_f, \mathbf{u}] \tag{2.68}$$

Thus we can write the functional $\mathcal{F}(\mathbf{u})$ extended to H_A as

$$F(\mathbf{u}) = \frac{1}{2}[\mathbf{u},\mathbf{u}] - [\mathbf{w}_f,\mathbf{u}] = \frac{1}{2}([\mathbf{u}-\mathbf{w}_f,\mathbf{u}-\mathbf{w}_f] - [\mathbf{w}_f,\mathbf{w}_f]) = \frac{1}{2}(|||\mathbf{u}-\mathbf{w}_f|||^2 - |||\mathbf{w}_f|||^2) \tag{2.69}$$

so that $\mathcal{F}(\mathbf{u})$ certainly achieves its minimum at $\mathbf{u} = \mathbf{w}_f \in H_A$.

Since D_A is dense in H_A, there must exist a sequence $\{\mathbf{u}_n\} \in D_A$ such that $\lim \mathbf{u}_n = \mathbf{w}_f$. The result (2.68) can therefore be written as

$$\lim_{m,n \to \infty} <\mathbf{f} - A\mathbf{u}_n, \mathbf{u}> = 0 \quad \forall\, \mathbf{u} \in D_A \tag{2.68$'$}$$

Now, if $\mathbf{w}_f \in D_A$, then clearly

$$\lim_{n\to\infty} \mathbf{u}_n = \mathbf{w}_f = \mathbf{s} \tag{2.70}$$

where $A\mathbf{s} = \mathbf{f}$ so that the minimization of the functional $\mathcal{F}(\mathbf{u})$ in H_A leads to the same minimum point $\mathbf{w}_f = \mathbf{s}$ as minimization in D_A.

If $\mathbf{w}_f \notin D_A$ there is no point in D_A such that $A\mathbf{w}_f = \mathbf{f}$. In that case we say that $\mathbf{w}_f$ is a generalized solution of the operator equation characterized by:

$$\lim_{n\to\infty} [\mathbf{u}_n, \mathbf{u}] = \lim_{n\to\infty} <A\mathbf{u}_n, \mathbf{u}> = <\mathbf{f}, \mathbf{u}> \tag{2.71}$$

In conclusion, $\forall\, \mathbf{f} \in H$ there exists a $\mathbf{w}_f \in H_A$ such that $\min \mathcal{F}(\mathbf{u}) = \mathcal{F}(\mathbf{w}_f)$.

2.7.1.2 The Special Problem of the Minimum of a Quadratic Functional

Let $B_2(\mathbf{u})$ be a homogeneous quadratic functional. We call it positive definite if it is defined on a dense set in an Hilbert space H and satisfies the inequality

$$B_2(\mathbf{u}) \geq \beta^2 \|\mathbf{u}\|_H^2 \tag{2.72}$$

Further, let $L(\mathbf{u})$ be a linear bounded functional whose domain of definition contains the domain of definition of $B_2(\mathbf{u})$. The problem of finding the minimum of the quadratic functional

$$Q_2(\mathbf{u}) = \frac{1}{2}B_2(\mathbf{u}) - L(\mathbf{u}) \tag{2.73}$$

is called "the special variational problem" by Mikhlin[17] and is basically solved in the same way as the ordinary previous problem.

Let us introduce a new Hilbert space H_B: its elements are the elements of the domain of definition of $Q_2(\mathbf{u})$ and the scalar product and norm are defined by

$$\{\mathbf{u},\mathbf{v}\} = B_{11}(\mathbf{u},\mathbf{v}) \qquad \{\mathbf{u},\mathbf{u}\} = \{\{\mathbf{u}\}\}^2 = B_{11}(\mathbf{u},\mathbf{u}) = B_2(\mathbf{u}) \tag{2.74}$$

where

$$B_{11}(\mathbf{u},\mathbf{v}) = [B_2(\mathbf{u} + \mathbf{v}) - B_2(\mathbf{u}) - B_2(\mathbf{v})] \tag{2.75}$$

If the space H_B is incomplete, it is completed with the usual procedure by using elements of the original space H and the inequality

$$\{\{\mathbf{u}\}\}^2 \geq \beta^2 \, \| \mathbf{u} \|_H^2 \tag{2.76}$$

holds in the completed space. Because of (2.76), the linear functional $L(\mathbf{u})$, bounded in H, is still bounded in the new space H_B:

$$| L(\mathbf{u}) | \leq k \, \| \mathbf{u} \|_H \leq \frac{k}{\beta} \{\{\mathbf{u}\}\} \tag{2.77}$$

Thus, by the representation theorem of linear bounded functionals,

$$L(\mathbf{u}) = \{\mathbf{u}, \mathbf{u}_0\} \tag{2.78}$$

where $\mathbf{u}_0$ is a definite element of H_B. Now

$$Q_2(\mathbf{u}) = \frac{1}{2}\{\mathbf{u}, \mathbf{u}\} - \{\mathbf{u}, \mathbf{u}_0\} = \frac{1}{2}\,[\{\{\mathbf{u} - \mathbf{u}_0\}\}^2 - \{\{\mathbf{u}_0\}\}^2] \tag{2.79}$$

This equation enables to extend $Q_2(\mathbf{u})$ to the whole of H_B: obviously $Q_2(\mathbf{u})$ achieves its global minimum on H_B for $\mathbf{u} = \mathbf{u}_0$ and this minimum equals $-\frac{1}{2}\{\{\mathbf{u}_0\}\}^2$

We can also obtain the variational condition whose solution gives $\mathbf{u}_0$. In fact, from the condition of minimum of $Q_2(\mathbf{u})$ at $\mathbf{u}_0$

$$Q_2(\mathbf{u}_0 + t\mathbf{v}) - Q_2(\mathbf{u}_0) \geq 0 \qquad \forall\, \mathbf{v} \in H_B \qquad \forall\, t \in R \tag{2.80}$$

we get

$$2t\{B_{11}(\mathbf{u}_0,\mathbf{v}) - L(\mathbf{v})\} + t^2 B_2(\mathbf{u}_0) \geq 0 \qquad \forall\, \mathbf{v} \in H_B \qquad \forall\, t \in R \tag{2.81}$$

Thus taking into account (2.72), we get the variational equation

$$B_{11}(\mathbf{u},\mathbf{v}) - L(\mathbf{v}) = 0 \qquad \forall\, \mathbf{v} \in H_B \tag{2.82}$$

whose unique solution gives $\mathbf{u}_0$. From (2.82) with $\mathbf{v} = \mathbf{u}_0$ we have $B_{11}(\mathbf{u}_0,\mathbf{u}_0) = L(\mathbf{u}_0)$ and we have min $Q_2(\mathbf{u}) = Q_2(\mathbf{u}_0) = {}^1\!/_2 B_2(\mathbf{u}_0) - L(\mathbf{u}_0) = -{}^1\!/_2 B_2(\mathbf{u}_0) = -{}^1\!/_2 \{\{\mathbf{u}_0\}\}^2$

2.8 CALCULUS OF OPERATORS AND FUNCTIONALS; EXTREME VALUES OF FUNCTIONALS

Stability analysis of continuous structural systems, via energy methods, involves finding the minimum of functionals on given sets.

The question whether a functional attains its minimum on a given set is a fundamental problem in functional analysis. In finite-dimensional spaces the well-known Weiestrass theorem, which states that a continuous function defined on a closed and bounded set has a maximum and a minimum, is of great utility.

The generalization of the Weierstrass theorem consists in the statement that a continuous mapping of a compact set M of a metric space X into R assumes a maximum and a minimum at some points of M. In the following discussion let X be a vector space, Y a normed space, and T a mapping defined on a domain $D \subset X$ and having range $R \subset Y$. If the normed space Y is the field of the real numbers R, the mapping T is a functional.

2.8.1 DIFFERENTIAL AND DERIVATIVES

2.8.1.1 Introductory Remarks

We begin by defining differentials and derivatives in functional spaces. Let us review what is meant by saying that the function $f: R \to R$ is differentiable at x_0 with derivative $f'(x_0)$. We assert the existence of the limit

$$\lim_{x \to x_0} \frac{f(x) - f(x_0)}{x - x_0}$$

and this limit, if it exists, is defined to be the derivative of at x_0. From the definition of the derivative it is seen that

$$\frac{f(x) - f(x_0)}{x - x_0} = f'(x_0) + \varepsilon$$

where $\lim_{(x - x_0) \to 0} \varepsilon = 0$. We can also write

$$[f(x) - f(x_0)] - f'(x_0)(x - x_0) = \varepsilon(x - x_0)$$

For small values of $(x - x_0)$ this difference is small. Geometrically in calculus we say that $\varepsilon(x - x_0)$ represents the portion of the increment $f(x) - f(x_0)$ cut off by the tangent line to the graph of f(x) at the point where the derivative is computated (Fig. 2.7). We therefore give the geometric description of the derivative of the function f(x). The function $f: R \to R$ is differentiable at x_0 if the graph of f has a unique tangent at x_0. We note that the tangent line to the graph of f at x_0 defines the graph of an affine linear map $y = mx + c$ and that this map

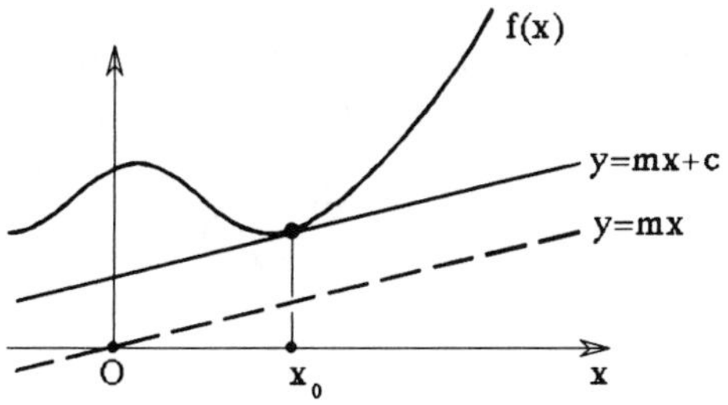

FIGURE 2.7

provides an affine linear approximation to f at x_0. We can define the derivative to be the linear map y = mx. (Of course, we can easily obtain the affine linear approximation from the derivative by noting that $c = f(x_0) - mx_0$.)

The idea of defining the derivative of a function as a linear approximation is fundamental to a proper understanding of the idea of a derivative. The analytic approach defines the derivative as a point in R, whereas the geometrical approach gives the derivative as an element of the space of the linear maps from R to R. These two different definitions can be reconciled by saying that the derivative $f'(x_0) \in R$, analytically evaluated, corresponds to the linear map $x \to f'(x_0)x$, that is, the geometrically defined derivative of f at x_0. All these concepts are basic to the definition of differentials and derivatives in functional spaces. After this remark let us define, first of all, the Gateaux differential.

2.8.1.2 The Gateaux and Fréchet Differentials

Let $\mathbf{x} \in D \subset X$ and let $\mathbf{h}$ be arbitrary in X. If the limit

$$DT(\mathbf{x};\mathbf{h}) = \lim_{\alpha \to 0} \left(\frac{1}{\alpha}\right) \left[T(\mathbf{x} + \alpha\mathbf{h}) - T(\mathbf{x})\right] \tag{2.83}$$

exists, it is called the Gateaux differential or the weak differential of T at $\mathbf{x}$ in the direction of the increment $\mathbf{h}$. The differential may exist for some $\mathbf{h}$ and fail to exist for others: if the limit exists for each $\mathbf{h} \in X$, the mapping T is said to be Gateaux differentiable at $\mathbf{x}$.

Note that the definition implies that $\mathbf{x} + \alpha\mathbf{h} \in D \subset X$ for some α sufficiently small. The limit is taken in the usual sense of the norm convergence in Y. The Gateaux differential is homogeneous in $\mathbf{h}$ in the sense that

$$DT(\mathbf{x};\ \lambda\mathbf{h}) = \lambda\ DT(\mathbf{x};\ \mathbf{h}) \tag{2.84}$$

but is not, in general, linear in $\mathbf{h}$.

For fixed $\mathbf{x} \in D \subset X$ and $\mathbf{h}$ regarded as variable, the Gateaux differential defines a mapping from X to Y. If, in particular, T is linear we have $DT(\mathbf{x};\ \mathbf{h}) = T(\mathbf{h})$. The most frequent application of this definition is in the case where Y is the real line and the transformation T reduces to a functional on X. Thus if f is a functional on X, the Gateaux differential of f, if it exists, is

$$Df(\mathbf{x};\ \mathbf{h}) = \left[\frac{\partial f(\mathbf{x} + \alpha\mathbf{h})}{\partial \alpha}\right]_{\alpha = 0} \tag{2.85}$$

and, for each fixed $\mathbf{x} \in X$, $Df(\mathbf{x};\ \mathbf{h})$ is a functional with respect to the variable $\mathbf{h} \in X$.

With the variational calculus notation, the Gateaux differential is also called the variation of the function f at x in the direction h and is written as

$$Df(\mathbf{x};\ \mathbf{h}) = \delta f(\mathbf{x};\ \mathbf{h}) \tag{2.84'}$$

Example 1. Let $X = R^n$ and let $f(\mathbf{x}) = f(x_1,x_2,...,x_n)$ be a function on R^n having continuous partial derivatives with respect to each variable x_j; thus, the Gateaux differential of f is

$$Df(\mathbf{x};\ \mathbf{h}) = \sum_{i=1}^{n} \frac{\partial f}{\partial x_i} h_i$$

Example 2. Let X = C[0,1] and let

$$f(x) = \int_0^1 g(x(t), t)\, dt$$

where it is assumed that the function derivative g_x exists and is continuous with respect to x and t. Then

$$Df(x; h) = \int_0^1 g_x(x(t), t)\, h(t)\, dt$$

The Gateaux differential generalizes the concept of directional derivative familiar in finite dimensional space. The existence of the Gateaux differential is a rather weak requirement, however, since its definition requires no norm on X; hence, properties of the Gateaux differential are not easily related to continuity. When X is normed, a more satisfactory definition is given by the Fréchet differential.

Let T be a mapping defined on an open domain D in a normed space X and having range in a normed space Y. If for a fixed $\mathbf{x} \in$ D and each $\mathbf{h} \in$ X there exists a DT($\mathbf{x}$; $\mathbf{h}$) $\in$ Y that is linear and continuous with respect to $\mathbf{h}$ and such that

$$\lim_{\|\mathbf{h}\| \to 0} \left(\frac{1}{\|\mathbf{h}\|} \right) \left\| T(\mathbf{x}+\mathbf{h}) - T(\mathbf{x}) - DT(\mathbf{x};\mathbf{h}) \right\| = 0 \tag{2.86}$$

then T is said Fréchet differentiable at $\mathbf{x}$ and DT($\mathbf{x}$; $\mathbf{h}$) is said to be the Fréchet differential of T at $\mathbf{x}$ with increment $\mathbf{h}$.

If T represents a functional f on X and the Fréchet differential of f exists at $\mathbf{x}$, we get

$$\lim_{\|\mathbf{h}\| \to 0} \left(\frac{1}{\|\mathbf{h}\|} \right) \left\| f(\mathbf{x}+\mathbf{h}) - f(\mathbf{x}) - Df(\mathbf{x};\mathbf{h}) \right\| = 0 \tag{2.87}$$

Thus we can write

$$f(\mathbf{x} + \mathbf{h}) = f(\mathbf{x}) + Df(\mathbf{x}; \mathbf{h}) + o(\|\mathbf{h}\|) \tag{2.88}$$

where, Df($\mathbf{x}$; $\mathbf{h}$) being linear and continuous with respect to $\mathbf{h}$, we have

$$|Df(\mathbf{x}; \mathbf{h})| \leq k\, \|\mathbf{h}\| \tag{2.89}$$

and

$$\lim_{\|\mathbf{h}\| \to 0} \frac{[f(\mathbf{x}+\mathbf{h}) - f(\mathbf{x})] - Df(\mathbf{x};\mathbf{h})}{\|\mathbf{h}\|} = \lim_{\|\mathbf{h}\| \to 0} \frac{o(\|\mathbf{h}\|)}{\|\mathbf{h}\|} = 0 \tag{2.90}$$

The Fréchet differential is thus the linear approximation to f at $\mathbf{x}$.

If the Fréchet differential exists, it is unique and the Gateaux and Fréchet differential are equal. Another important property is given by the following proposition: if the mapping T defined on an open set D in X has a Fréchet differential at $\mathbf{x}$, then T is continuous at $\mathbf{x}$.

We give now the definition of derivative. Let the transformation T defined on an open domain D $\subset$ X is Fréchet differentiable throughout D. At a fixed point $\mathbf{x} \in$ D the Fréchet differential DT($\mathbf{x}$; $\mathbf{h}$) is, by definition, of the form

$$DT(\mathbf{x}; \mathbf{h}) = T_x \mathbf{h} \tag{2.91}$$

where T_x is a bounded linear operator from X to Y. Thus, as $\mathbf{x}$ varies over D the correspondence $\mathbf{x} \to T_x$ defines a mapping from D into the normed linear space B(X,Y) of the linear bounded operators from X to Y; this mapping is called the Fréchet derivative T' of T. Thus we have, by definition,

$$DT(\mathbf{x}; \mathbf{h}) = T'(\mathbf{x})\mathbf{h} \tag{2.92}$$

In the special case where the first transformation is a functional f on the space X, we have

$$Df(\mathbf{x}; \mathbf{h}) = f'(\mathbf{x})\mathbf{h} \tag{2.93}$$

or, with a different notation,

$$Df(\mathbf{x}; \mathbf{h}) = \nabla_x f(\mathbf{x})\mathbf{h} \tag{2.93'}$$

where $f'(\mathbf{x})$ or $\nabla_x f(\mathbf{x})$ is a mapping from D into the normed linear space X* of the linear bounded functionals from X to R. The element $f'(\mathbf{x})$ is called the gradient of f at **x** and is sometimes denoted $\nabla f(\mathbf{x})$ rather than $f'(\mathbf{x})$. We sometimes write $< f'(\mathbf{x}), \mathbf{h} >$ for $Df(\mathbf{x}; \mathbf{h})$ since it is bounded and linear in **h**.

2.8.2 CALCULUS OF OPERATORS AND FUNCTIONALS

Much of the theory of ordinary derivatives can be generalized to Fréchet derivatives. For instance, the implicit function theorem and Taylor series have very satisfactory extensions. So we can give a very useful inequality that replaces the mean value theorem for ordinary functions.

Let T be Fréchet differentiable on an open domain D. Let $\mathbf{x} \in D$ and suppose that $\mathbf{x} + \alpha\mathbf{h} \in D$ for all α, $0 \le \alpha \le 1$. Then

$$\| T(\mathbf{x}+\mathbf{h}) - T(\mathbf{x}) \| \le \| \mathbf{h} \| \sup_{0<\alpha<1} \| T'(\mathbf{x}+\alpha\mathbf{h}) \| \tag{2.94}$$

We can extend the previous definitions of the Gateaux and Fréchet differentials to differentials of higher order. Thus we say that $D^2f(\mathbf{x}; \mathbf{u}, \mathbf{v})$ is the second Gateaux differential of the functional $f(\mathbf{x})$ at **x** in D and with increments **u** and $\mathbf{v} \in X$, where

$$D^2f(\mathbf{x}; \mathbf{u}, \mathbf{v}) = \left[\frac{\partial Df(\mathbf{x} + \alpha\mathbf{v}; \mathbf{u})}{\partial\alpha}\right]_{\alpha=0} = \left[\frac{\partial Df(\mathbf{x} + \alpha\mathbf{u}; \mathbf{v})}{\partial\alpha}\right]_{\alpha=0} \tag{2.95}$$

If $f(\mathbf{x})$ is twice Fréchet differentiable at **x**, on the analogy of (2.73), we write

$$f(\mathbf{x} + \mathbf{h}) = f(\mathbf{x}) + Df(\mathbf{x}; \mathbf{h}) + 1/2D^2f(\mathbf{x}; \mathbf{h},\mathbf{h}) + o(\|\mathbf{h}\|^2) \tag{2.96}$$

where, $Df(\mathbf{x}; \mathbf{h})$ and $D^2f(\mathbf{x}; \mathbf{h})$ being linear and continuous with respect to **h**, we have

$$|Df(\mathbf{x}; \mathbf{h})| \le k\,\|\mathbf{h}\| \qquad |D^2f(\mathbf{x}; \mathbf{h})| \le k\,\|\mathbf{h}\|^2 \tag{2.97}$$

$$\lim_{\|\mathbf{h}\|\to 0} \left(\frac{1}{\|\mathbf{h}\|^2}\right) o\left(\|\mathbf{h}\|^2\right) = 0 \tag{2.97'}$$

A generalization of (2.96) is the Taylor formula:

$$f(\mathbf{x} + \mathbf{h}) = f(\mathbf{x}) + Df(\mathbf{x}; \mathbf{h}) + \frac{1}{2!} D^2f(\mathbf{x}; \mathbf{h}, \mathbf{h}) + \cdots + \frac{1}{n!} D^nf(\mathbf{x}; \mathbf{h},\mathbf{h},\ldots,\mathbf{h}) + o(\|\mathbf{h}\|^n) \tag{2.98}$$

or,

$$f(\mathbf{x}+\mathbf{h}) = f(\mathbf{x}) + f_1(\mathbf{x}; \mathbf{h}) + f_2(\mathbf{x}; \mathbf{h}, \mathbf{h}) + \ldots + f_n(\mathbf{x}; \mathbf{h}, \mathbf{h},\ldots,\mathbf{h}) + o\left(\|\mathbf{h}\|^n\right) \tag{2.98'}$$

where

$$f_1(\mathbf{x}; \mathbf{h}) = \frac{1}{1!} Df(\mathbf{x}; \mathbf{h}), \quad f_2(\mathbf{x}; \mathbf{h}, \mathbf{h}) = \frac{1}{2!} D^2f(\mathbf{x}; \mathbf{h}, \mathbf{h})$$

$$\ldots f_n(\mathbf{x}; \mathbf{h}, \mathbf{h},\ldots,\mathbf{h}) = \frac{1}{n!} D^nf(\mathbf{x}; \mathbf{h}, \mathbf{h},\ldots,\mathbf{h}) \tag{2.98''}$$

The following further notations are very frequent and will be employed in the text

$$D^n f(\mathbf{x};\, \mathbf{h}, \mathbf{h}, \ldots, \mathbf{h}) = D^n f(\mathbf{x}_i\, \mathbf{h}'') = D^n f(\mathbf{x};\, \mathbf{h})$$

$$f_n(\mathbf{x};\, \mathbf{h}, \mathbf{h}, \ldots, \mathbf{h}) = f_n(\mathbf{x};\, \mathbf{h}^n) = f_n(\mathbf{x};\, \mathbf{h}) \qquad (2.98''')$$

If T: X → Y is Fréchet differentiable on an open domain D the derivative maps D into B(X,Y) and may itself be Fréchet differentiable on a subset $D_1 \subset D$. In this case the Fréchet derivative of T' is called the second Fréchet derivative of T and is denoted by T''.

So the following inequality can be proved in a manner paralleling that of the previous preposition (2.94): Let T be Fréchet differentiable on an open domain D. Let $\mathbf{x} \in D$ and suppose that $\mathbf{x} + \alpha\mathbf{h} \in D$ for all α, $0 \le \alpha \le 1$. Then

$$\| T(\mathbf{x}+\mathbf{h}) - T(\mathbf{x}) - T'(\mathbf{x})\mathbf{h} \| \le \frac{1}{2} \| \mathbf{h} \|^2 \sup_{0<\alpha<1} \| T''(\mathbf{x}+\alpha\mathbf{h}) \| \qquad (2.99)$$

2.8.3 MAXIMA AND MINIMA

It is simple to apply the concepts of the Gateaux and Fréchet differentials to the problem of minimizing or maximizing a functional on a linear space. The technique leads quite naturally to the bases of the calculus of variations (where, in fact, the abstract concepts of differentials originated). We can therefore extend the familiar technique of minimizing a function of a single variable by ordinary calculus to a similar technique based on more general differentials. In this way we obtain analogs of the classical necessary conditions for local extrema.

Let f be a real-valued functional defined on a normed space X with norm || ||. A point $\mathbf{x}_0 \in X$ is said to be a weak relative minimum of f on X if there is an open ball $S_\rho = \{\mathbf{x} \in X: \|\mathbf{x}\| \le \rho\} \subset X$ centered on $\mathbf{x}_0$ such that $f(\mathbf{x}_0) \le f(\mathbf{x})$ for all $\mathbf{x} \in S_\rho$. The point $\mathbf{x}_0$ is said a strict relative minimum of f on X if there exists an open ball $S_\rho \subset X$ centered on $\mathbf{x}_0$ such that

$$\inf_{\partial S_\varepsilon} f(\mathbf{x}) > f(\mathbf{x}_0) \quad \forall\, \mathbf{x} \in \partial S_\varepsilon \qquad (2.100)$$

with $\partial S_\varepsilon = \{\mathbf{x} \in X : \|\mathbf{x} - \mathbf{x}_0\| = \varepsilon\}$ and for any $\varepsilon > 0$ such that $0 < \varepsilon < \rho$. Relative maxima are defined similarly.

Hence we can recall the following fundamental theorem of great utility in the theory of elastic stability: let the real-valued functional f have a Gateaux differential on a vector space X. A necessary condition for f to have an extremum at $\mathbf{x}_0 \in X$ is that $Df(\mathbf{x};\, \mathbf{h}) = 0$ for all $\mathbf{h} \in \mathbf{X}$.

For every $\mathbf{h} \in X$, the function $f(\mathbf{x}_0 + \alpha\mathbf{h})$ of the real variable α must achieve an extremum at α = 0. Thus, by ordinary calculus,

$$\left[\frac{df(x_0 + \alpha\mathbf{h})}{d\alpha}\right]_{\alpha=0} = Df(\mathbf{x};\, \mathbf{h}) = 0 \qquad (2.101)$$

A point at which $Df(\mathbf{x};\, \mathbf{h}) = 0$ for all $\mathbf{h}$ is called a stationary point; hence, this theorem merely states that extrema occur at stationary points.

We can give now an important sufficient condition for the existence of a strict relative minimum or a strict relative maximum of a functional f. Suppose that f: X → R is Fréchet differentiable on X and twice Fréchet differentiable at $\mathbf{x}_0$ with $Df(\mathbf{x}_0;\, \mathbf{h}) = 0$ for all $\mathbf{h} \in X$. A sufficient condition for the point $\mathbf{x}_0$ to be a strict relative minimum for f is that there exists a $c > 0$ such that

$$D^2 f(\mathbf{x}_0;\, \mathbf{h},\mathbf{h}) \ge c\|\mathbf{h}\|^2 \quad \forall\, h \in X \qquad (2.102)$$

where $D^2f(\mathbf{x}_0; \mathbf{h},\mathbf{h})$ is quadratic and continuous with respect to $\mathbf{h}$. In fact, from the Taylor theorem (2.98) we get

$$f(\mathbf{x}_0+\mathbf{h}) = f(\mathbf{x}_0) + \frac{1}{2}D^2f(\mathbf{x}_0; \mathbf{h},\mathbf{h}) + r(\mathbf{h}) \tag{2.98$'$}$$

where

$$\lim_{\|\mathbf{h}\|\to 0} \frac{r(\mathbf{h})}{\|\mathbf{h}\|^2} = 0 \tag{2.103}$$

Now if $D^2f(\mathbf{x}_0; \mathbf{h},\mathbf{h}) \geq c\|\mathbf{h}\|^2\ \forall\ \mathbf{h} \in X$, it follows that

$$f(\mathbf{x}_0+\mathbf{h}) - f(\mathbf{x}_0) \geq \frac{c}{2}\|\mathbf{h}\|^2 + r(\mathbf{h}) \tag{2.104}$$

and so, for $\mathbf{h} \neq \mathbf{0}$,

$$\frac{f(\mathbf{x}_0+\mathbf{h}) - f(\mathbf{x}_0)}{\|\mathbf{h}\|^2} \geq \frac{c}{2} + \frac{r(\|\mathbf{h}\|)}{\|\mathbf{h}\|^2} \geq \frac{c}{2} - \frac{|r(\|\mathbf{h}\|)|}{\|\mathbf{h}\|^2} \tag{2.104$'$}$$

Choose the radius ρ of the ball $S_\rho = \{h : \|h\| \leq \rho\}$ such that $|r(h)| / \|\mathbf{h}\|^2 \leq c/4$, $\mathbf{h} \neq \mathbf{0} \in S_\rho$. Then, for any $\mathbf{h}$ such that $\mathbf{h} \neq \mathbf{0} \in S_\rho$

$$\frac{f(\mathbf{x}_0+\mathbf{h}) - f(\mathbf{x}_0)}{\|\mathbf{h}\|^2} \geq \frac{c}{2} - \frac{|r(\|\mathbf{h}\|)|}{\|\mathbf{h}\|^2} \geq \frac{c}{2} - \frac{c}{4} = \frac{c}{4} \tag{2.104$''$}$$

It follows immediately that

$$f(\mathbf{x}_0+\mathbf{h}) - f(\mathbf{x}_0) \geq \frac{c}{4}\|\mathbf{h}\|^2 \quad \forall\ \mathbf{h} \in S_\rho \tag{2.105}$$

Hence $\mathbf{x}_0$ is a strict relative minimum for f.

The requirement of the twice strong differentiability at $\mathbf{x}_0$ for f can be weakened assuming only the twice strong semi-differentiability of f at $\mathbf{x}_0$. In fact, let us assume that, in place of condition (2.103) (but always with the condition that $|D^2f(\mathbf{x}_0; \mathbf{h},\mathbf{h})| \leq k\ \|\mathbf{h}\|_H^2$, we have that, for any $\varepsilon > 0$ there exists a ball S_δ centered at $\mathbf{x}_0$ such that

$$-\varepsilon < \frac{r(\mathbf{h})}{\|\mathbf{h}\|^2} \quad \forall\ \mathbf{h} \neq \mathbf{0} \in S_\delta \tag{2.106}$$

Thus from (2.98′), (2.102), and (2.106) we get $f(x_0 + \mathbf{h}) - f(\mathbf{x}_0) \geq \left(\frac{c}{2} - \varepsilon\right)\|\mathbf{h}\|^2$. Choose $\varepsilon > 0$ such that $\left(\frac{c}{2} - \varepsilon\right) > 0$ and we find again condition (2.105) $\forall\ \mathbf{h} \in S_\delta$.

2.9 GENERALIZED DERIVATIVES AND SOBOLEV SPACES

Let Ω be some open bounded region in r-dimensional space, and let $\partial\Omega$ be its boundary. By a bounded strip in the region Ω we mean the set of points of Ω whose distance to the boundary $\partial\Omega$ is no greater than some given number δ; this number is called the width of the bounded strip. A region Ω' is said to be a subregion of Ω if all points of Ω' belong to Ω and is said to be an interior subregion of Ω if there exists a bounded strip in the region Ω that does not contain points of the subregion Ω'.

By Φ_m we denote the set of real functions $\phi(\mathbf{x})$ that are m times continuously differentiable in Ω and vanish in a bounded strip, generally different for each function, in this region. Thus we say that $\phi(x)$ has compact support in Ω because there exists a compact set $M \in \Omega$ such that $\phi(x) = 0$ in $\Omega - M$. Let $k = (k_1, k_2, \ldots, k_r)$ be a vector with integral components and let

$$|k| = \sum_{i=1}^{r} |k_i| \tag{2.107}$$

thus we define, with a multiindex notation, the symbol D^k to mean

$$D^k = \frac{\partial^{|k|}}{d\,x_1^{k_1}\,\partial x_2^{k_2} \ldots \partial x_r^{k_r}} \tag{2.108}$$

Thus, at first we assume that in the closed region $\overline{\Omega} = \Omega \cup \partial\Omega$ the real function $u(\mathbf{x})$ has continuous derivatives $D^k u$ for any k such that $|k| \leq m$. Let $\phi(\mathbf{x})$ be any function in the set Φ_m. On the surface $\partial\Omega$ the function $\phi(\mathbf{x})$ and all its derivatives vanish, so integration by parts yields the formula

$$\int_\Omega u D^k \phi \, d\mathbf{x} = (-1)^{|k|} \int_\Omega \phi D^k u \, d\mathbf{x} \tag{2.109}$$

Now, let $u(\mathbf{x})$ be some function that is summable in any interior subregion of Ω and assume that there exists a function $w(\mathbf{x})$ that is also summable in any interior subregion of Ω and is such that for any function $\phi(\mathbf{x}) \in \Phi_k$ we have the identity

$$\int_\Omega u D^k \phi \, d\mathbf{x} = (-1)^{|k|} \int_\Omega \phi \, w \, d\mathbf{x} \tag{2.110}$$

Then $w(\mathbf{x})$ is called the generalized kth derivative with respect to $(x_1, x_2, \ldots, x_r)$ of the function $u(\mathbf{x})$ in the region Ω. A generalized derivative is denoted by the usual symbol

$$w(\mathbf{x}) = D^k u \tag{2.111}$$

For definition we assume $D^0 u = u$.

We say that a region is star-shaped with respect to a given point if any ray beginning at this point intersects the boundary only once. For example, any convex region is star-shaped with respect to any of its points. In this section we will consider only finite regions that can be represented in the form of the union of a finite number of overlapping subregions that are each star-shaped with respect to any point in some ball.

Let a function $u(\mathbf{x})$ be summable in a region Ω with all generalized derivatives $D^k u$ for all $|k| \leq m$ in Ω. Then $u(\mathbf{x})$ has all their possible generalized derivatives from zero up to the order m, and they are summable in Ω.

The Sobolev spaces $H^{(m)}(\Omega)$ consist of all functions that in Ω have all their generalized derivatives $D^k u \in L_2(\Omega)$ for all $|k| \leq m$ in Ω. Both the functions and their derivatives up to the order m are thus square integrable in Ω. The norm of this space is given by the formula

$$\|\mathbf{u}\|^{(m)} = \left(\int_\Omega \sum_{0 \leq |k| \leq m} \left(D^k u\right)^2 dx \right)^{1/2} = \left(\sum_{0 \leq |k| \leq m} \|D^k u\|^2 \right)^{1/2} \tag{2.112}$$

and with the corresponding inner product

$$< u, v >^{(m)} = \int_{\Omega} \sum_{0 \le |k| \le m} \left(D^k u\right)\left(D^k v\right) dx = \sum_{0 \le |k| \le m} < D^k u, D^k v >_{L_2(\Omega)} \tag{2.113}$$

where the sum is taken over all possible combinations of indices such that $0 \le |k| \le m$. It is clear from the definition of $H^{(m)}(\Omega)$ that if $m_2 > m_1 > 0$, then

$$H^{(m_2)}(\Omega) \subset H^{(m_1)}(\Omega) \subset H^{(0)}(\Omega) = L_2(\Omega) \tag{2.114}$$

The embedding of $H^{(m_2)}(\Omega)$ in $H^{(m_1)}(\Omega)$ is compact, i.e., any set bounded in $H^{(m_2)}(\Omega)$ is compact in $H^{(m_1)}(\Omega)$. Another important result available for Sobolev spaces points out the properties of the element functions of these spaces.

For functions of a single variable the existence of the derivative implies continuity, whereas for functions of several variables this is not the case. A conjecture also is that if for a given underlying space R^r derivatives of sufficiently high order are square integrable, then this order might be strong enough to imply continuity of the function and perhaps continuity of its derivatives up to an order s. This is the import of the Sobolev's embedding theorem: if $u(x) \in H^{(m)}(\Omega)$ where Ω is an open set in R^r, then $u \in C^{(s)}(\Omega)$ provided

$$m > r/2 + s \tag{2.115}$$

and the Sobolev inequality

$$\|u\|_{C(\Omega)} \le M \|u\|^{(m)} \tag{2.116}$$

holds where $C(\Omega)$ denotes the space of functions that are continuous in Ω, M is a constant that does not depend on the function u(x), and

$$\|u\|_{C(\Omega)} \le \max_{\Omega} |u| \tag{2.117}$$

is the uniform norm. Further we have that the bounded set in $H^{(m)}(\Omega)$ is compact in $C(\Omega)$.

This theorem is of fundamental importance. All the smoothness properties of the functions $u(x) \in H^{(m)}(\Omega)$ can be clearly pointed out. In fact, if $u(x) \in H^{(m)}(\Omega)$, u will have generalized derivatives up to the order m. Thus, if we consider the generalized derivatives D^1u of order 1, they will have generalized up to the order $m - 1$; i.e., D^1u will belong to the space $H^{(m-1)}(\Omega)$. If we again apply the Sobolev theorem, we can deduce that if we have $m - 1 > r/2$, the first derivatives will be continuous and

$$\max_{\Omega} | D^1 u | \le C \| D^1 u \|^{(m-1)} \le C\left(\| D^1 u \|^2 + \| D^2 u \|^2 + \ldots + \| D^m u \|^2\right)$$

$$\le C\left(\| u \|^2 + \| D^1 u \|^2 + \| D^2 u \|^2 + \ldots + \| D^m u \|^2\right) = C \| D^1 u \|^{(m)} \tag{2.118}$$

This procedure can be further applied and we deduce that all the functions of the space $H^{(m)}(\Omega)$ are continuous in Ω) with the derivatives up to order s such that condition (2.115) holds, i.e. up to order

$$s < m - r/2 \tag{2.119}$$

and

$$\| u \|_{C^{(s)}(\Omega)} \le M \| u \|^{(m)} \tag{2.120}$$

where $C^{(s)}$ is the space of s-times continuously differentiable functions on Ω with norm

$$\| u \|_{C^{(s)}(\Omega)} = \sum_{0 \le |t| \le s} \max_{\Omega} | D^{(t)} u | \tag{2.121}$$

It is easy to recognize that the energy spaces analyzed in Section. 2.7 are special Sobolev spaces. Every definite positive operator defines an energy norm: it can be shown that the energy norm and the appropriate Sobolev norms are equivalent.

We apply these results to analyze the properties of the functions of some Sobolev spaces and of the corresponding energy spaces. Let us consider first of all the elastic beam and the corresponding linear differential equation

$$EI \frac{d^4 v}{dz^4} = p \tag{2.122}$$

with corresponding linear boundary conditions. The corresponding Sobolev space is the space $H^{(2)}$ [0,L], i.e., the space of the functions v(z) that in the closed interval [0,L] have all their generalized derivatives from zero up to the order two. Both the functions v(z) and their derivatives up to the order two are square integrable in [0,L]. The norm of this space is

$$\|u\|^{(2)} = \left(\int_\Omega (u^2)\, dx + \int_\Omega (D^1 u)^2\, dx + \int_\Omega (D^2 u)^2\, dx \right)^{1/2} \tag{2.123}$$

The corresponding energy space H_A is the space of all the functions v(z) defined in the closed interval [0,L] with generalized derivatives up to order two; both the functions and their derivatives up to order two are square integrable over [0,L]. In the space H_A is defined the norm

$$|||\, v \,||| = \left(\frac{1}{2} EI \int_0^L \left(D^2 v\right)^2 dz \right)^{1/2} \tag{2.124}$$

It is easy to show that

$$|||\, v \,||| \ge k_1 \| v \|^{(2)} \tag{2.125}$$

A set of displacements bounded in H_A[0,L] is also bounded in $H^{(2)}$[0,L]. Consider now the space $H^{(1)}$[0,L], i.e., the space of all the functions v(z) that have all their generalized derivatives from zero up to order 1 square integrable in [0,L]. The embedding of $H^{(2)}$[0,L] in $H^{(1)}$[0,L] is compact; i.e., any set bounded in $H^{(2)}$[0,L] is compact in $H^{(1)}$[0,L].

Furthermore, according to (2.119), the maximum order of the derivatives in the norm of $H^{(2)}$[0,L] is m = 2 and the dimension of the region Ω in which the functions of this space are defined (i.e., the interval [0,L]) is r = 1. Thus the functions v(z) with their first derivatives v'(z) are continuous in the interval [0,L] and

$$\max_{[0.1]} |v(z)| \le c_1 \|v\|^{(2)} \qquad \max_{[0.1]} |v'(z)| \le c_2 \|v\|^{(2)} \tag{2.126}$$

Let us consider now the case of elastic plates. In this case the energy norm is given by

$$\|w\|^{(2)} = \frac{D}{2} \int_\Omega \left\{ \left(\nabla^2 w\right)^2 + 2(1-\nu) \left[\left(\frac{\partial^2 w}{\partial x \partial y} \right)^2 - \frac{\partial^2 w \partial^2 w}{\partial x^2 \partial y^2} \right] \right\} d\Omega \tag{2.127}$$

In this case we have m = 2 and r = 2. Thus s < 1. The displacement functions w(x,y) belong to the space $H^{(2)}(\Omega)$ and have norm

$$|||\, w\, |||^{(2)} = \left(\int_\Omega (w)^2 d\Omega + \int_\Omega \left(D^1 w\right)^2 d\Omega + \int_\Omega \left(D^2 w\right)^2 d\Omega \right)^{1/2} \tag{2.128}$$

They are thus continuous in Ω and

$$\max_{(\Omega)} |w(x,y)| \le c_1 \, ||w||^{(2)} \tag{2.129}$$

Let us now consider the tridimensional elastic body. The energy norm is given by

$$|||\, \mathbf{u}\, |||^2 = \frac{1}{2}\int_\Omega \sigma_{ij}^{(1)}\ \varepsilon_{ij}^{(1)}\ d\Omega \tag{2.130}$$

where

$$\varepsilon_{ij}^{(1)} = \frac{1}{2}\int_\Omega \left(u_{i,j} + u_{j,i}\right) \tag{2.131}$$

is the first-order strain tensor and $\sigma_{ij}^{(1)}$ the corresponding stress tensor, given by

$$\sigma_{ij}^{(1)} = C_{ijpq}\ \varepsilon_{pq}^{(1)} \tag{2.132}$$

The displacement functions **u**(x,y,z) belong to the space $H^{(1)}(\Omega)$ and have norm

$$||\, u\, ||^{(1)} = \left(\int_\Omega (\mathbf{u}\cdot\mathbf{u})^2 d\Omega + \int_\Omega \left(D^1\mathbf{u}\cdot D^1\mathbf{u}\right) d\Omega \right)^{1/2} \tag{2.133}$$

In this case m = 1 and r = 3. Thus the displacement functions of the space $H^{(1)}(\Omega)$ of the elastic tridimensional body can be discontinuous functions.

REFERENCES

1. Brown, A. L. and Page, A., Elements of Functional Analysis, Van Nostrand Reinhold, London, 1970.
2. Boccara, N., Analyse Functionnelle: Une Introduction pour Physiciens, Ellipses, Paris, 1984.
3. Courant, R. and Hilbert, D., Methods of Mathematical Physics, Vol. I, II, Interscience, New York, 3rd ed., 1961.
4. Curtain, R. F. and Pritchard, A. J., Functional Analysis in Modern Applied Mathematics, Academic Press, London, 1977.
5. Fichera, G., Existence theorems in elasticity, Handbuch der Physik, Band VI a/2, Springer-Verlag, Berlin, 1976.
6. Field, M. J., Differential Calculus and its Applications, Van Nostrand Reinhold, New York, 1976.
7. Fiorenza, R., Elementi di Analisi funzionale ed Applicazioni. Liguori, Napoli, 1991.
8. Halmos, P., Introduction to Hilbert Spaces, Chelsea, New York, 1964.
9. Kolmogorov, A. N. and Fomin, S. V., Elementos de la Teorìa de Functiones y del Analysis Functional, MIR, Moscow, 1972.
10. Kreyzig, E., Introductory Functional Analysis with Applications, John Wiley & Sons, New York, 1978.
11. Lions, J. L., Cours d'Analyse Numérique, Ecole Polytechnique, Promotion 1971, Paris, 1973.
12. Liusternik, L. A., and Sobolev, V., Elements of Functional Analysis, Frederick Ungar, New York, 1961.
13. Luenberger, D. G., Optimization by Vector Space methods, John Wiley & Sons, New York, 1969.
14. Marti, J. T., Introduction to Sobolev Spaces and Finite Element Solution Of Elliptic Boundary Value Problems, Academic Press, London, 1986.
15. Mason, J., Methods of Functional Analysis for Application in Solid Mechanics, Elsevier, Amsterdam, 1985.

16. Mikhlin, S. G.,Variational Methods in Mathematical Physics, Pergamon Press, Oxford, 1964.
17. Mikhlin, S. G., The Problem of the Minimum of a Quadratic Functional, Holden-Day, San Francisco, 1965.
18. Milne, R. D. Applied Functional Analysis: An Introductory Treatment, Pitman Advanced Publ. Program, Boston, 1980.
19. Oden, J. T., Applied Functional Analysis, Prentice-Hall, Englewood Cliffs, NJ, 1979.
20. Reddy, J. N. and Rasmussen, M. L., Advanced Engineering Analysis, John Wiley & Sons, New York, 1982.
21. Reddy, J. N., Applied Functional Analysis and Variational Methods in Engineering, McGraw-Hill, New York, 1986.
22. Riesz, F. and SZ. Nagy, B., Functional Analysis, Frederick Ungar, New York, 1955.
23. Schechter, M., Principles of Functional Analysis, Academic Press, New York, 1971.
24. Smirnov, V. I., A Course of Higher Mathematics, Vol. 5 Integration and Functional Analysis, Pergamon Press, New York, 1964.
25. Sobolev, S. L., Applications of Functional Analysis in Mathematical Physics, Transl. of Math. Monographs, Vol. 7, American Mathematical Society, Providence, RI, 1963.
26. Vainberg, M. M. and Trenogin, V. A., Theory of Branching of Solutions of Non-linear Equations, Noordhoff Int. Publ., Leyden, 1974.
27. Vainberg, M. M., Variational Methods for the Study of Non-linear Operators, Holden-Day, San Francisco, 1964.
28. Yosyda, K., Functional Analysis, John Wiley & Sons, New York, 1979.

Chapter 3

STABILITY ANALYSIS OF CONTINUOUS ELASTIC SYSTEMS

3.1 INTRODUCTION

The aim of this chapter is to present a new formulation of the theory of equilibrium stability of infinite-dimensional elastic structural systems loaded by conservative forces. In a first general statement of the theory, the norm of the space of the configurations of the system will not be explicitly chosen and stability criteria will be only formally given. On the other hand, the requirement of stability of the unstressed initial state in the context of the energy method implies that the definition of stability properly has to be given by assuming the "energy norm" as the measure of distances among configurations. In this framework, stability, as well as instability, is analyzed using the second differential energy criterion. Moreover, the undecided case of stability at the critical state will be examined.

3.2 ANALYSIS OF STABILITY

3.2.1 DEFINITION OF STABILITY

Let us consider a continuous elastic system T. Let Ω be the region of the space occupied by T at the reference configuration C, frequently assumed coinciding with the stress free configuration C_i. An admissible deformation of T can be represented by the displacement field $\mathbf{u}(\mathbf{x})$, $\mathbf{x} \in \Omega$, satisfying suitable boundary conditions. The space of all the admissible displacement fields $\mathbf{u}(\mathbf{x})$ is denoted by $H(\Omega)$, assumed to be a Banach space with the norm $||\mathbf{u}||_H$, not yet specified. For simplicity, we will also indicate this norm with $||\mathbf{u}||$.

Let us consider now a motion of T, defined at any instant t by the pair

$$\mathbf{m}(\mathbf{x},t) = \begin{bmatrix} \mathbf{u}(\mathbf{x},t) \\ \dot{\mathbf{u}}(\mathbf{x},t) \end{bmatrix} \tag{3.1}$$

where $\dot{\mathbf{u}}(\mathbf{x},t)$ is the velocity field; the $\dot{\mathbf{u}}(\mathbf{x},t)$ is assumed to belong to the space $L_2(\Omega)$. The norm of $\dot{\mathbf{u}}(\mathbf{x},t)$ can be thus represented by the kinetic energy

$$||\dot{\mathbf{u}}||^2 = \mathcal{T}(\dot{\mathbf{u}}) = \frac{1}{2}\int_\Omega \rho\dot{\mathbf{u}} \cdot \dot{\mathbf{u}}\, d\Omega \tag{3.2}$$

where ρ is the mass density. The space H_m of the states of motion will thus be the product space $H(\Omega) \times L_2(\Omega)$ with norm

$$||\mathbf{m}||^2 = ||\mathbf{u}||_H^2 + \mathcal{T}(\dot{\mathbf{u}}) = ||\mathbf{u}||^2 + \mathcal{T}(\dot{\mathbf{u}}) \tag{3.3}$$

Consider now an equilibrium configuration C_0 of T loaded by a given distribution of conservative forces. Let $\mathbf{u}_0$ be the corresponding displacement field that defines C_0 with respect to the initial unstressed configuration C_i.

We focus our attention on the behavior of the system T disturbed at the equilibrium configuration C_0. Next, let us impart to the system T and at time t = 0 the disturbance

$$\mathbf{m}(\mathbf{x},0) = \begin{bmatrix} \mathbf{u}(\mathbf{x},0) \\ \dot{\mathbf{u}}(\mathbf{x},0) \end{bmatrix} \tag{3.4}$$

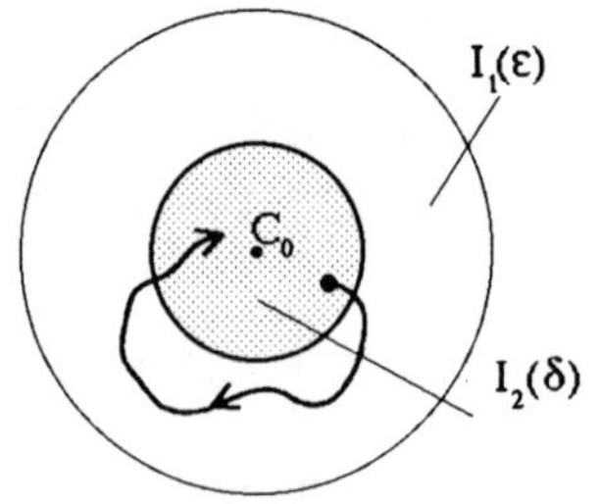

FIGURE 3.1

The subsequent disturbed motion of T is

$$\mathbf{m}(\mathbf{x},t) = \begin{bmatrix} \mathbf{u}(\mathbf{x},t) \\ \dot{\mathbf{u}}(\mathbf{x},t) \end{bmatrix} \tag{3.5}$$

where $\mathbf{u}(\mathbf{x},t)$ is now the additional displacement that moves T from C_0 to another configuration C and $\dot{\mathbf{u}}(\mathbf{x},t)$ is the corresponding velocity field. According to the classical Liapounov[1] definition of stability the equilibrium configuration C_0 of T (Fig. 3.1) is stable if, for each spherical neighborhood $I_1(\varepsilon)$ of C_0 in H_m, it is possible to find a corresponding spherical neighborhood $I_2(\delta)$ such that every initial disturbance that starts within $I_2(\delta)$ yields a disturbed motion always contained in $I_1(\varepsilon)$.

In short, the configuration C_0 is stable if

$$\forall\, \varepsilon > 0, \exists\, \delta > 0 : \|\mathbf{m}(\mathbf{x},0)\| < \delta \Rightarrow \|\mathbf{m}(\mathbf{x},t)\| < \varepsilon\, \forall\, t < 0 \tag{3.6}$$

3.2.2 THE ENERGY CRITERION OF STABILITY

The structural system is elastic and the external loads are conservative. Hence the potential energy of all the external and internal forces acting on the system does exist: it will be a nonlinear functional $\mathcal{E}(\mathbf{u})$ which we assume to be defined in the space $H(\Omega)$ of the admissible displacement fields $\mathbf{u}(\mathbf{x})$. It consists of a line integral, a surface integral, a volume integral, or a sum of such integrals, whose integrals all depend on the displacement vector $\mathbf{u}$ and its spatial derivatives.

The system T, in equilibrium at the fixed configuration C_0, is moved from C_0 to another configuration C, and $\mathbf{u}$ is the additional displacement field that takes T from C_0 to C. In the passage of T from C_0 to C the corresponding increment of the potential energy $\mathcal{E}(\mathbf{u})$ is

$$\Delta\mathcal{E}(\mathbf{u}) = \mathcal{E}(C) - \mathcal{E}(C_0) \tag{3.7}$$

For simplicity this increment of the potential energy $\Delta\mathcal{E}(\mathbf{u})$ will be denoted by $\mathcal{E}(\mathbf{u})$; by definition, $\mathcal{E}(C_0) = 0$.

According to the principle of virtual works, at the equilibrium configuration C_0 the following variational condition holds

$$D\mathcal{E}(\mathbf{0}; \delta\mathbf{u}) = 0 \qquad \forall\, \delta\mathbf{u} \in H(\Omega) \tag{3.8}$$

where $D\mathcal{E}(\mathbf{0}; \delta\mathbf{u})$ is the Gateaux differential of $\mathcal{E}(\mathbf{u})$ at C_0 in the direction $\delta\mathbf{u}$.

Next, in order to overcome the difficulties pointed out in Section 1.2, let us analyze the Lagrange-Dirichlet theorem in its extension from discrete to continuous systems. For an elastic infinite-dimensional system, it has been proven that the equilibrium solution is stable in the sense of Liapounov when the condition of a minimum of the potential energy is replaced by that one of a "potential well". Equivalent definitions of a "potential well" have been given or used by various authors, such as Gurtin,[2] Koiter,[3] Coleman,[4] Hughes, Kato, and Marsden,[5]

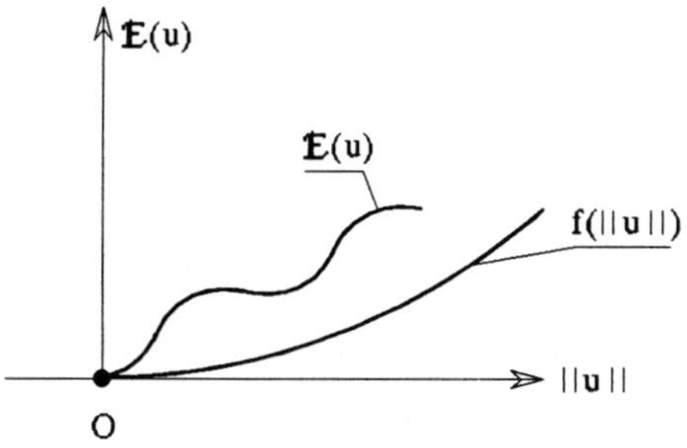

FIGURE 3.2

Knops,[6] and Como and Grimaldi.[17] The extension of the Lagrange-Dirichlet theorem can be given in the following way: a conservative system is stable at an equilibrium configuration C_0 in $H(\Omega)$ if

1. the potential energy functional $\mathcal{E}(\mathbf{u})$ is continuous at C_0, i.e.,

$$\lim_{\|\mathbf{u}\|_H \to 0} \mathcal{E}(\mathbf{u}) = 0 \tag{3.9}$$

2. we can find a number $\bar{\rho} > 0$ such that in the spherical neighborhood

$$S_{\bar{\rho}} = \{\mathbf{u} \in H(\Omega) : 0 \leq \|\mathbf{u}\|_H \leq \bar{\rho}\} \tag{3.10}$$

of the origin $\mathbf{0} = C_0$, there exists a function $f(\|\mathbf{u}\|_H)$ of the displacement norm $\|\mathbf{u}\|_H$ such that

a. $f(\|\mathbf{u}\|_H) > 0$ for $\mathbf{u} \neq \mathbf{0}$, $\mathbf{u} \in S_{\bar{\rho}}$ $f(\|\mathbf{u}\|_H) = 0 \Leftrightarrow \mathbf{u} = \mathbf{0}$ (3.11)

b. $f(\|\mathbf{u}_2\|) > f(\|\mathbf{u}_1\|)$ if $\|\mathbf{u}_2\| > \|\mathbf{u}_1\|$ and $\mathbf{u}_1, \mathbf{u}_2 \in S_{\bar{\rho}}$ (3.12)

c. $\mathcal{E}(\mathbf{u}) \geq f(\|\mathbf{u}\|)$ $\mathbf{u} \in S_{\bar{\rho}}$ (3.13)

i.e., the potential energy is bounded below by a positive strictly increasing function $f(\|\mathbf{u}\|_H)$ of the norm $\|\mathbf{u}\|_H$ in $S_{\bar{\rho}}$.

Condition 2 corresponds to the existence of a "potential well" for the energy functional $\mathcal{E}(\mathbf{u})$ at the origin $\mathbf{u} = \mathbf{0}$ (Fig. 3.2). An equivalent condition is the assumption that the infima of $\mathcal{E}(\mathbf{u})$ on the spheres ∂S_ρ, $0 < \rho < \bar{\rho}$, centered at C_0 are monotonically increasing with distance from C_0. It is worthwhile, for the subsequent analysis, to recall the main lines of the proof of this extension of the Lagrange-Dirichlet theorem.

First let us define in H_m the total energy $E_T(\mathbf{m})$ of the system T as the functional

$$E_T(\mathbf{m}) = \mathcal{E}(\mathbf{u}) + \mathcal{T}(\dot{\mathbf{u}}) \tag{3.14}$$

We will show, according to assumption 2, that, like $\mathcal{E}(\mathbf{u})$ in H, the total energy functional $E_T(\mathbf{m})$, defined in the space H_m, also has positive infima over the spherical boundaries of the origin $\mathbf{0} = C_0$; i.e.,

$$\inf_{\partial S_\varepsilon} E_T(\mathbf{m}) > 0 \quad \forall\, \varepsilon : 0 < \varepsilon < \bar{\rho} \tag{3.15}$$

where

$$\partial S_\varepsilon = \{\mathbf{m} \in H_m : \|\mathbf{m}\| = \varepsilon,\ \varepsilon < \bar{\rho}\} \tag{3.16}$$

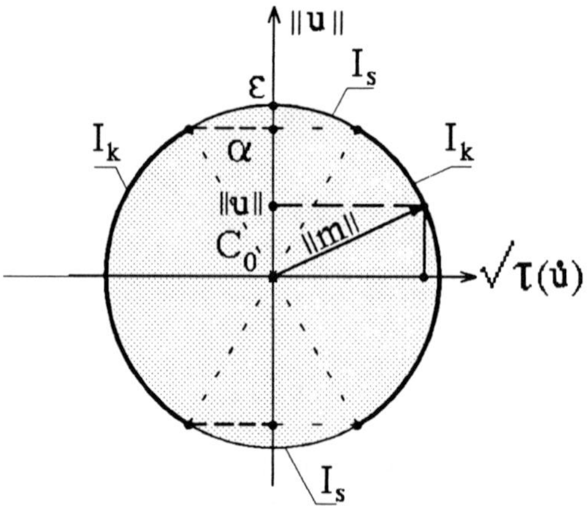

FIGURE 3.3

is the boundary of a spherical neighborhood, belonging to $S_{\bar{\rho}}$, of the origin O in the space H_m. Let us partition ∂S_ε as

$$\partial S_\varepsilon = I_k \cup I_s \tag{3.17}$$

where

$$I_k = \left\{ \mathbf{m} = \begin{bmatrix} \mathbf{u} \\ \dot{\mathbf{u}} \end{bmatrix} \in \partial S_\varepsilon : 0 \le \|\mathbf{u}\| < \alpha < \varepsilon < \bar{\rho} \right\} \tag{3.18}$$

$$I_s = \left\{ \mathbf{m} = \begin{bmatrix} \mathbf{u} \\ \dot{\mathbf{u}} \end{bmatrix} \in \partial S_\varepsilon : \alpha \le \|\mathbf{u}\| \le \varepsilon < \bar{\rho} \right\} \tag{3.19}$$

The sets I_k and I_s respectively correspond to states of motion of norm ε with dominant contents of kinetic or of strain energy (Fig. 3.3). For any state of motion $\mathbf{m} = \begin{bmatrix} \mathbf{u} \\ \dot{\mathbf{u}} \end{bmatrix} \in I_k$ we have, according to (3.3) and (3.18)

$$\varepsilon^2 \ge \mathcal{T}(\dot{\mathbf{u}}) = \varepsilon^2 - \|\mathbf{u}\|^2 > \left(\varepsilon^2 - \alpha^2\right) > 0 \tag{3.20}$$

and, accordingly,

$$E_T(\mathbf{m}) \ge \mathcal{T}(\dot{\mathbf{u}}) > \left(\varepsilon^2 - \alpha^2\right) > 0, \; \mathbf{m} \in I_k \tag{3.21}$$

Further, for $\mathbf{m} = \begin{bmatrix} \mathbf{u} \\ \dot{\mathbf{u}} \end{bmatrix} \in I_s$, where

$$0 < \mathcal{T}(\dot{\mathbf{u}}) = \varepsilon^2 - \|\mathbf{u}\|^2 \le \left(\varepsilon^2 - \alpha^2\right) \tag{3.22}$$

because

$$\alpha^2 \le \|\mathbf{u}\|^2 \le \varepsilon^2 \tag{3.23}$$

Thus, when $\mathbf{m} \in I_s$ the displacement $\mathbf{u}$, (i.e., the component of $\mathbf{m}$ in H) belongs to the annular region A (Fig. 3.4)

$$A = \{\mathbf{u} \in H : \alpha \le \|\mathbf{u}\| \le \varepsilon, \; \varepsilon < \bar{\rho}\} \tag{3.24}$$

But for $\mathbf{u} \in A \subset S_{\bar{\rho}}$, according to assumption

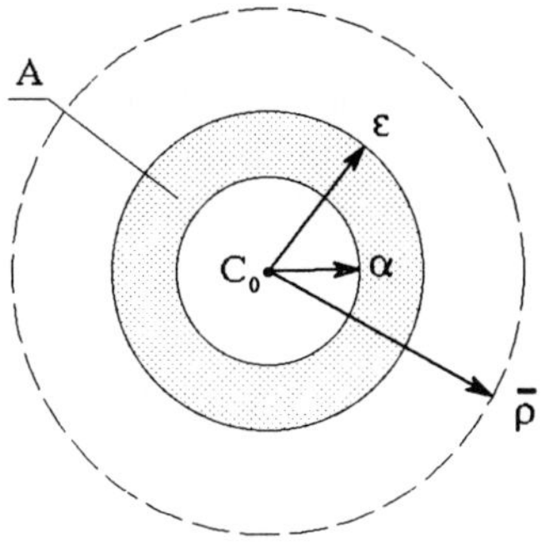

FIGURE 3.4

$$\mathcal{E}(\mathbf{u}) \geq f(\|\mathbf{u}\|) \geq f(\alpha) > 0 \tag{3.25}$$

Thus

$$E_T(\mathbf{m}) \geq \mathcal{E}(\mathbf{u}) \geq f(\alpha) > 0 \qquad \mathbf{m} \in I_s \tag{3.26}$$

and $E_T(\mathbf{m})$ is also bounded below in I_s. Then, taking into account (3.17), (3.21), and (3.26), condition (3.15) is satisfied.

Further, taking into account (3.3), the continuity of $\mathcal{E}(\mathbf{u})$ at $C_0 = \mathbf{0} \in H(\Omega)$ implies continuity of $E_T(\mathbf{m})$ at $C_0 = \mathbf{0} \in H_m$. Hence,

$$\lim_{\|\mathbf{m}\|H_m \to 0} E_T(\mathbf{m}) = E_T(\mathbf{0}) = 0 \tag{3.27}$$

Now, given any $\varepsilon : 0 < \varepsilon < \bar{\rho}$ let $E_T^*(\varepsilon)$ be:

$$E_T^*(\varepsilon) = \inf_{\partial S_\varepsilon} E_T(\mathbf{m}) > 0 \tag{3.28}$$

For every ε let $0 < \theta < E_T^*(\varepsilon)$. Thus, according to continuity of $E_T(\mathbf{m})$ at 0 we can write

$$\forall\, \varepsilon > 0, \forall\, \theta : 0 < \theta < E_T^*(\varepsilon), \exists\, \delta > 0 : \|\mathbf{m}\| < \delta \Rightarrow E_T(\mathbf{m}) < \theta < E_T^*(\varepsilon) \tag{3.29}$$

Hence, if the disturbance $\mathbf{m}(\mathbf{x},t)$ initiates at the point $\mathbf{m}(\mathbf{x},0)$ contained inside the spherical neighborhood S_δ, i.e., such that $\|\mathbf{m}(\mathbf{x},0)\| < \delta$, we have $E_T(\mathbf{m}(0)) < \theta < E_T^*(\varepsilon)$. But the total energy $E_T(\mathbf{m})$ is constant during the motion, so, for every time $t > 0$, $E_T(\mathbf{m}(t)) < \theta < E_T^*(\varepsilon)$.

Thus the point representing the state of motion of T in H_m can never reach the boundary ∂S_ε on which $E_T(\mathbf{m}) \geq E_T^*(\varepsilon)$. The disturbed motion will therefore always be contained within the neighborhood S_ε.The equilibrium is thus stable at C_0 and the theorem is proven.

Condition (3.13) represents a stronger version of condition (1.56) that, together with continuity of $\mathcal{E}(\mathbf{u})$ in the whole neighborhood of C_0, implies stability of C_0 for discrete elastic systems.

The weaker condition

$$\exists\, \bar{\rho} > 0 : \mathbf{u} \in H, 0 < \|\mathbf{u}\|_H < \bar{\rho} \Rightarrow \mathcal{E}(\mathbf{u}) > 0 \tag{3.30}$$

is unable to extend condition (1.56) to continuous elastic systems. In fact, the sphere ∂S_ε is not a compact set in the infinite-dimensional space H. Thus, in spite of the assumed continuity of $\mathcal{E}(\mathbf{u})$ in a neighborhood S_ρ of O in H, the minimum of $\mathcal{E}(\mathbf{u})$ could not exist in the set ∂S_ε and, although $\mathcal{E}(\mathbf{u}) > 0\ \forall\, \mathbf{u} \in \partial S_\rho$, the greatest lower bound of $\mathcal{E}(\mathbf{u})$ on ∂S_ρ could be zero. In any case, without the additional condition (3.13) the infimum of $E_T(\mathbf{m})$ over ∂S_ε could be zero. Condition (3.13), in place of (3.30), with (3.9) is thus sufficient to formulate the energy criterion for infinite-dimensional elastic systems.

Conditions (3.9) and (3.13), assumed in this section, are implicitly contained inside the conditions 1 and 2 of Section 1.3. The continuous function $\mathcal{E}(\mathbf{u})$, the potential energy of elastic discrete systems, in fact achieves its minimum on the compact sphere ∂S_ρ centered at C_0. Because of the continuity of $\mathcal{E}(\mathbf{u})$ in a whole neighborhood of C_0, there exists a ball $S_{\rho'}$ centered at C_0 of radius ρ' such that inside this ball the minima of $\mathcal{E}(\mathbf{u})$ will monotonically increase with the distance $\|\mathbf{u}\|$ from C_0.

3.2.3 THE SECOND DIFFERENTIAL ENERGY CRITERION

According to definitions 2 given in the previous section, the function

$$f(\|\mathbf{u}\|) = \gamma \|\mathbf{u}\|^\mu \quad \mu > 1, \gamma > 0 \tag{3.31}$$

positive and strictly increasing with $\|\mathbf{u}\|$, can be used to extend the Lagrange-Dirichlet theorem to continuous elastic systems. After this preliminary statement, examine the validity of the stability test, recalled in Chapter 1, involving the second differential

$$\mathcal{E}_2(\mathbf{u}) = \frac{1}{2!} D^2\mathcal{E}(0; \mathbf{u}) \tag{3.32}$$

of the potential energy functional $\mathcal{E}(\mathbf{u})$, evaluated at C_0 along the displacement field $\mathbf{u}$. If, with a suitable choice of the norm $\|\mathbf{u}\|$,

$$\mathcal{E}_2(\mathbf{u}) \geq \gamma_1 \|\mathbf{u}\|^2 \Rightarrow \mathcal{E}(\mathbf{u}) \geq \gamma_2 \|\mathbf{u}\|^2 \quad \gamma_1 > 0, \gamma_2 > 0 \tag{3.33}$$

condition (3.13) is satisfied and with (3.9a) we have stability at C_0. Consequently the following criterion of elastic stability can be given: Let $\mathcal{E}(\mathbf{u})$ be twice Fréchet differentiable in H at the equilibrium configuration C_0 and let ∂S^* be the boundary of the unit sphere in the space H. Thus the elastic system T is stable at C_0 if

$$\omega = \inf_{\partial S^*} \mathcal{E}_2(\mathbf{u}^*) > 0 \tag{3.34}$$

with $\mathbf{u}^* \in \partial S^*$. Equivalently there is stability at C_0 if

$$\omega = \inf_{H-\{\mathbf{0}\}} \frac{\mathcal{E}_2(\mathbf{u})}{\|\mathbf{u}\|_H^2} > 0 \tag{3.35}$$

or

$$\mathcal{E}_2(\mathbf{u}) > \omega \|\mathbf{u}\|_H^2, \quad \omega > 0 \tag{3.36}$$

It is useful to give a sketch of the proof of this important statement. According to Taylor's theorem, we have, in fact

$$\mathcal{E}(\mathbf{u}) = \mathcal{E}_1(\mathbf{u}) + \mathcal{E}_2(\mathbf{u}) + o(\|\mathbf{u}\|_H^2) \tag{3.37}$$

where, because of the assumed strong (Fréchet) differentiability of $\mathcal{E}(\mathbf{u})$ at $\mathbf{0}$, we have

$$\mathcal{E}_2(\mathbf{u}) \leq k \|\mathbf{u}\|_H^2 \tag{3.38}$$

and

$$\lim_{\|\mathbf{u}\| \to 0} \frac{o(\|\mathbf{u}\|_H^2)}{\|\mathbf{u}\|_H^2} = 0 \tag{3.39}$$

Equilibrium of the configuration C_0 yields, according to (3.8),

$$\mathcal{E}(\mathbf{u}) = \mathcal{E}_2(\mathbf{u}) + o\left(\|\mathbf{u}\|_H^2\right) \tag{3.40}$$

Hence, if u^* is an element of ∂S^*, that is, such that $\|\mathbf{u}^*\|_H = 1$, with $\mathbf{u}^* = \mathbf{u} / \|\mathbf{u}\|$ we get

$$\mathcal{E}(\mathbf{u}) = \|\mathbf{u}\|_H^2 \left[\mathcal{E}_2(\mathbf{u}^*) + \frac{o\left(\|\mathbf{u}\|_H^2\right)}{\|\mathbf{u}\|_H^2} \right] \tag{3.41}$$

and according to (3.34)

$$\mathcal{E}(\mathbf{u}) \geq \|\mathbf{u}\|_H^2 \left[\omega + \frac{o\left(\|\mathbf{u}\|_H^2\right)}{\|\mathbf{u}\|_H^2} \right] \geq \|\mathbf{u}\|_H^2 \left[\omega - \frac{\left|o\left(\|\mathbf{u}\|_H^2\right)\right|}{\|\mathbf{u}\|_H^2} \right] \tag{3.41$'$}$$

Taking into account (3.39): $\forall\, \varepsilon > 0\ \exists\, \rho > 0 : \mathbf{u}\, \varepsilon\, \mathbf{H}$,

$$0 < \|\mathbf{u}\|_H < \rho \Rightarrow \frac{\left|o(\|\mathbf{u}\|_H^2\right|}{\|\mathbf{u}\|_H^2} < \varepsilon \tag{3.41$''$}$$

But we can choose ε such that $\omega - \varepsilon > 0$. Thus;

$$\exists\, \rho > 0 : 0 < \|\mathbf{u}\|_H < \rho \Rightarrow \mathcal{E}(\mathbf{u}) > (\omega - \varepsilon)\, \|\mathbf{u}\|_H^2 > 0 \tag{3.42}$$

and the proof is completed.

To summarize, for infinite-dimensional elastic systems the second differential criterion implies equilibrium stability at C_0 if

1. $\mathcal{E}(\mathbf{u})$ is at least twice strong differentiable at C_0.
2. Condition (3.34) is satisfied.

The condition $\mathcal{E}_2(\mathbf{u}) \geq 0$ is only necessary for stability, so the Hadamard condition[7] is only a necessary condition of stability.

We want to point out now, using the semi-differentiability concept,[8] that we can obtain a condition, weaker than (3.39) that, always coupled to (3.34), is able to imply the existence of a potential well of $\mathcal{E}(\mathbf{u})$ at the equilibrium state C_0 and thus stability.

Let us assume conditions (3.34), (3.38), and, in place of (3.39), the weaker condition, that for every $\varepsilon > 0$, there exists a spherical neighborhood S_δ of C_0 such that

$$-\varepsilon < \frac{o\left(\|\mathbf{u}\|_H^2\right)}{\|\mathbf{u}\|_H^2} \qquad \forall\, \mathbf{u} : 0 < \|\mathbf{u}\|_H < \delta \tag{3.43}$$

In fact, from conditions (3.41) and (3.43), and by taking ε such that $\omega - \varepsilon > 0$, we again get

$$\mathcal{E}(\mathbf{u}) \geq (\omega - \varepsilon)\, \|\mathbf{u}\|_H^2 \qquad \forall\, \mathbf{u} : \|\mathbf{u}\|_H < \delta \tag{3.42$'$}$$

which yields stability. Thus conditions (3.34), (3.38), coupled to the twice semi-differentiability of $\mathcal{E}(\mathbf{u})$ in place of (3.39), imply stability of C_0.

For infinite-dimensional elastic structures the existence of a "potential well" at the equilibrium state is sufficient to stability. Of course, the problem arises of determining under which conditions the potential energy $\mathcal{E}(\mathbf{u})$ posseses a "potential well" or is twice strong differentiable or semi-differentiable. Before trying to solve this problem, another question, due to the norm dependance of stability definition for infinite dimensional systems, must also be analyzed.

3.3 THE REQUIREMENT OF STABILITY AT THE UNSTRESSED STATE; THE CHOICE OF THE ENERGY NORM

The complete solution of the stability problem of a continuous system can be obtained, from a mathematical point of view, only by determining the sets of norms that imply stability or instability. From a physical point of view, on the other hand, any class of stability problems of continuous systems requires the formulation of a mechanical model of the phenomenon and, consequently, the choice of a suitable norm.

In the wide class of buckling and snapping problems in structural mechanics, the unstable state is generally attained during a loading process, when the system finds a way to elude the loading action. Unstable equilibriums are, in fact, always strictly connected to the action of external loads or to a state of internal stresses.

Let us focus our attention on the unstressed or "natural" configuration of the system. A physically consistent theory of stability should satisfy the requirement of stability of the *unstressed state* of the system.

It is well known, on the other hand, that the "natural" configuration of a continuous elastic system can itself be unstable with respect to some choices of the norm. There are many interesting examples in the literature.[9,10,11] So, for instance, taking as measure of the initial disturbance the corresponding energy increment and, as a measure of the consequent disturbed motion, the local values of the displacement fields (and of their gradients) together with the displacement velocities, the free configuration of a sphere of elastic material turns out to be unstable. This result can be applied to any sufficiently small internal spherical region of an elastic body. In bidimensional or tridimensional elastic wave propagation, in fact, momentary and local concentrations of energy can occur. An interesting example of instability of a beam at its natural state was also shown in Section 1.2.

The unstressed rectilinear configuration of an elastic rod is in fact unstable if we choose as measure of the distance between the flexural perturbed motions **m**(z,t) and the initial rectilinear configuration of the beam each one of the norms

$$\|\mathbf{m}\|_1 = \left[\int_0^L v^2 dz\right]^{1/2} + \left[\int_0^L \dot{v}^2 dz\right]^{1/2} \tag{3.44}$$

$$\|\mathbf{m}\|_2 = \max_{z\in[0,L]} |v(z)| + \max_{z\in[0,L]} |\dot{v}(z)| \tag{3.45}$$

i.e., the L_2 norm and the uniform norm. In both cases, in fact, unbounded amounts of strain energy can be imparted to the rod with very "small" disturbances (of course, "small" in the sense of the chosen norms (3.44) or (3.45)). On the contrary, the "natural" rectilinear configuration of the bar is stable with respect to the energy norm, i.e., the norm which involves the second derivatives of the displacement functions.

Summing up all the previous considerations, we can affirm that the main difficulties contained in the statement of the energy stability criteria of elastic continuous structural systems are related to the following arguments:

1. The norm dependence of the stability concept for continuous systems demands the choice of a suitable norm satisfying some essential physical requirements. Among them, according to Como and Grimaldi,[12] the *stability of the unstressed configuration* can be considered a crucial starting point in formulating a theory of stability of continuous structural systems.

2. The assumption of smoothness — for instance the n times strong differentiability ($n \geq 2$) — of potential energy is generally verified for the elastic discrete systems. On the contrary, for continuous elastic systems the smoothness of the energy functional depends on the chosen norm.
3. It is very difficult to have a potential energy functional that is sufficiently smooth with respect to norms satisfying the requirement of the stability of the unstressed state.

Let us attempt to obtain now more information from the condition 1. Hence, we try to apply the second variation energy criterion stated in Section 3.2.2 to continuous elastic systems in their *unstressed states.*

The second differential of the potential energy $\mathcal{E}(\mathbf{u})$ at the "natural" configuration C_i of the system is

$$\left[\mathcal{E}_2(\mathbf{u})\right]_{\text{Nat}} = \frac{1}{2}\int_\Omega s_{ij}\, e_{ij}\, d\Omega = \frac{1}{2}|||\mathbf{u}|||^2 \tag{3.46}$$

where $|||\mathbf{u}|||$ is the "energy" norm of the admissible displacement field $\mathbf{u}(\mathbf{x})$.

Thus, in order to apply the energy criterion of the second differential $\mathcal{E}_2(\mathbf{u})$, the potential energy functional $\mathcal{E}(\mathbf{u})$ has to be at least twice Fréchet differentiable at the "natural" configuration C_i. Thus the functional $\mathcal{E}_2(\mathbf{u})$ has to be quadratic and continuous with respect to the chosen norm $||\mathbf{u}||_H$. Then we have

$$\exists k > 0 : \left[\mathcal{E}_2(\mathbf{u})\right]_{\text{Nat}} \leq k\,||\mathbf{u}||_H^2 \quad \forall\ \mathbf{u} \in H(\Omega) \tag{3.47}$$

or, by means of (3.46)

$$\exists k > 0 : |||\mathbf{u}|||^2 \leq k\,||\mathbf{u}||_H^2 \ \forall\ \mathbf{u} \in H(\Omega) \tag{3.48}$$

i.e., the norm $||\ ||_H$ in $H(\Omega)$ *cannot be weaker* than the "energy" norm $|||\mathbf{u}|||$. Thus, for instance, we cannot use the uniform norm (2.6) or the L_2 norm (2.7′) as measures of distance among configurations of the infinite-dimensional rod. On the contrary, for the elastic beam, to satisfy (3.48) with $v \in C^{(2)}$ we could use the norm

$$||\mathbf{v}||_H = \max_{z\in[0,L]} |\mathbf{v}(z)| + \max_{z\in[0,L]} |\mathbf{v}'(z)| + \max_{z\in[0,L]} |\mathbf{v}''(z)| \tag{3.48′}$$

After this first result, let us go back now to the condition of stability at the stress-free state.

To have stability at the unstressed state of the system it is thus necessary, according to (3.34), that

$$\inf_{\mathbf{u}\in H(\Omega),\mathbf{u}\neq\mathbf{0}} \frac{\left[\mathcal{E}_2(\mathbf{u})\right]_{\text{Nat}}}{||\mathbf{u}||_H^2} = \frac{1}{2} \inf_{\mathbf{u}\in H(\Omega),\mathbf{u}\neq\mathbf{0}} \frac{|||\mathbf{u}|||^2}{||\mathbf{u}||_H^2} = \omega_1 > 0 \tag{3.49}$$

where $||u||_H$ is the chosen norm in the space $H(\Omega)$ of the admissible displacement fields. But condition (3.49) means

$$|||\mathbf{u}||| \geq (2\omega_1)^{1/2}\ ||\mathbf{u}||_H \tag{3.50}$$

i.e., the norm $||\mathbf{u}||_H$ of $H(\Omega)$ *cannot be stronger* than the energy norm $|||\mathbf{u}|||$. Thus, according to Como and Grimaldi,[12] we deduce that the second differential energy criterion, in its classical form, entails stability of the unstresssed configuration of the continuous elastic structures only if the norm in $H(\Omega)$ is equivalent to the energy norm in the sense that

$$k_1 |||\mathbf{u}||| \leq ||\mathbf{u}||_H \leq k_2 |||\mathbf{u}||| \quad k_1, k_2 \geq 0 \tag{3.51}$$

Condition (3.51) cannot be satisfied by choosing the L_2 norm, the uniform norm or norms like (3.45′). The use of uniform norm like (3.45′) — on the contrary, often assumed — is therefore unacceptable to formulate energy criteria of stability for infinite dimensional elastic systems.

On the other hand, the only requirement of stability at the unstressed configuration does not necessarily require choosing of the energy norm. On the contrary, the requirement of stability at the unstressed state, together with the use of the energy criterion and the consequent assumption of a smooth potential energy functional, requires choosing of energy norm.

We can also use the concept of semi-differentiability, discussed in Sections 2.8.3 and 3.2.2, to obtain a condition, weaker than the twice strong differentiability at C_0 of $\mathcal{E}(\mathbf{u})$, able to imply stability of the unstressed state. In this case, always with conditions (3.48) and (3.50), i.e., always choosing the energy norm, only the twice semi-differentiability of $\mathcal{E}(\mathbf{u})$ is required, i.e., the (3.43) in place of the (3.39).

In the next chapter, studying the smoothness properties of the potential energy functional of various structural systems such as beams, plates, and shells, it will be shown that the requirement of strong differentiability of $\mathcal{E}(\mathbf{u})$ is frequently met in the energy spaces of the admissible displacement fields.

3.4. STABILITY ANALYSIS WITH RESPECT TO THE ENERGY NORM

3.4.1 PRELIMINARY DEFINITIONS AND ASSUMPTIONS

Let us consider a structural elastic system T. Let Ω be the region, with a sufficiently regular boundary $\partial\Omega$, occupied by T at the unstressed initial configuration C_i. The positive-definite linear elastic "stiffness" operator A, which controls the response of the structure in the passage from C_i to the close configuration C, is associated to the system.

Also let C_0 be the fundamental configuration of the system T in equilibrium under the action of the given loads and let

$$\mathbf{u}(\mathbf{P}) \qquad \mathbf{P} \in \Omega$$

be the field that defines the additional displacement of T from C_0 to a nearby configuration C.

The space M of these additional displacements can be represented by all the functions in Ω, continuous with all their derivatives up to the order n and satisfying suitable linear boundary conditions. We associate to this space M the energy product (2.54) and the energy norm (2.55) by using the "stiffness" operator A of the system just recalled. This inner product space M, dense in the space $L_2(\Omega)$, is not complete. It can be completed according to the procedure that we have recalled in Section 2.7.1; the completed space is the energy space H_A.

Let us consider now the functional $\mathcal{E}(\mathbf{u})$, i.e., the additional potential energy in the displacement of the elastic structural system from configuration C_0 to the nearby configuration C. The functional $\mathcal{E}(\mathbf{u})$ is defined in a suitable neighborhood of C_0, i.e., in the spherical set $S_\rho = \{\mathbf{u} \in M : |||\mathbf{u}||| \leq \rho\}$ of the space M. We will admit that $\mathcal{E}(\mathbf{u})$ can be extended to the corresponding neighborhood S_ρ of the energy space H_A.

Further, we assume that the functional $\mathcal{E}(\mathbf{u})$ is n times Fréchet ($n \geq 2$) differentiable at the fundamental configuration C_0 in the energy space H_A.

3.4.2 THE FORMULATION OF THE ENERGY CRITERIA BY USING THE ENERGY NORM

Let us formulate the energy criteria of stability by assuming the energy norm, with the additional potential energy functional properly extended to H_A. The space of the states of motions of the structure is now identified as

$$H_{Am} = H_A \times L_2 \tag{3.52}$$

Thus, the stability of the equilibrium configuration C_0 is examined with respect to the norm of the space H_{Am} where, according to definition (3.6),

$$\|\mathbf{m}\|^2 = \left\| \begin{matrix} \mathbf{u} \\ \dot{\mathbf{u}} \end{matrix} \right\|^2 = |||\mathbf{u}|||^2 + \mathcal{T}(\dot{\mathbf{u}}) \tag{3.53}$$

Let C_0 be the fixed equilibrium configuration the stability of which is sought. To reformulate in the "energy" space the above discussed Lagrange-Dirichlet theorem, we will assume that:

1. The potential energy functional is continuous at C_0 with respect to the energy norm of T, i.e.,

$$\lim_{|||\mathbf{u}|||\to 0} \mathcal{E}(\mathbf{u}) = 0 \tag{3.54}$$

   where $\mathbf{u}$ represents any additional displacement that moves T from C_0 to C in the neighborhood of C_0
2. There exists a spherical neighborhood $S_{\bar{\rho}}$ of C_0 and a function $f(|||\mathbf{u}|||)$ of the energy norm $|||\mathbf{u}|||$ such that

$$\text{a. } f(|||\mathbf{u}|||) > 0 \Leftrightarrow \mathbf{u} \neq \mathbf{0} \qquad f(|||\mathbf{u}|||) = \mathbf{0} \Leftrightarrow \mathbf{u} = 0 \tag{3.55}$$

$$\text{b. } f(|||\mathbf{u}_2|||) > f(|||\mathbf{u}_1|||) \qquad |||\mathbf{u}_2||| > |||\mathbf{u}_1||| \qquad \mathbf{u}_1, \mathbf{u}_2 \in S_{\bar{\rho}} \tag{3.55'}$$

$$\text{c. } \mathcal{E}(\mathbf{u}) \geq f(|||\mathbf{u}|||) \qquad \mathbf{u} \in S_{\bar{\rho}} \tag{3.55''}$$

i.e., the potential energy functional is bounded below by $f(|||\mathbf{u}|||)$ in $S_{\bar{\rho}}$.

Thus we have stability at C_0, i.e., condition (3.6) has to be satisfied with respect to the norm (3.53) of H_{Am}. Likewise, the criterion of stability involving the second differential can be immediately re-established.

Let $\mathcal{E}(\mathbf{u})$ be at least twice the Fréchet differentiable at $C_0 \in H_A$ and let

$$\partial S^* = \{\mathbf{u} \in H_A : |||\mathbf{u}||| = 1\} \tag{3.56}$$

The elastic system is stable at the equilibrium configuration C_0 if

$$\omega = \inf_{\partial S^*} \mathcal{E}_2(\mathbf{u}) > 0 \tag{3.57}$$

or, with a different notation,

$$\mathcal{E}_2(\mathbf{u}) > \omega\, |||\mathbf{u}|||^2 \qquad \omega > 0 \quad \forall\, \mathbf{u} \in H_A - \{\mathbf{0}\} \tag{3.57'}$$

3.4.2.1 Variational Formulation of the Second Differential Energy Criterion

Let us assume now that

$$\omega_1 = \inf_{\partial S^*} \mathcal{E}_2(\mathbf{u}) = \min_{\partial S^*} \mathcal{E}_2(\mathbf{u}) \tag{3.58}$$

i.e.,

$$\exists\, \mathbf{u}_1^* \in \partial S^* : \mathcal{E}_2(\mathbf{u}_1^*) = \min_{\partial S^*} \mathcal{E}_2(\mathbf{u}) \tag{3.58'}$$

This assumption is verified in many structural problems and will be discussed later. With (3.58) the stability criterion (3.57) becomes a workable condition for practical applications and gives rise to the following variational formulation: if the functional $\mathcal{E}_2(\mathbf{u})$ attains its minimum over ∂S^*, namely

$$\exists\, \mathbf{u}_1^* \in \partial S^* : \mathcal{E}_2(\mathbf{u}_1^*) = \min_{\partial S^*} \mathcal{E}_2(\mathbf{u}^*) = \omega_1 \tag{3.59}$$

or

$$\mathcal{E}_2(\mathbf{u}_1) = \min_{H_A - \{\mathbf{0}\}} \frac{\mathcal{E}_2(\mathbf{u})}{|||\mathbf{u}|||^2} = \omega_1 \tag{3.59'}$$

where $\mathbf{u}_1^* = \mathbf{u}_1/|||\mathbf{u}_1|||$, thus ω_1 and $\mathbf{u}_1$ are the smallest eigenvalue and the corresponding eigenelement of the variational problem

$$\mathcal{E}_{11}(\mathbf{u},\mathbf{v}) - 2\omega[\mathbf{u},\mathbf{v}] = 0 \qquad \forall\, \mathbf{v} \in H_A \tag{3.60}$$

The integrand of $\mathcal{E}_2(\mathbf{u})$ is a complete homogeneous function of order two in the argument $\mathbf{u}$. Thus we have

$$\mathcal{E}_2(\mathbf{u} + t\mathbf{v}) = \mathcal{E}_2(\mathbf{u}) + t\, \mathcal{E}_{11}(\mathbf{u},\mathbf{v}) + t^2\, \mathcal{E}_2(\mathbf{v}) \tag{3.61}$$

where $\mathcal{E}_{11}(\mathbf{u},\mathbf{v})$, homogeneous of the first order in $\mathbf{u}$ and $\mathbf{v}$, is the integral of the sum of all terms obtained through development of the integrand of $\mathcal{E}_2(\mathbf{u} + \mathbf{v})$. The following relation also holds

$$\mathcal{E}_{11}(\mathbf{u},\mathbf{u}) = 2\, \mathcal{E}_2(\mathbf{u}) \tag{3.61'}$$

Thus, if we take an arbitrary $\mathbf{v} \in H_A$ and an arbitrary real number t, we have

$$\frac{\mathcal{E}_2(\mathbf{u}_1 + t\mathbf{v})}{|||\mathbf{u}_1 + t\mathbf{v}|||^2} \geq \omega_1 \Rightarrow t^2\left(\mathcal{E}_2(\mathbf{v}) - \omega_1\, |||\mathbf{v}|||^2\right) + t\left(\mathcal{E}_{11}(\mathbf{u}_1,\mathbf{v}) - 2\omega_1[\mathbf{u}_1,\mathbf{v}]\right) \geq 0, \forall\, \mathbf{v} \in H_A, \forall\, t \in R \tag{3.62}$$

and (3.60) follows. Thus ω_1 and $\mathbf{u}_1$ are eigensolutions of the variational problem (3.60). Further, let $\mathbf{u}'$ and ω' be eigenvalue and eigensolutions of problem (3.60). Taking $\mathbf{u}' = \mathbf{v}$ from (3.60) we get

$$\omega' = \frac{\mathcal{E}_2(\mathbf{u}')}{|||\mathbf{u}'|||^2} \tag{3.63}$$

Since $\omega_1 = \min \mathcal{E}_2(\mathbf{u})$, over ∂S^*, ω_1 is the smallest eigenvalue of the problem (3.60).

3.4.2.2 The Existence of the Eigen-solutions of the Variational Problem of the Second Differential Criterion

In order to apply variational condition (3.60) it is necessary to prove the existence of the element $\mathbf{u}_1$, which gives the minimum of $\mathcal{E}_2(\mathbf{u})$ over ∂S^*. To overcome this difficulty we observe first of all that in most structural problems the second differential $\mathcal{E}_2(\mathbf{u})$ takes the form

$$\mathcal{E}_2(\mathbf{u}) = \frac{1}{2}\, |||\mathbf{u}|||^2 - \lambda\, \mathcal{L}_2(\mathbf{u}) \tag{3.64}$$

where λ is the load multiplier and $\mathcal{L}_2(\mathbf{u})$ is a quadratic functional describing the destabilizing effect of the loads. Further, for many structural models, it is easy to verify that $\mathcal{E}_2(\mathbf{u})$, and therefore $\mathcal{L}_2(\mathbf{u})$, are continuous in H_A. Moreover, the stability criterion (3.57) becomes

$$\inf_{H_A-\{\mathbf{0}\}}\left[1-2\lambda\frac{\mathcal{L}_2(\mathbf{u})}{|||\mathbf{u}|||^2}\right]>0 \tag{3.65}$$

The existence of an element $\mathbf{u}_1 \in H_A$ such that

$$\inf_{H_A-\{\mathbf{0}\}}\frac{\mathcal{E}_2(\mathbf{u})}{|||\mathbf{u}|||^2}=\frac{\mathcal{E}_2(\mathbf{u}_1)}{|||\mathbf{u}_1|||^2}=\max_{H_A-\{\mathbf{0}\}}\frac{\mathcal{E}_2(\mathbf{u})}{|||\mathbf{u}|||^2} \tag{3.66}$$

is equivalent to the existence of an element $\mathbf{u}_1 \in H_A$ such that

$$\sup_{H_A-\{\mathbf{0}\}}\frac{\mathcal{L}_2(\mathbf{u})}{|||\mathbf{u}|||^2}=\frac{\mathcal{L}_2(\mathbf{u}_1)}{|||\mathbf{u}_1|||^2}=\max_{H_A-\{\mathbf{0}\}}\frac{\mathcal{L}_2(\mathbf{u})}{|||\mathbf{u}|||^2} \tag{3.67}$$

The following theorem can be helpful to prove the existence of the element $\mathbf{u}_1$. If for any sequence $\{\mathbf{u}_n\}_{n\in N}$ of displacement fields bounded in H_A it is possible to pick out a subsequence $\{\Phi_k\}_{k\in N}$ such that

$$\lim_{k,m\to\infty}\mathcal{L}_2(\Phi_k-\Phi_m)=0 \tag{3.68}$$

then an element of ∂S^* exists that maximizes (or minimizes) the functional $\mathcal{L}_2(\mathbf{u})$ over ∂S^*, and the differential $\mathcal{E}_2(\mathbf{u})$ of $\mathcal{E}(\mathbf{u})$ attains its infimum or its supremum over ∂S^*.

In fact, it will be shown that the sequence $\{\Phi_k\}_{k\in N}$ of elements of ∂S^* that maximizes (or minimizes) the functional $\mathcal{L}_2(\mathbf{u})$ if according to (3.68), is Cauchy in the sense of $\mathcal{L}_2(\mathbf{u})$, converges in the energy space to an element $\mathbf{u}_1^*$ of ∂S^*. Thus, the functional $\mathcal{L}_2(\mathbf{u})$, because of its continuity in H_A, will attain its supremum (or its infimum) at $\mathbf{u}_1^*$.

The proof of this theorem, given in,[12] goes back to the classical proof of existence of the smallest eigenvalue of a positive definite operator.[13] Since the functional $\mathcal{L}_2(\mathbf{u})$ is continuous in H_A it is bounded on ∂S^*, i.e., $\forall\ \mathbf{u} \in H_A$

$$|\mathcal{L}_2(\mathbf{u})| \le \alpha\, |||\mathbf{u}|||^2 \quad \alpha > 0 \tag{3.69}$$

or equivalently

$$|\mathcal{L}_2(\mathbf{u}^*)| \le \alpha \qquad \alpha > 0\ \mathbf{u}^* \in \partial S^* \tag{3.69$'$}$$

Thus, we can put

$$l_i = \inf_{\partial S^*}\mathcal{L}_2(\mathbf{u}) \quad l_s = \sup_{\partial S^*}\mathcal{L}_2(\mathbf{u}) \tag{3.70}$$

Let $\{\mathbf{w}_n^*\}_{n\in N}$ be a sequence of elements of ∂S^* maximizing $\mathcal{L}_2(\mathbf{u})$, i.e., such that

$$\lim_{n\to\infty}\mathcal{L}_2(\mathbf{w}_n^*)=l_s \tag{3.71}$$

The sequence $\{\mathbf{w}_n^*\}_{n\in N}$, because of (3.56), is bounded in H_A. Thus, according to the assumptions, it is possible to pick out a subsequence $\{\Phi_k\}_{k\in N}$ satisfying conditions (3.68). Hence, the elements Φ_k possess the following properties:

$$\text{a.}\quad |||\Phi_k||| = 1 \tag{3.72}$$

$$\text{b.}\quad \lim_{k\to\infty} \mathcal{L}_2(\Phi_k) = 1_s \tag{3.73}$$

$$\text{c.}\quad \lim_{k,m\to\infty} \mathcal{L}_2(\Phi_k - \Phi_m) = 0 \tag{3.74}$$

Now take an arbitrary element $\mathbf{v} \in H_A$ such that $|||\mathbf{v}||| < C =$ constant. Let t be an arbitrary real number. Thus, the element $\Phi_k + t\mathbf{v}$ is in general different from zero. Hence, we can write, $\forall$ $t \in R$

$$\begin{gathered} \frac{\mathcal{L}_2(\Phi_k + t\mathbf{v})}{|||\Phi_k + t\mathbf{v}|||^2} \le 1_s \Leftrightarrow t^2\left(\mathcal{L}_2(\mathbf{v}) - 1_s\,|||\mathbf{v}|||^2\right) \\ + t\left(\mathcal{L}_{11}(\Phi_k,\mathbf{v}) - 21_s[\Phi_k,\mathbf{v}]\right) + \left(\mathcal{L}_2(\Phi_k) - 1_s\right) \le 0 \end{gathered} \tag{3.75}$$

The quadratic expression in the variable t in inequality (3.75) does not change sign for any value of t: thus its discriminant is not positive, i.e., taking account of (3.69)

$$\left(\mathcal{L}_{11}(\Phi_K,\mathbf{v}) - 21_s[\Phi_K,\mathbf{v}]\right)^2 \le 4\left(\mathcal{L}_2(\mathbf{v}) - 1_s\,|||\mathbf{v}|||^2\right)\left(\mathcal{L}_2(\Phi_K) - 1_s\right) \tag{3.75'}$$

From (3.75′) and (b) it follows that

$$\lim_{k\to\infty} \left\{\mathcal{L}_{11}(\Phi_k, \mathbf{v}) - 21_s\,[\Phi_k, \mathbf{v}]\right\} = 0 \tag{3.76}$$

This condition holds uniformly if $|||\mathbf{v}||| < C$, where C is a constant that does not depend upon $\mathbf{v}$. Let assume now

$$\mathbf{v} = \Phi_k - \Phi_m \tag{3.77}$$

where the number m is arbitrary. Then $|||\mathbf{v}||| = |||\Phi_k - \Phi_m||| \le 2$.

In fact, by taking into account that $|||\Phi_k||| = 1$, we have, $\forall$ k, m $\in$ N, $|||\Phi_k - \Phi_m|||^2 = [\Phi_k - \Phi_m, \Phi_k - \Phi_m] = 2 - 2[\Phi_k, \Phi_m] \le 2 + 2|[\Phi_k, \Phi_m]| \le 2 + 2|||\Phi_k|||\ |||\Phi_m||| \le 4$. It follows from (3.76)

$$\lim_{k\to\infty} \left\{\mathcal{L}_{11}(\Phi_k, \Phi_k - \Phi_m) - 21_s\,[\Phi_k, \Phi_k - \Phi_m]\right\} = 0 \tag{3.78}$$

Numbers k and m are equivalent. Interchanging them, we have

$$\lim_{m\to\infty} \left\{\mathcal{L}_{11}(\Phi_m, \Phi_m - \Phi_k) - 21_s\,[\Phi_m, \Phi_m - \Phi_k]\right\} = 0 \tag{3.79}$$

Summing (3.79) and (3.78) gives

$$\lim_{k,m\to\infty} \left\{\mathcal{L}_2(\Phi_k - \Phi_m) - 1_s\,|||\Phi_k - \Phi_m|||^2\right\} = 0 \tag{3.80}$$

because, according to (3.62′), $\mathcal{L}_{11}(\Phi_k - \Phi_m, \Phi_k - \Phi_m) = 2\mathcal{L}_2(\Phi_k - \Phi_m)$. Thus, taking into account assumption (c), we get

$$\lim_{k,m\to\infty} |||\Phi_k - \Phi_m|||^2 = 0 \tag{3.81}$$

The space H_A is complete and therefore there exists an element $\mathbf{u}_1 \in H_A : |||\Phi_k - \mathbf{u}_1||| \to 0$. Since $\forall k \in N$

$$|||\Phi_k - \mathbf{u}_1||| \geq \left| |||\Phi_k||| - |||\mathbf{u}_1||| \right| = \left| 1 - |||\mathbf{u}_1||| \right| \tag{3.82}$$

we deduce that $|||\mathbf{u}_1||| = 1$, i.e., $\mathbf{u}_1 \in \partial S^*$. Moreover, $\mathcal{L}_2(\mathbf{u})$ is continuous in H_A and therefore

$$\lim_{k,m\to\infty} \mathcal{L}_2(\Phi_k) = \mathcal{L}_2(\mathbf{u}_1) = 1_s \tag{3.83}$$

It follows that an element $\mathbf{u}_1$, such that

$$\mathbf{u}_1 \in \partial S^* : \mathcal{L}_2(\mathbf{u}_1) = \sup_{\partial S^*} \mathcal{L}_2(\mathbf{u}) \tag{3.84}$$

does exist. The same result holds for the infimum of $\mathcal{L}_2(\mathbf{u})$ on ∂S^*.

In the next chapter we will show that the energy norms of structural models such as beams, plates, and shells, are equivalent to the norms of the corresponding Sobolev space $H^{(2)}$. At the same time it will also be shown that the functional $\mathcal{L}_2(\mathbf{u})$ is upperbounded by the norm of the corresponding Sobolev space $H^{(1)}$. Thus, for most structural systems the following inequalities hold:

$$|||\mathbf{u}||| \geq \chi_1 |||\mathbf{u}|||_{H^{(2)}} \tag{3.85}$$

$$\left|\mathcal{L}_2(\mathbf{u})\right| \leq \chi_2 \, |||\mathbf{u}|||^2_{H^{(1)}} \tag{3.86}$$

Further, every set of displacement fields bounded in energy, i.e., in $H_A(\Omega)$, is also bounded in $H^{(2)}(\Omega)$. It is known, on the other hand, according to the Sobolev embedding theorem recalled in Section 2.9, that every set bounded in $H^{(2)}(\Omega)$ is compact in $H^{(1)}(\Omega)$. Therefore the assumptions of the existence theorem are verified.

3.4.2.3 Critical State: The Koiter Condition of Stability at the Critical State

If we have:

$$\inf_{\partial S^*} \mathcal{E}_2(\mathbf{u}) = 0 \tag{3.87}$$

the equilibrium of T is at the critical state. In such a situation the stability criterion (3.57) fails and stability of this position of neutral equilibrium is undecided.

Let us assume, according to the previous analysis, the existence of a displacement field $\mathbf{u}_c^*$ such that:

$$\mathcal{E}_2(\mathbf{u}_c^*) = \min_{\partial S^*} \mathcal{E}_2(\mathbf{u}) = 0 \tag{3.88}$$

At the critical state the second differential of the total potential energy becomes positive semi-definite. Thus the second differential is zero at $\mathbf{u}_c^*$ but positive at any other displacement $\mathbf{u}^* \in \partial S^*$. For simplicity, assume uniqueness of the critical displacement $\mathbf{u}_c^*$. From (3.60) the critical displacement $\mathbf{u}_c^*$ is then the solution of the classical variational problem

$$\mathcal{E}_{11}(\mathbf{u}_c^*, \mathbf{v}) = 0 \quad \forall\, \mathbf{v} \in H_A \tag{3.89}$$

representing the neutral equilibrium, as discussed at the Chapter 1. Further, we will limit our analysis to the case of a potential energy functional at least four times Fréchet differentiable.

Necessary conditions to the existence of a minimum of $\mathcal{E}(\mathbf{u})$ or of a potential well at the critical state are

$$\mathcal{E}_3(\mathbf{u}_c^*) = 0 : \mathcal{E}_4(\mathbf{u}_c^*) \geq 0 \tag{3.90}$$

In fact, at the critical state, taking $\mathbf{u}_c = \alpha\mathbf{u}_c^*$, we get

$$\mathcal{E}(\mathbf{u}_c) = \mathcal{E}_3(\mathbf{u}_c) + o(|||\mathbf{u}_c|||^3) = \alpha^3\left[\mathcal{E}_3(\mathbf{u}_c^*) + \frac{o(\alpha^3)}{\alpha^3}\right] \tag{3.91}$$

where

$$\lim_{\alpha\to 0}\frac{o(\alpha^3)}{\alpha^3} = 0 \tag{3.92}$$

Consequently, if α is sufficiently small, the sign of $\mathcal{E}(\mathbf{u}_c)$ equals the sign of $\alpha^3\mathcal{E}_3(\mathbf{u}_c^*)$; hence, the existence of the potential well is $\mathcal{E}_3(\mathbf{u}_c^*) = 0$. Thus we can write

$$\mathcal{E}(\mathbf{u}_c) = \alpha^4\left[\mathcal{E}_4(\mathbf{u}_c^*) + \frac{o(\alpha^4)}{\alpha^4}\right] \tag{3.93}$$

At the critical state, when the system moves along the critical direction $\mathbf{u}_c^*$, it is also necessary that $\mathcal{E}_4(\mathbf{u}_c^*) \geq 0$. But the conditions

$$\mathcal{E}_3(\mathbf{u}_c^*) = 0 \qquad \mathcal{E}_4(\mathbf{u}_c^*) > 0 \tag{3.90'}$$

are not sufficient for stability at the critical state.

Let us now analyze the behavior of $\mathcal{E}(\mathbf{u})$ when the system moves from C_0 along a direction $\mathbf{u}^* \neq \mathbf{u}_c^*$ with $\mathbf{u} = \alpha\mathbf{u}^*$. We get

$$\mathcal{E}(\mathbf{u}) = \alpha^2\left[\mathcal{E}_2(\mathbf{u}^*) + \frac{o(\alpha^2)}{\alpha^2}\right] \tag{3.94}$$

Thus, $\forall\, \mathbf{u}^* \neq \mathbf{u}_c^*$ (Fig. 3.5)

$$\exists\ \bar{\alpha} > 0 : |\alpha| < \bar{\alpha} \Rightarrow \mathcal{E}(\alpha\mathbf{u}^*) > 0 \tag{3.95}$$

The functional $\mathcal{E}(\mathbf{u}) = \mathcal{E}(\alpha\mathbf{u}^*)$ is positive over the segment $\{\mathbf{u} = \alpha\mathbf{u}^* : 0 < |\alpha| \leq \bar{\alpha}\}$ for every direction of displacement $\mathbf{u}^* \neq \mathbf{u}_c^*$ and where

$$\bar{\alpha} = \bar{\alpha}(\mathbf{u}^*) \tag{3.96}$$

However, taking into account that

$$\min_{\partial S^*} \mathcal{E}_2(\mathbf{u}^*) = \mathcal{E}_2(\mathbf{u}_c^*) = 0 \tag{3.97}$$

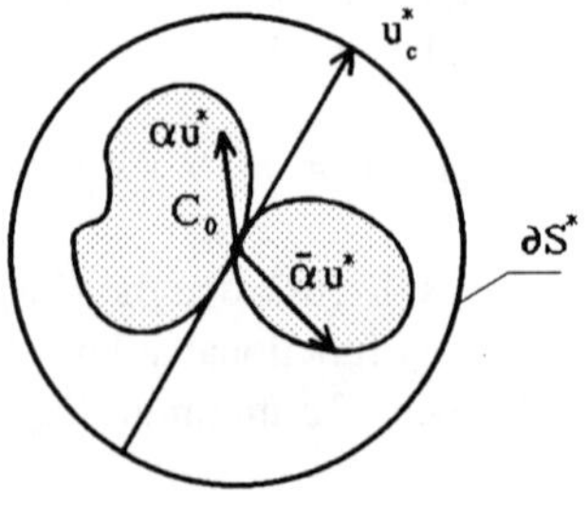

FIGURE 3.5

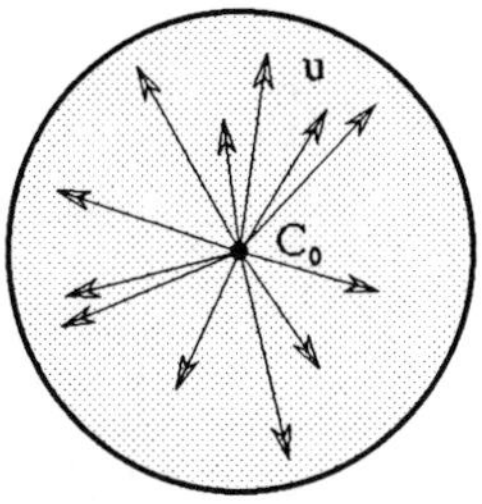

FIGURE 3.6

we can have, in spite of (3.90′),

$$\inf_{\partial S^*} \bar{\alpha}(\mathbf{u}^*) = 0 \tag{3.98}$$

and no longer are we acquainted with the existence of a neighborhood $S_{\bar{\rho}}$ of C_0 where $\mathcal{E}(\mathbf{u}) \geq \gamma\, |||\mathbf{u}|||^{\mu}$ $\mu > 1$, $\gamma > 0$. In conclusion, equilibrium can be unstable at the critical state, even though, along the critical displacement $\mathbf{u}_c^*$, $\mathcal{E}_3(\mathbf{u}_c^*) = 0$, $\mathcal{E}_4(\mathbf{u}_c^*) > 0$. The fact that conditions (3.90′) are only necessary but not sufficient to the existence of the potential well shows that the equilibrium at C_0 may be unstable even though $\mathcal{E}(\mathbf{u}) > 0$ for every additional displacement field (Fig. 3.6).

On the other hand, a condition sufficient to entail stability of the critical state is given by the following Koiter theorem:[14] the equilibrium at the critical state is stable if the following conditions hold

$$\mathcal{E}_3(\mathbf{u}_c^*) = 0 \qquad \mathcal{E}_4(\mathbf{u}_c^*) - \mathcal{E}_2(\Phi_1) > 0 \tag{3.99}$$

where Φ_1, orthogonal in H_A to $\mathbf{u}_c^*$ in the sense that

$$[\mathbf{u}_c^*, \Phi_1] = 0 \tag{3.100}$$

is the unique solution of the variational problem

$$\mathcal{E}_{11}(\Phi, \mathbf{v}) + \mathcal{E}_{21}(\mathbf{u}_c^*, \mathbf{v}) = 0 \qquad \forall\, \mathbf{v} \in H_A \tag{3.101}$$

To prove the theorem, let us investigate the behavior of the potential energy functional $\mathcal{E}(\mathbf{u})$ along displacements having direction close to the critical direction $\mathbf{u}_c^*$. Define the following set of finite additional displacement fields:

$$C_1 = \left\{\mathbf{u} = \alpha\left(\mathbf{u}_c^* + \beta\Phi^*\right) : |\alpha| \leq \alpha_0,\ |\beta| \leq \beta_0,\ \Phi^* \in \partial S_1^*\right\} \tag{3.102}$$

where

$$\partial S_1^* = \left\{\mathbf{u}^* \in \partial S^* : \left[\mathbf{u}^*, \mathbf{u}_c^*\right] = 0\right\} \tag{3.103}$$

The set C_1 represents a conical region with axis along the direction $\mathbf{u}_c^*$ (Fig. 3.7). The quantities α_0 and β_0 represent respectively the length and the width of this region.

Let us evaluate the functional $\mathcal{E}(\mathbf{u})$ along a displacement $\mathbf{u}$ belonging to the conical region C_1. We have

$$\mathcal{E}(\mathbf{u}) = \mathcal{E}_2(\alpha\mathbf{u}_c^* + \alpha\beta\Phi^*) + \mathcal{E}_3(\alpha\mathbf{u}_c^* + \alpha\beta\Phi^*) + \mathcal{E}_4(\alpha\mathbf{u}_c^* + \alpha\beta\Phi^*) + o|||\mathbf{u}|||^4 \tag{3.104}$$

To develop the various terms of (3.104) it is useful to take into account that the integrands of $\mathcal{E}_m(\mathbf{u})$ ($m = 2,3,\ldots$) are complete homogeneous functions of order m in the argument $\mathbf{u}$ and its derivatives; thus the following expansion holds

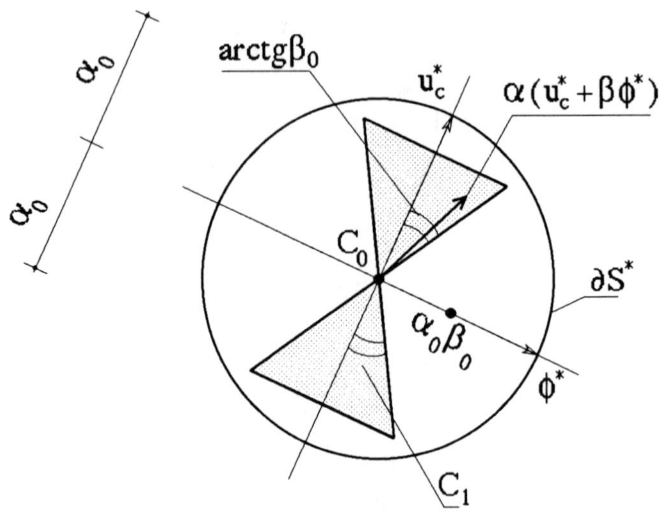

FIGURE 3.7

$$\mathcal{E}_m(\mathbf{u}+\mathbf{v}) = \mathcal{E}_m(\mathbf{u}) + \mathcal{E}_{m-1,1}(\mathbf{u},\mathbf{v}) + \mathcal{E}_{m-2,2}(\mathbf{u},\mathbf{v}) + \ldots + \mathcal{E}_{1,m-1}(\mathbf{u},\mathbf{v}) + \mathcal{E}_m(\mathbf{v}) \tag{3.105}$$

In this expression $\mathcal{E}_{m-n,n}(\mathbf{u},\mathbf{v})$ is the integral of all terms that are obtained through development of the integrand of $\mathcal{E}_m(\mathbf{u}+\mathbf{v})$. $\mathcal{E}_{m-n,n}(\mathbf{u},\mathbf{v})$ is homogeneous of the nth order in the function **v** and its derivatives and homogeneous of the (m–n)th order in the function **u** and its derivatives. The integrals $\mathcal{E}_{mn}$ and $\mathcal{E}_{nm}$ are interchanged when **u** and **v** are interchanged. The following result also holds

$$\mathcal{E}_{m-n,n}(\mathbf{u},\mathbf{u}) = \mathcal{E}_{n,m-n}(\mathbf{u},\mathbf{u}) = \binom{m}{m-n}\mathcal{E}_m(\mathbf{u}) = \binom{m}{n}\mathcal{E}_m(\mathbf{u}) \tag{3.106}$$

in which $\binom{m}{m-n} = \binom{m}{n}$ represents the binomial coefficient

$$\binom{m}{n} = \frac{m(m-1)(m-2)\ldots(m-n+1)}{n!} = \frac{m!}{n!(m-n)!} = \binom{m}{m-n} \tag{3.107}$$

Thus, for instance,

$$\mathcal{E}_3(\mathbf{u}+\mathbf{v}) = \mathcal{E}_3(\mathbf{u}) + \mathcal{E}_{21}(\mathbf{u},\mathbf{v}) + \mathcal{E}_{12}(\mathbf{u},\mathbf{v}) + \mathcal{E}_3(\mathbf{v}) \tag{3.105$'$}$$

$$\mathcal{E}_4(\mathbf{u}+\mathbf{v}) = \mathcal{E}_4(\mathbf{u}) + \mathcal{E}_{31}(\mathbf{u},\mathbf{v}) + \mathcal{E}_{22}(\mathbf{u},\mathbf{v}) + \mathcal{E}_{13}(\mathbf{u},\mathbf{v}) + \mathcal{E}_4(\mathbf{v}) \tag{3.105$''$}$$

and where

$$\mathcal{E}_{21}(\mathbf{u},\mathbf{u}) = \mathcal{E}_{12}(\mathbf{u},\mathbf{u}) = 3\ \mathcal{E}_3(\mathbf{u}) \tag{3.106$'$}$$

$$\mathcal{E}_{13}(\mathbf{u},\mathbf{u}) = \mathcal{E}_{31}(\mathbf{u},\mathbf{u}) = 4\ \mathcal{E}_4(\mathbf{u}) \qquad \mathcal{E}_{22}(\mathbf{u},\mathbf{u}) = 6\ \mathcal{E}_4(\mathbf{u}) \tag{3.106$''$}$$

Thus, taking into account that $\mathcal{E}_{11}(\mathbf{u}_c^*,\mathbf{v}) = 0$, according to (3.89), the potential energy can be expressed in the form

$$\mathcal{E}(\mathbf{u}) = \alpha^2\beta^2\mathcal{E}_2(\Phi^*) + \alpha^3\left[\beta\mathcal{E}_{21}(\mathbf{u}_c^*,\Phi^*) + \beta^2\mathcal{E}_{12}(\mathbf{u}_c^*,\Phi^*) + \beta^3\mathcal{E}_3(\Phi^*)\right] + \alpha^4$$

$$+\left[\mathcal{E}_4(\mathbf{u}_c^*) + \beta\mathcal{E}_{31}(\mathbf{u}_c^*,\Phi^*) + \beta^2\mathcal{E}_{22}(\mathbf{u}_c^*,\Phi^*) + \beta^3\mathcal{E}_{13}(\mathbf{u}_c^*,\Phi^*) + \beta^4\mathcal{E}_4(\Phi^*)\right] + o(|||\mathbf{u}|||^4) \tag{3.104$'$}$$

$\forall\ \mathbf{u} \in C_1$. According to the assumption of the uniqueness of the minimizing element $\mathbf{u}_c^*$ of $\mathcal{E}_2(\mathbf{u}^*)$, we observe that the functional $\mathcal{E}_2(\mathbf{u})$ is positive definite in $H_A^{(1)}$, i.e.,

$$\mathcal{E}_2(\mathbf{u}) \geq \gamma |||\mathbf{u}|||^2 \qquad \gamma > 0,\ \forall\, \mathbf{u} \in H_A^{(1)} \tag{3.108}$$

where $H_A^{(1)}$ is the subspace of H_A defined as

$$H_A^{(1)} = \{\mathbf{u} \in H_A : [\mathbf{u}_c, \mathbf{u}] = 0\} \tag{3.109}$$

Further, the functionals $\mathcal{E}_{12}(\mathbf{u}_c^*, \Phi^*)$ and $\mathcal{E}_3(\Phi^*)$ are continuous with respect to Φ^* in $H_A^{(1)}$, and therefore we have, taking also in account that $|||\Phi^*||| = 1$:

$$\left|\mathcal{E}_{12}\left(\mathbf{u}_c^*, \Phi^*\right)\right| \leq k_{12}\ |||\Phi^*|||^2 \leq \frac{k_{12}}{\gamma}\ \mathcal{E}_2\left(\Phi^*\right) \tag{3.110}$$

$$\left|\mathcal{E}_3\left(\Phi^*\right)\right| \leq k_3\ |||\Phi^*|||^3 \leq \frac{k_3}{k_3\ |||\Phi^*|||^2\ (\gamma)}\ \mathcal{E}_2\left(\Phi^*\right) \tag{3.110$'$}$$

Therefore we obtain from (3.110) and (3.110′)

$$\left|\alpha^3 \beta^2 \mathcal{E}_{12}\left(\mathbf{u}_c^*, \Phi^*\right)\right| \leq \left|\alpha^3\right| \beta^2 k_{12}\ |||\Phi^*|||^2 \leq \alpha^2 \beta^2 |\alpha| \frac{k_{12}}{(\gamma)}\ \mathcal{E}_2\left(\Phi^*\right) \tag{3.111}$$

$$\left|\alpha^3 \beta^3 \mathcal{E}_3\left(\Phi^*\right)\right| \leq \left|\alpha^3\right| \left|\beta^3\right| k_3\ |||\Phi^*|||^3 \leq \alpha^2 \beta^2 |\alpha\beta| \frac{k_3}{(\gamma)}\ \mathcal{E}_2\left(\Phi^*\right) \tag{3.112}$$

Bearing in mind also that the functionals $\mathcal{E}_{31}(\mathbf{u})$, $\mathcal{E}_{22}(\mathbf{u})$, $\mathcal{E}_{13}(\mathbf{u})$, $\mathcal{E}_4(\mathbf{u})$, that appear in the last term at second member of (3.104′), are continuous in $H_A^{(1)}$ and therefore bounded over the unit sphere ∂S_1^*, we have

$$\left|\mathcal{E}_{rs}\left(\mathbf{u}_c^*, \Phi^*\right)\right| \leq k_{rs}\ |||\Phi^*|||^s = k_{rs}\ |||\Phi^*|||^2 \leq \frac{k_{rs}}{(\gamma)} \mathcal{E}_2\left(\Phi^*\right) \tag{3.113}$$

where r = 0,1,2,3,4, s = 0,1,2,3,4 but such that r + s = 4. Thus we get, $\forall\, \mathbf{u} \in C_1$;

$$\begin{aligned}
\mathcal{E}(\mathbf{u}) &\geq \alpha^2\beta^2\ \mathcal{E}\left(\Phi^*\right) + \alpha^3\beta\ \mathcal{E}_{21}\left(\mathbf{u}_c^*, \Phi^*\right) + \alpha^4\ \mathcal{E}_4\left(\mathbf{u}_c^*\right) - \alpha^2\beta^2\left|\alpha \mathcal{E}_{12}\left(\mathbf{u}_c^*, \Phi^*\right)\right| \\
&\quad -\alpha^4\left|\beta\ \mathcal{E}_{31}\left(\mathbf{u}_c^*, \Phi^*\right)\right| - \alpha^2\beta^2\left|\alpha\beta\ \mathcal{E}_3\left(\Phi^*\right)\right| - \alpha^4\beta^2\left|\mathcal{E}_{22}\left(\mathbf{u}_c^*, \Phi^*\right)\right| \\
&\quad -\alpha^4\beta^2\left|\beta\ \mathcal{E}_{13}\left(\mathbf{u}_c^*, \Phi^*\right)\right| - \alpha^4\beta^4\left|\mathcal{E}_4\left(\Phi^*\right)\right| - \left|o\left(|||\mathbf{u}|||^4\right)\right| \\
&\geq \alpha^2\beta^2\ \mathcal{E}_2\left(\Phi^*\right) + \alpha^3\beta\ \mathcal{E}_{21}\left(\mathbf{u}_c^*, \Phi^*\right) + \alpha^4 \mathcal{E}_4\left(\mathbf{u}_c^*\right) - \alpha^2\beta^2|\alpha| k_{12}\ |||\Phi^*|||^2 \\
&\quad -\alpha^4|\beta| k_{31}\ |||\Phi^*||| - \alpha^2\beta^2|\alpha\beta| k_3\ |||\Phi^*|||^3 - \alpha^4\beta^2\ k_{22}\ |||\Phi^*|||^2 \\
&\quad -\alpha^4\beta^2|\beta| k_{13}\ |||\Phi^*|||^3 - \alpha^4\beta^4\ k_4\ |||\Phi^*|||^4 - \left|o\left(|||\mathbf{u}|||^4\right)\right| \\
&\geq \left\{\alpha^2\beta^2\ \mathcal{E}_2\left(\Phi^*\right)(1-c) + \alpha^3\beta \mathcal{E}_{21}\left(\mathbf{u}_c^*, \Phi^*\right) + \alpha^4 \mathcal{E}_4\left(\mathbf{u}_c^*\right)\right\} - \alpha^4\left|\beta k_{31}\right| - \left|o\left(|||\mathbf{u}|||^4\right)\right|
\end{aligned} \tag{3.114}$$

where

$$c = \{|\alpha_0|\, k_{12} + |\alpha_0\beta_0|\, k_3 + \alpha_0^2 k_{22} + \alpha_0^2 |\beta_0|\, k_{13} + \alpha_0^2 \beta_0^2 k_4\}/\gamma \tag{3.115}$$

is a quantity that can be made < 1 by suitably contracting the dimensions of the cone C_1. On the other hand

$$|||\mathbf{u}|||^4 = |||\alpha(\mathbf{u}_c^* + \beta\Phi^*)|||^4 = \alpha^4(1+\beta^2)^2 \tag{3.116}$$

and we obtain, $\forall \mathbf{u} \in C_1$

$$\frac{\mathcal{E}(\mathbf{u})}{|||\mathbf{u}|||^4} \geq \frac{1}{(1+\beta^2)^2}\left[\left(\frac{\beta}{\alpha}\right)^2 (1-c)\mathcal{E}_2(\Phi^*) + \frac{\beta}{\alpha}\mathcal{E}_{21}(\mathbf{u}_c^*, \Phi^*) + \mathcal{E}_4(\mathbf{u}_c^*)\right] - \frac{O(\beta_0)}{(1+\beta^2)^2} - \frac{o\,|||\mathbf{u}|||^4}{|||\mathbf{u}|||^4} \tag{3.117}$$

with

$$O(\beta_0) = k_{31}\,|\beta_0| \tag{3.115'}$$

Let us examine now the functional — which we call the Koiter quadratic functional — that appears in the square brackets of inequality (3.117)

$$\overline{\mathcal{K}}_2(\Phi^*) = \left[\left(\frac{\beta}{\alpha}\right)^2 \mathcal{E}_2(\Phi^*)(1-c) + \frac{\beta}{\alpha}\mathcal{E}_{21}(\mathbf{u}_c^*, \Phi^*) + \mathcal{E}_4(\mathbf{u}_c^*)\right] \tag{3.118}$$

This functional, if we put

$$\Phi = \frac{\beta}{\alpha}\Phi^* \tag{3.119}$$

can be written as

$$\overline{\mathcal{K}}_2(\Phi) = \left[\mathcal{E}_2(\Phi)(1-c) + \mathcal{E}_{21}(\mathbf{u}_c^*, \Phi) + \mathcal{E}_4(\mathbf{u}_c^*)\right] \quad \forall \Phi \in H_A^{(1)} \tag{3.118'}$$

and (3.117) becomes

$$\frac{\mathcal{E}(\mathbf{u})}{|||\mathbf{u}|||^4} \geq \frac{1}{(1+\beta^2)^2}\left(\overline{\mathcal{K}}_2(\Phi) - O(\beta_0)\right) - \frac{o\,|||\mathbf{u}|||^4}{|||\mathbf{u}|||^4} \tag{3.117'}$$

Now Φ belongs to the subspace $H_A^{(1)}$ defined by (3.109) and $\mathcal{E}_2(\Phi)$ is positive definite in $H_A^{(1)}$. At the same time the linear functional $\mathcal{E}_{21}(\mathbf{u}_c^*, \Phi)$ is continuous and therefore bounded in $H_A^{(1)}$. Thus $\overline{\mathcal{K}}_2(\Phi)$ will admit a minimum in $H_A^{(1)}$. The search for this minimum can in fact be worked out as in Section 2.7.1.2. We can introduce a new Hilbert space H_K whose elements are elements of $H_A^{(1)}$ but with scalar product and norm defined by

$$\{\Phi, \psi\} = \mathcal{E}_{11}(\Phi, \psi) \quad \{\{\Phi\}\}^2 = \{\Phi, \Phi\} = 2\mathcal{E}_2(\Phi) \tag{3.120}$$

because $\mathcal{E}_{11}(\Phi, \Phi) = 2\,\mathcal{E}_2(\Phi)$. The following inequalities

$$\{\{\Phi\}\} \geq \gamma|||\Phi||| \tag{3.121}$$

$$\left|\mathcal{E}_{21}\left(\mathbf{u}_c^*,\Phi\right)\right| \le k_{21}\,|||\Phi||| \le \frac{k_{21}}{\gamma}\{\{\Phi\}\}\;\hat{\Phi}_1 \tag{3.122}$$

hold $\forall\,\Phi \in H_A^{(1)}$. Thus, by the representation theorem of linear bounded functionals in the Hilbert space H_K, we can write

$$\mathcal{E}_{21}\left(\mathbf{u}_c^*,\Phi\right) = \left\{\Phi,\ \hat{\Phi}_1\right\} \tag{3.123}$$

where $\hat{\Phi}_1$ is a definite element of H_K. Hence from (3.118′) the Koiter functional $\overline{\mathcal{K}}_2(\Phi)$ becomes

$$\overline{\mathcal{K}}_2(\Phi) = \frac{1}{2}(1-c)\{\Phi,\Phi\} + \left\{\Phi,\ \hat{\Phi}_1\right\} + \mathcal{E}_4\left(\mathbf{u}_c^*\right) = (1-c)\left[\frac{1}{2}\{\Phi,\Phi\} - \left\{\Phi,\overline{\Phi}_1\right\}\right] + \mathcal{E}_4\left(\mathbf{u}_c^*\right) \tag{3.124}$$

where

$$\overline{\Phi}_1 = -\frac{\hat{\Phi}}{(1-c)} \tag{3.125}$$

Thus

$$\overline{\mathcal{K}}_2(\Phi) = \frac{1}{2}(1-c)\left[\left\{\left\{\Phi - \overline{\Phi}_1\right\}\right\}^2 - \left\{\left\{\overline{\Phi}_1\right\}\right\}^2\right] + \mathcal{E}_4\left(\mathbf{u}_c^*\right) \tag{3.126}$$

and $\overline{\mathcal{K}}_2(\Phi)$ can be extended to the whole H_K. Moreover, $\overline{\mathcal{K}}_2(\Phi)$ achieves its global minimum on H_K at $\Phi = \overline{\Phi}_1$. This minimum is

$$\min_{H_K}\overline{\mathcal{K}}_2(\Phi) = \overline{\mathcal{K}}_2\left(\overline{\Phi}_1\right) = \mathcal{E}_4\left(\mathbf{u}_c^*\right) - \frac{1}{2}(1-c)\left\{\left\{\overline{\Phi}_1\right\}\right\}^2 = \mathcal{E}_4\left(\mathbf{u}_c^*\right) - (1-c)\mathcal{E}_2\left(\overline{\Phi}_1\right) \tag{3.127}$$

To evaluate the element $\Phi = \overline{\Phi}_1$ at which the functional $\overline{\mathcal{K}}_2(\Phi)$ achieves its minimum, we can use a variational approach. In fact, we have

$$\overline{\mathcal{K}}_2\left(\overline{\Phi}_1 + t\boldsymbol{\eta}\right) - \overline{\mathcal{K}}_2\left(\overline{\Phi}_1\right) \ge 0\ \forall\, t \in R,\ \forall\,\boldsymbol{\eta} \in H_A^{(1)} \tag{3.128}$$

or equivalently

$$t^2(1-c)\mathcal{E}_2(\boldsymbol{\eta}) + t\{(1-c)\mathcal{E}_{11}(\Phi,\boldsymbol{\eta}) + \mathcal{E}_{21}(\mathbf{u}_c^*,\boldsymbol{\eta})\} \ge 0 \tag{3.128′}$$

to be satisfied $\forall\, t \in R$, $\forall\,\boldsymbol{\eta} \in H_A^{(1)}$. Thus in the neighborhood of $\overline{\Phi}_1$ the dominant term of the first member of (3.128′) equals zero, i.e.,

$$(1-c)\mathcal{E}_{11}(\Phi,\eta) + \mathcal{E}_{21}(\mathbf{u}_c^*,\eta) = 0\ \forall\,\boldsymbol{\eta} \in H_A^{(1)} \tag{3.129}$$

Let us take $\Phi = \overline{\Phi}_1$ and $\eta = \overline{\Phi}_1$. Thus from condition (3.129) we have

$$\mathcal{E}_{21}(\mathbf{u}_c^*,\overline{\Phi}_1) = -(1-c)\mathcal{E}_{11}(\overline{\Phi}_1,\overline{\Phi}_1) = -2(1-c)\mathcal{E}_2(\overline{\Phi}_1) \tag{3.129′}$$

and we find again the value $\overline{\mathcal{K}}_2(\overline{\Phi}_1)$ given by (3.127). Then we can consider the auxiliary functional

$$\mathcal{K}_2(\Phi) = \mathcal{E}_2(\Phi) + \mathcal{E}_{21}(\mathbf{u}_c^*, \Phi) + \mathcal{E}_4(\mathbf{u}_c^*) \quad \forall \Phi \in H_A^{(1)} \tag{3.118''}$$

We observe that, if μ denotes the value of $\mathcal{K}_2(\Phi) - \mathcal{E}_4(\mathbf{u}_c^*)$ for some displacement Φ, then the functional $\overline{\mathcal{K}}_2(\Phi) - \mathcal{E}_4(\mathbf{u}_c^*)$ evaluated in $\Phi/(1-c)$ takes the value $\mu/(1-c)$. Then the minimum of $\overline{\mathcal{K}}_2(\Phi) - \mathcal{E}_4(\mathbf{u}_c^*)$ is given by the minimum of $\mathcal{K}_2(\Phi) - \mathcal{E}_4(\mathbf{u}_c^*)$ divided by $(1-c)$. In turn, the element $\Phi_1 = (1-c)\overline{\Phi}_1$ that satisfies the condition

$$\mathcal{E}_{11}(\Phi,\eta) + \mathcal{E}_{21}(u_c^*,\eta) = 0 \qquad \forall\, \eta \in H_A^{(1)} \tag{3.130}$$

gives the minimum of $\mathcal{K}_2(\Phi)$. Thus

$$\min_{H_K} \mathcal{K}_2(\Phi) = \mathcal{E}_4(\mathbf{u}_c^*) - \mathcal{E}_2(\Phi_1) \tag{3.131}$$

and

$$\min_{H_K} \overline{\mathcal{K}}_2(\Phi) = \frac{1}{1-c} \min_{H_K} \mathcal{K}_2 - \frac{c}{1-c} \mathcal{E}_4(\mathbf{u}_c^*) \tag{3.132}$$

With a suitable contraction of the dimensions α_0 and β_0 of the conical region C_1, i.e., with a suitable contraction of the quantity c, it is possible to choose a $\bar{c} > 0$, i.e., a conical region $\overline{C}_1$ such that, for every $c > 0$, $0 < c < \bar{c}$

$$\min_{H_K} \mathcal{K}_2(\Phi) > 0 \Rightarrow \min_{H_K} \overline{\mathcal{K}}_2(\Phi) > 0 \tag{3.133}$$

These results are fundamental to the analysis of stability at the critical state. In fact from inequality (3.117′)

$$\frac{\mathcal{E}(\mathbf{u})}{|||\mathbf{u}|||^4} \geq \frac{1}{(1+\beta_0^2)^2}\left[\min_{H_K} \overline{\mathcal{K}}_2(\Phi) - O(\beta_0)\right] - \frac{\left|o(|||\mathbf{u}|||^4)\right|}{|||\mathbf{u}|||^4} \tag{3.117''}$$

$\forall\, \mathbf{u} \in C_1$. Thus, taking account of (3.131) and (3.132), if we have

$$\mathcal{E}_4(\mathbf{u}_c^*) - \mathcal{E}_2(\Phi_1) > 0 \tag{3.134}$$

then, with a suitable contraction of the conical region C_1, from eq. (3.117″) we get

$$\mathcal{E}(\mathbf{u}) \geq \gamma_1\, |||\mathbf{u}|||^4 \quad \gamma_1 > 0 \;\forall\, \mathbf{u} \in C_1 \tag{3.135}$$

In brief, if condition (3.134) is satisfied, it is possible to find a conical set C_1 and a positive quantity γ_1 such that condition (3.135) holds. Let us consider now the set C_2 (Fig. 3.8)

$$C_2 = \left\{\mathbf{u} = \alpha(\Phi^* + \varepsilon\mathbf{u}_c^*) : |\alpha| < \alpha_0 |\varepsilon| < \varepsilon_0\right\},\ \varepsilon_0 = \frac{1}{\beta_0}, \Phi^* \in H_A^{(1)} \tag{3.136}$$

In the conical set C_2 we have

$$\mathcal{E}(\mathbf{u}) = \mathcal{E}_2(\alpha\Phi^* + \alpha\varepsilon\mathbf{u}_c^*) + o(|||\mathbf{u}|||^2) \geq \alpha^2\, \mathcal{E}_2(\Phi^*) - \left|o(|||\mathbf{u}|||^2)\right| \tag{3.137}$$

and

$$\frac{\mathcal{E}(\mathbf{u})}{|||\mathbf{u}|||^2} \geq \frac{1}{1+\varepsilon_0^2}\, \mathcal{E}_2(\Phi^*) - \frac{\left|o(|||\mathbf{u}|||^2)\right|}{|||\mathbf{u}|||^2} \tag{3.138}$$

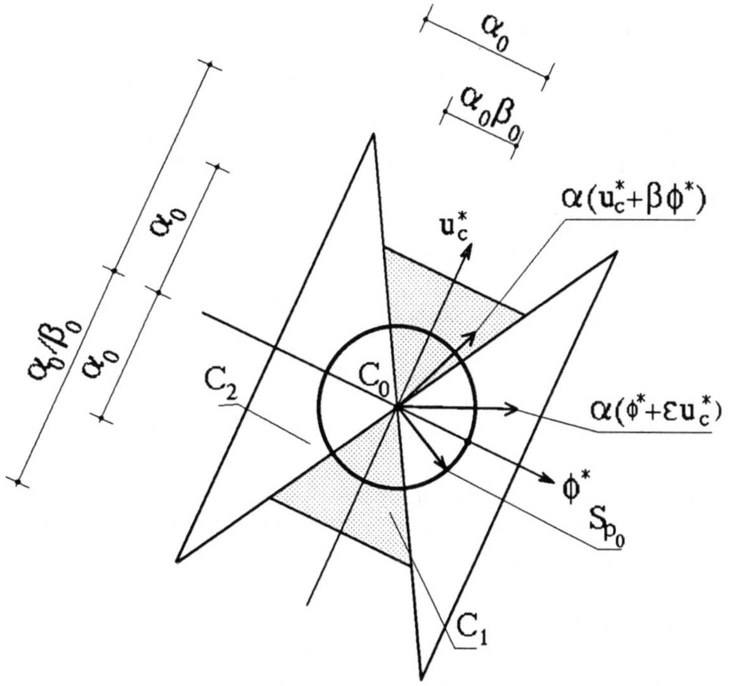

FIGURE 3.8

But on the set

$$\Phi S_A^{*(1)} = \left\{ \mathbf{u} \in H_A^{(1)} : |||\mathbf{u}||| = 1 \right\} \tag{3.139}$$

we can admit that the minimum of $\mathcal{E}_2(\Phi^*)$ exists and that this minimum is positive. Let $\tilde{\Phi}^*$ be the element of $\Phi S_A^{*(1)}$ where this minimum is attained. Thus for a suitable value of α_0, we can evaluate a positive constant γ_2 such that

$$\frac{\mathcal{E}(\mathbf{u})}{|||\mathbf{u}|||^2} \geq \frac{1}{1+\varepsilon_0^2} \mathcal{E}_2\left(\tilde{\Phi}^*\right) - \frac{\left|o\left(|||\mathbf{u}|||^2\right)\right|}{|||\mathbf{u}|||^2} \geq \gamma_2 > 0, \quad \forall \mathbf{u} \in C_2 \tag{3.140}$$

Taking into account (3.135) and (3.140), we can conclude (Fig. 3.8) that there exists a spherical neighborhood S_{ρ_0} of C_0 with $0 < \rho_0 < 1$ such that

$$\mathcal{E}(\mathbf{u}) > \gamma\, |||\mathbf{u}|||^4 \qquad \forall\, \mathbf{u} \in S_{\rho_0} \tag{3.141}$$

where

$$\gamma = \min\,(\gamma_1, \gamma_2) \tag{3.142}$$

because $|||\mathbf{u}|||^2 > |||\mathbf{u}|||^4$, $\mathbf{u} \in S_{\rho_0}$. Thus, the critical state is stable.

We point out now that condition (3.130) can be released by the constraint condition to which the element Φ_1 is subjected, i.e., to satisfy the variational eq. (3.130) for every $\eta \in H_A^{(1)}$.

Let us consider, on the contrary, a generic additional displacement $\mathbf{v} \in H_A$. This displacement can be composed as

$$\mathbf{v} = \alpha \mathbf{u}_c^* + \boldsymbol{\eta} \qquad \mathbf{v} \in H_A, \boldsymbol{\eta} \in H_A^{(1)} \tag{3.143}$$

Substitution of (3.143) into the variational condition (3.130) yields

$$\mathcal{E}_{11}(\boldsymbol{\Phi},\mathbf{v}) + \mathcal{E}_{21}(\mathbf{u}_c^*,\mathbf{v}) = \mathcal{E}_{11}(\boldsymbol{\Phi}, \alpha\mathbf{u}_c^* + \boldsymbol{\eta}) + \mathcal{E}_{21}(\mathbf{u}_c^*, \alpha\mathbf{u}_c^* + \boldsymbol{\eta})$$
$$= \alpha\mathcal{E}_{11}(\boldsymbol{\Phi},\mathbf{u}_c^*) + \mathcal{E}_{11}(\boldsymbol{\Phi},\boldsymbol{\eta}) + \alpha\mathcal{E}_{21}(\mathbf{u}_c^*,\mathbf{u}_c^*) + \mathcal{E}_{21}(\mathbf{u}_c^*,\boldsymbol{\eta}) = 0 \tag{3.144}$$

Because, according to (3.89) and (3.106′), $\mathcal{E}_{11}(\boldsymbol{\Phi}, \mathbf{u}_c^*) = 0$, $\mathcal{E}_{21}(\mathbf{u}_c^*, \mathbf{u}_c^*) = 3\ \mathcal{E}_3(\mathbf{u}_c^*) = 0$ and we again find condition (3.130). Thus, $\boldsymbol{\Phi}_1$ is a solution of (3.100) and of the released variational equation

$$\mathcal{E}_{11}(\boldsymbol{\Phi},\mathbf{v}) + \mathcal{E}_{21}(\mathbf{u}_c^*,\mathbf{v}) = 0 \qquad \mathbf{v} \in H_A \tag{3.145}$$

equivalent to a set of differential equations and boundary conditions. Only when

$$\boldsymbol{\Phi}_1 = \mathbf{0} \tag{3.146}$$

conditions (3.99) degenerate into (3.90′). Only in this case conditions (3.90′) are sufficient to stability.

In conclusion we point out that the proof has shown that conditions (3.99), (3.100) and (3.101) are sufficient to assure the existence of a potential well at C_0 and, consequently, the stability of C_0 at the critical state.

3.4.2.4 Sufficient Conditions of Instability

In the previous sections we analyzed a sufficient condition of stability involving the existence of a positive infimum of the second differential $\mathcal{E}_2(\mathbf{u})$ on the spherical boundary of S^*. To complete this analysis, we extend to continuous elastic structures the classical Liapounov *instability* criterion, still involving the second differential of the potential energy.

With the assumption of $\mathcal{E}(\mathbf{u})$ twice Fréchet differentiable at $\mathbf{0}$, it is in fact possible to prove that the equilibrium is unstable if there exists an additional displacement $\mathbf{u}_c^*$ along which the second differential $\mathcal{E}_2(\mathbf{u})$ is negative. For instance, if the system is the elastic beam of Fig.1. 12, in equilibrium in the vertical position under the axial dead load N, we will assume $N > N_c$, where N_c is the critical Euler load of the beam. Thus, deforming the beam along the buckling mode v_c^*, we will have

$$\mathcal{E}_2(v_c^*) = \omega < 0 \tag{3.147}$$

After these preliminary remarks, at the inital time t = 0 let us impart to the system T, in equilibrium at C_0, a disturbance $\mathbf{m}(0)$ represented by an additional deformation corresponding to the first buckling mode of the system

$$\mathbf{m}(0) = \begin{bmatrix} C\mathbf{u}_c^* \\ 0 \end{bmatrix} \qquad C \in R \tag{3.148}$$

i.e., let us move T along the buckling displacement $\mathbf{u}_c$

$$\mathbf{u}_c = C\mathbf{u}_c^* \tag{3.149}$$

The corresponding total energy of the system, immediately after the imparted disturbance, i.e., at time t = 0, is given by

$$E_T(0) = \mathcal{E}_2(\mathbf{u}_c) + o\left(|||\mathbf{u}_c|||^2\right) = C^2\omega + o\left(|||\mathbf{u}|||^2\right) \tag{3.150}$$

Taking into account (3.147) and sufficiently restricting the value |C|, we can move T along the direction (3.149) so that the initial value $E_T(0)$ of the total energy is negative. Thus, we have

$$\operatorname{sgn} E_T(0) = \operatorname{sgn} C^2\omega < 0 \tag{3.151}$$

Now let us analyze the evolution of the disturbed motion of T and introduce the measure

$$F(t) = \int_\Omega \mu \mathbf{u}(t) \cdot \mathbf{u}(t)\, d\Omega \tag{3.152}$$

where μ is the mass density and $\mathbf{u}(t)$ is the additional displacement of T at time t during the evolution of the disturbed motion

$$\mathbf{m}(t) = \begin{bmatrix} \mathbf{u}(t) \\ \dot{\mathbf{u}}(t) \end{bmatrix} \tag{3.153}$$

According to the previous assumptions, $\mathbf{u}(t) \in H_A$, $\dot{\mathbf{u}}(t) \in L_2$, and the motion $\mathbf{m}(t)$ will belong to the product space $H_m = H_A \times L_2$. Thus the norm of $\mathbf{m}(t)$ will be

$$|||\mathbf{m}(t)|||_{H_m} = \left(|||\mathbf{u}|||^2 + ||\dot{\mathbf{u}}||^2_{L_2} \right)^{1/2} \tag{3.154}$$

Taking into account (2.49), we observe that

$$||\mathbf{m}(t)||_{H_m} \geq |||\mathbf{u}(t)|||^2 \geq \gamma ||\mathbf{u}(t)||^2 \geq \gamma_1 F(t) \quad \gamma, \gamma_1 > 0 \tag{3.155}$$

where, for simplicity, we have used the notation $||\mathbf{u}(t)||_{L_2} = ||\mathbf{u}(t)||$. By differentiating (3.152) we obtain

$$\dot{F}(t) = 2\int_\Omega \mu \mathbf{u} \cdot \dot{\mathbf{u}}\, d\Omega \tag{3.156}$$

$$\ddot{F}(t) = 2\int_\Omega \mu (\dot{\mathbf{u}} \cdot \dot{\mathbf{u}} + \mathbf{u} \cdot \ddot{\mathbf{u}})\, d\Omega \tag{3.157}$$

Equation (3.157) can be very useful to the analysis of the perturbed motion (3.153). The disturbed motion of T, in fact, will satisfy the virtual work equation that, taking into account also the inertial forces, gives

$$\int_\Omega \mu \ddot{\mathbf{u}} \cdot \delta \mathbf{v}\, d\Omega + \mathcal{E}_{11}(\mathbf{u}, \delta \mathbf{v}) + \ldots = 0 \tag{3.158}$$

to be satisfied for every admissible virtual displacement $\delta\mathbf{v}$. The symbol ... denotes terms having higher order of smallness, relative to displacements and velocities, than the terms written out earlier. Thus, from (3.158), taking $\delta\mathbf{v} = \mathbf{u}(t)$, we get

$$2\mathcal{E}_2(\mathbf{u}) + \ldots = -\int_\Omega \mu \ddot{\mathbf{u}} \cdot \mathbf{u}\, d\Omega \tag{3.159}$$

Substituting (3.159) in (3.157) gives

$$\ddot{F}(t) = 2\int_\Omega \mu \dot{\mathbf{u}}(t) \cdot \dot{\mathbf{u}}(t) d\Omega - 4\mathcal{E}_2(\mathbf{u}(t)) + \ldots = 4\left[\mathcal{T}(\dot{\mathbf{u}}(t)) - \mathcal{E}_2(\mathbf{u}(t))\right] + \ldots \tag{3.160}$$

The total energy of the system on the other hand is given by

$$E_T(\mathbf{m}(t)) = \mathcal{T}(\dot{\mathbf{u}}(t)) + \mathcal{E}_2(\mathbf{u}(t)) + \ldots \tag{3.161}$$

Thus, substitution of (3.161) into (3.160) gives

$$\ddot{F}(t) = 8\mathcal{T}(\dot{\mathbf{u}}(t)) - 4E_T(\mathbf{m}(t)) + \ldots \tag{3.162}$$

Since the total energy is constant during the motion, from (3.162) and (3.150) we get

$$\ddot{F}(t) = 8\mathcal{T}(\dot{\mathbf{u}}(t)) - 4\,E_T(\mathbf{m}(0)) + \ldots = 8\mathcal{T}(\dot{\mathbf{u}}(t)) - 4\,C^2\omega + \ldots \tag{3.163}$$

Consequently the following inequality holds

$$\ddot{F}(t) \geq 4C^2\,|\omega| + \ldots \tag{3.164}$$

Then, by integration, we have

$$F(t) \geq\; 2C^2(|\omega| + \ldots)\,t^2 + A_1 t + A_2 \tag{3.165}$$

where A_1 and A_2 are arbitrary constants. Thus, according to (3.155), the disturbing motion $\mathbf{m}(t)$ has a norm in H_m such that

$$\|\mathbf{m}(t)\|_{H_m} \geq |||\mathbf{u}(t)||| \geq \gamma\|\mathbf{u}(t)\| \geq \gamma_1 F(t) \geq \gamma_1\left[2C^2(|\omega| + \ldots)t^2 + A_1 t + A_2\right] \tag{3.166}$$

Consider a spherical neighborhood S_ρ of $\mathbf{0}$ with radius $\rho > 0$ sufficiently small such that

$$(|\omega| + \ldots) > 0 \tag{3.167}$$

For every $\varepsilon > 0 : 0 < \varepsilon < \rho$ and for every choice of A_1, A_2 and of the amplitude C of the initial disturbance such that $E_T(\mathbf{0}) < 0$, there will exist a sufficiently large time $\bar{t}$ such that,

$$\gamma_1\left[2C^2(|\omega| + \ldots)\bar{t}^{\,2} - |A_1|\,\bar{t} - |A_2|\right] > \varepsilon^2 \tag{3.168}$$

and (Fig. 3.9)

$$\|\mathbf{m}(t)\|_{H_m} > \varepsilon \;\; \forall\, t \geq \bar{t} \tag{3.169}$$

Hence instability occurs if condition (3.147) is satisfied. In conclusion, we can affirm that, with

$$\omega = \min_{\delta S^*} \mathcal{E}_2(\mathbf{u}^*) \tag{3.170}$$

we have stability if $\omega > 0$ and instability if $\omega < 0$.

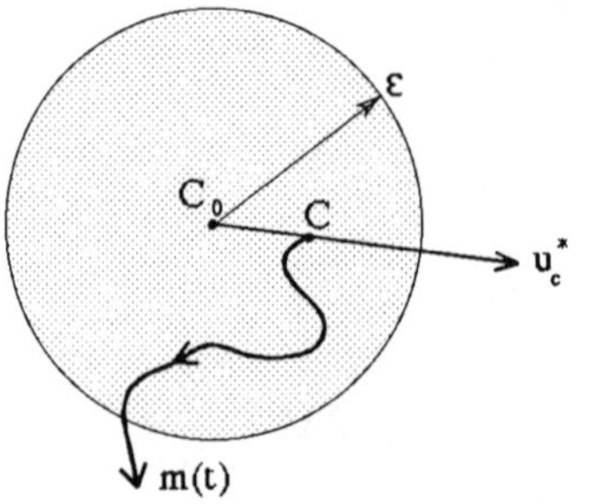

FIGURE 3.9

REFERENCES

1. Liapounov, A. M., Problème general de la stabilité du mouvement. Societé Mathématique de Kharkow. Translated by Davaux, E., Ann. Fac. Sci. Toulouse, 2nd Ser. 9, Reprinted at Princeton: Princeton University Press (Ann. of Math.), 1949.
2. Gurtin, M. E., Thermodynamics and stability. Arch. Rational. Mech. Anal. 59, 1975.
3. Koiter, W. T., The energy criterion of stability for continuous elastic bodies, I and II, Proc. Koninkl. Ned. Akad. Wetenschap B 68, 1965.
4. Coleman, B. D., The energy criterion for stability in continuum thermodynamics; Rend. Sem. Mat. Fis. Milano 43, 1973.
5. Hughes, T. J. R., Kato, T., and Marsden, J. E., Well posed quasi-linear second order hyperbolic systems with applications to non linear elasticity and general relativity, Arch. Rational Mech. Anal. 63, 1977.
6. Knops, R. J., On potential wells and stability in non linear elasticity, Math. Proc. Camb. Phil. Soc., 84, 1978.
7. Hadamard, J., Leçons sur la Propagation des Ondes et les Equations de l'Hydrodynamique. Paris, 1903. Reprinted in New York, Chelsea, 1949.
8. Panagiotopoulos, P. D., Inequality Problems in Mechanics and Applications. Birkhäuser, Boston, 1985.
9. Shield, R. T., and Green, A. E., On certain methods in the stability theory of continuous systems, Arch. Rational Mech. Anal. 12, 354–360, 1963.
10. John, F., Formation of singularities in one dimensional nonlinear wave propagation, Comm. Pure Appl. Math. 27, 377–405, 1974.
11. Knops, R. J. and Wilkes, E. W., Theory of elastic stability. Handbuch der Physik, Band Vla/3, Springer-Verlag, Berlin-Heidelberg New York, 1973.
12. Como, M. and Grimaldi, A., Stability, buckling and postbuckling of elastic structures. Meccanica 14, 1977.
13. Mikhlin, S. G., Variational Methods in Mathematical Physics, Pergamon Press, Oxford, 1964.
14. Koiter, W. T., Over de stabiliteit von het elastisch enerwicht, Thesis. Delft, H. J. Paris, Amsterdam 1945, English Transl. NASA YY-F-10, 833, 1967.
15. Koiter, W. T., A basic open problem in the theory of elastic stability, Paper presented at the Joint IUTAM/IMU Symposium on Applications of Methods of Functional Analysis to Problems of Mechanics, Marseille, 1975.
16. Knops, R. J., Instability and ill-posed Cauchy problems in elasticity. In Mechanics of Solids, ed. by H. G. Hopkins and M. J. Sewell, Pergamon Press, Oxford and New York, 1982.
17. Como, M., and Grimaldi, A., On the existence of potential wells in stability analysis of beams and shells, Proc. STAMM 94, Lisbon, July 1994, in press.

REFERENCES

Chapter 4

EQUILIBRIUM STABILITY OF CONTINUOUS ELASTIC STRUCTURES

4.1 INTRODUCTION

The aim of this chapter is to investigate the equilibrium stability of various structural elements such as beams, plates, and shells, and as far as possible, the general elastic continuum in the framework of the previously stated theory.

First of all some fundamental smoothness properties of the potential energy functionals of various structural elements in their energy spaces are carefully investigated. Then the equilibrium stability of the axially loaded inextensible elastic rod, according to the classical model of the "Elastica", is studied in detail, and both plane flexural buckling and stability are analyzed. Subsequently the problem is extended in the tridimensional space considering the combined torsional-flexural deformations of the rod. Buckling and stability problems of shafts under constant torque and of deep beams under transversal loads are then worked out as particular examples. Some basic stability problems of plates and shells are then studied extensively by analyzing special properties of their strain energy functionals into the corresponding energy spaces. Finally, the energy space of the elastic continuum, together with the convergences of the strain and stress tensors, is examined. The still open problem of the statement of the energy criterion in tridimensional elasticity is discussed in detail at the end of the chapter.

4.2 BUCKLING AND STABILITY ANALYSIS OF THE UNIDIMENSIONAL BEAM MODEL

4.2.1 THE "ELASTICA"

4.2.1.1 Kinematics of the Model

In treating bending and extension of elastic rods the following common assumptions, (see, for instance, Love[1]) are used:

1. Cross sections remain plane.
2. Cross sections remain orthogonal to the deformed axis.
3. Cross sections don't get deformed in their plane.

In this framework a fundamental role is played by the " Elastica ". Let us first consider pure bending of an inextensible rod. With reference to Fig. 4.1, let

$$\mathbf{f}(z) = v(z)\mathbf{j} + w(z)\mathbf{k} \tag{4.1}$$

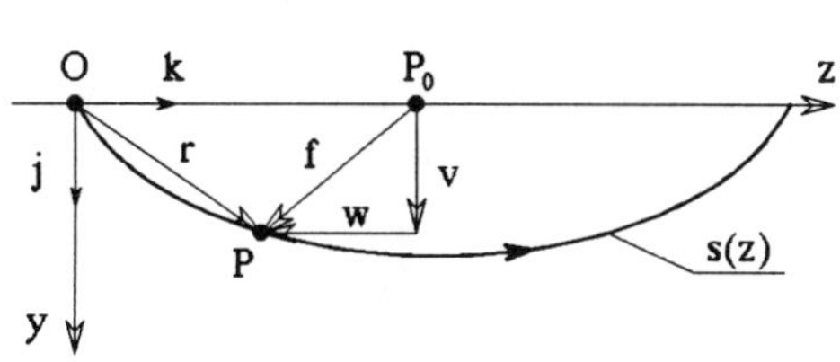

FIGURE 4.1

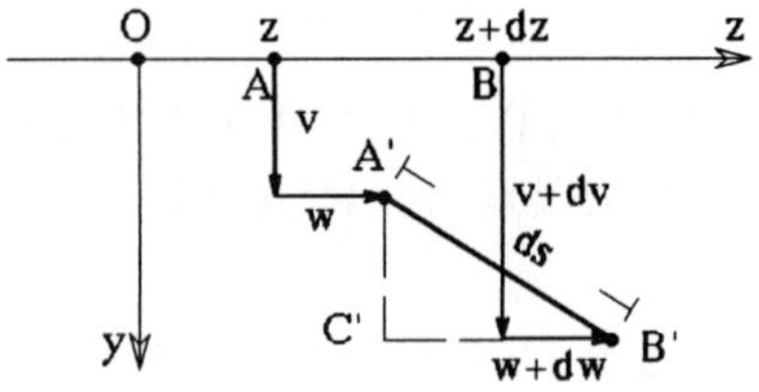

FIGURE 4.2

be the displacement vector of a generic point P_o on the beam axis. In (4.1) **j** and **k** are unit vectors on the fixed axes Oz and Oy, and v(z) and w(z) are the displacement components along the y and z directions. We can begin to assume that v(z), w(z) $\in C^n[0,L]$. Moreover, let

$$s = s(z) \quad z \in [0,L] \tag{4.2}$$

be the curvilinear abscissa, a function of z, along the deformed axis of the rod. Let P be a point on the deformed axis; the position of P is thus defined by the vector function

$$\mathbf{r}[s(z)] = z(s)\mathbf{k} + \mathbf{f}(s) \tag{4.3}$$

From direct inspection of Fig. 4.2 we get

$$\overline{A'C'} = dv \quad \overline{C'B'} = z + dz + w + dw - z - w = dz + dw \tag{4.4}$$

Thus

$$ds^2 = dv^2 + (dz + dw)^2 \tag{4.5}$$

or

$$\left(\frac{ds}{dz}\right)^2 = v'^2 + (1 + w')^2 \tag{4.6}$$

The inextensibility condition

$$\frac{ds}{dz} = 1 \tag{4.7}$$

taking into account (4.6), yields

$$w' \le 0 \quad v'^2(z) \le 1 \quad z \in [0, L] \tag{4.8}$$

and we get

$$w' = \sqrt{1 - v'^2} - 1 \tag{4.9}$$

Equations (4.9) and (4.8) represent the constraint conditions between the functions w(z) and v(z) due to the inextensibility of the beam axis. From (4.3), the vector

$$\mathbf{T} = \frac{d\mathbf{r}}{ds} \qquad 4.10)$$

tangent to the deformed axis of the rod is also given by

$$\mathbf{T} = \mathbf{k} + \frac{d\mathbf{f}}{ds} \tag{4.11}$$

Thus, the vector function

$$\frac{d\mathbf{T}}{ds} = \frac{d^2\mathbf{f}}{ds^2} \tag{4.12}$$

gives a measure of the flexure of the rod. The modulus of this last vector represents, in fact, the bending curvature of the rod axis. We have

$$\left|\frac{1}{\rho}\right| = \left|\frac{d\mathbf{T}}{ds}\right| = \left|\frac{d^2\mathbf{f}}{ds^2}\right| \tag{4.13}$$

On the other hand, using the notation

$$(\dot{\ }) = \frac{d}{ds}, \quad (\ )' = \frac{d}{dz} \tag{4.14}$$

and taking into account (4.1) and (4.7), we get

$$\frac{d^2\mathbf{f}}{ds^2} = \ddot{v}\mathbf{j} + \ddot{w}\mathbf{k} = v''\mathbf{j} + w''\mathbf{k} \tag{4.15}$$

and the curvature $1/\rho$ of the rod axis is

$$\left|\frac{1}{\rho}\right| = \left|\frac{d^2\mathbf{f}}{ds^2}\right| = \sqrt{v''^2 + w''^2} \tag{4.16}$$

On the other hand, from (4.9) we get

$$w'' = -\frac{v'v''}{\sqrt{1-v'^2}} \tag{4.17}$$

and finally, assuming positive the curvature of the bent rod of Fig. 4.1 (i.e., assuming $1/\rho$ positive or negative according to whether the upper side of the region in the plane yz bounded above by the curve v(z) is concave or convex), we have

$$\frac{1}{\rho} = -\frac{v''}{\sqrt{1-v'^2}} \tag{4.18}$$

For z varying in the interval [0,L], every function v(z), with second derivative v"(z) finite and first derivative v'(z) satisfying condition (4.8), represents a flexural deformation of the inextensible axis of the rod.

4.2.1.2 The Energy Space of the Displacement Functions of the Rod

According to the linear theory of plane bending of the elastic rod, the "energy product" and the "energy norm" are

$$[v_1, v_1] = \int_0^L EIv_1''v_2''\,dz \qquad |||v|||^2 = \int_0^L EIv''^2\,dz \tag{4.19}$$

Thus, the "energy space" H_A of the displacements functions v(z) is the set of functions with second derivatives square integrable on the interval [0,L] and satisfying linear homogeneous boundary conditions. These last conditions exclude any rigid displacement of the rod and involve v(z) or the first derivative v'(z). The corresponding Sobolev space is $H^{(2)}[0,L]$. According to the Sobolev embedding theorem — discussed in Section 2.9 — the following inequalities hold:

$$\exists\, k_1 > 0 : \left(\int_0^L v^2 \, dz \right)^{1/2} = \|v\| \le k_1 \, |||v|||; \; \exists\, k_2 > 0 : \left(\int_0^L v'^2 \, dz \right)^{1/2} = \|v'\| \le k_2 \, |||v||| \tag{4.20}$$

$$\exists\, k_3 > 0 : \max_{z \in [0,L]} |v(z)| \le k_3 \, |||v|||; \; \exists\, k_4 > 0 : \max_{z \in [0,L]} |v'(z)| \le k_4 \, |||v||| \tag{4.21}$$

The functions v(z) of the space H_A are therefore continuous with their first derivatives. Of course, if $\{v_n\}_{n \in N}$ is a sequence of displacements converging in energy norm to the element v_0, i.e.,

$$|||v_n - v_o||| \to 0 \tag{4.22}$$

in virtue of the foregoing inequalities, sequences as $\{v_n\}_{n \in N}$ and $\{v_n'\}_{n \in N}$ converge uniformly to $v_0(z)$ and $v_0'(z)$. It can be easily proven that if the sequence $\{v_n\}_{n \in N}$ of displacements converges in energy to the element $v_0(z)$, i.e., according to (4.22), we also get

$$\left\| \frac{1}{\rho_n} - \frac{1}{\rho_o} \right\| = \left(\int_0^L \left(\frac{1}{\rho_n} - \frac{1}{\rho_o} \right)^2 dz \right)^{1/2} \to 0 \tag{4.23}$$

If we assume that the elastic constitutive equation $M = M(\frac{1}{\rho})$ satisfies the following inequality

$$\left| M\left(\frac{1}{\rho_1}\right) - M\left(\frac{1}{\rho_2}\right) \right| \le k \left| \frac{1}{\rho_1} - \frac{1}{\rho_2} \right| \tag{4.24}$$

as, for instance, in the linear model where

$$M = EI \frac{1}{\rho} \tag{4.25}$$

from (4.22) we have

$$\|M_n - M_o\| \to 0 \tag{4.26}$$

Thus, in the nonlinear case also, with assumption (4.24), convergence in energy norm of displacement functions implies convergence in the mean of the corresponding curvatures and bending moments.

4.2.1.3 The Energy Functional of the Rod

Let us consider now an elastic rod under an axial dead load N (Fig. 4.3). The strain energy functional $\mathcal{V}(u)$ of the inexestensible rod is a quadratic functional of the curvature $1/\rho(z)$,

$$\mathcal{V}(v) = \frac{EI}{2} \int_0^L \left(\frac{1}{\rho(z)} \right)^2 dz = \frac{EI}{2} \int_0^L \frac{v''^2}{1 - v'^2} \, dz \tag{4.27}$$

The potential energy of the axial load N is

$$\mathcal{U}(v) = N \int_0^L (\sqrt{1 - v'^2} - 1) \, dz \tag{4.28}$$

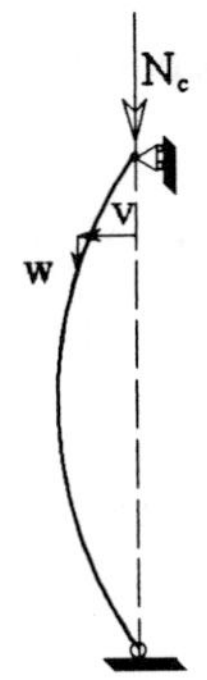

FIGURE 4.3

According to (4.7) functionals $\mathcal{V}(v)$ and $\mathcal{U}(v)$ can be defined for any displacement field belonging to the set

$$S = \{\ v(z) : v'^2 < \delta < 1\} \qquad \forall\ z \in [0,L] \tag{4.29}$$

This set, taking into account inequality (4.21), is represented by a spherical neighborhood of the origin O in H_A. Hence,

$$S = \{v \in H_A : |||v||| < \delta'\} \tag{4.30}$$

Thus, the potential energy functional

$$\mathcal{E}(v) = \frac{EI}{2}\int_0^L \frac{v''^2}{1-v'^2}\,dz + N\int_0^L \left(\sqrt{1-v'^2}-1\right) dz \tag{4.31}$$

is defined in a suitable spherical neighborhood of the origin O in H_A.

We evaluate the first two weak differentials of $\mathcal{E}(v)$ at the point v_1 of the set S and along the direction v_2. Thus, we have

$$D\mathcal{E}(v_1, v_2) = \left[\frac{\partial \mathcal{E}(v_1+\beta v_2)}{\partial \beta}\right]_{\beta=0} \tag{4.32}$$

and

$$D\mathcal{E}(v_1, v_2) = EI\int_0^L \frac{v_1''v_2''(1-v_1'^2) + v_1'v_2'v_1''^2}{(1-v_1'^2)^2}\,dz - N\int_0^L \frac{v_1'v_2'}{\sqrt{1-v_1'^2}} \tag{4.33}$$

Hence at $v_1 = 0$ we get

$$D\mathcal{E}(0,v) = 0 \tag{4.34}$$

Let us indicate with

$$D^{(n)}\mathcal{E}(v_1;\ v,\ v, \ldots\ v) = D^{(n)}\mathcal{E}(v_1;\ v^n) = \left[\frac{d^n\mathcal{E}(v_1+\alpha v)}{d\alpha^n}\right]_{\alpha=0} \tag{4.35}$$

the nth differential of $\mathcal{E}(\mathbf{v})$ at $\mathbf{v}_1$ in the direction v. Thus, for instance, we have

$$D^2\mathcal{E}(v_1; v, v) = EI\int_0^L \left[\frac{v''^2}{1-v_1'^2} + \frac{v_1''^2 v'^2 + 4v_1' v_1'' v' v''}{\left(1-v_1'^2\right)^2} + \frac{4v_1'^2 v_1''^2 v'^2}{\left(1-v_1'^2\right)^3}\right] dz$$

$$-N\int_0^L \left[\frac{v'^2}{\sqrt{1-v_1'^2}} + \frac{v_1'^2 v'^2}{\left(1-v_1'^2\right)^{3/2}}\right] dz \tag{4.36}$$

and

$$D^2\mathcal{E}(0; v_2, v_3) = EI\int_0^L v_2'' v_3''\, dz - N\int_0^L v_2'' v_3''\, dz \tag{4.37}$$

and so on. According to the foregoing definitions, we can write

$$D\mathcal{E}(0; v) = \left[\frac{d\mathcal{E}(\alpha v)}{d\alpha}\right]_{\alpha=0} \quad D^2\mathcal{E}(0; v, v) = \left[\frac{d^2\mathcal{E}(\alpha v)}{d\alpha^2}\right]_{\alpha=0} \tag{4.38}$$

$$D^3\mathcal{E}(0; v, v, v) = \left[\frac{d^3\mathcal{E}(\alpha v)}{d\alpha^3}\right]_{\alpha=0} \quad D^4\mathcal{E}(0; v, v, v, v) = \left[\frac{d^4\mathcal{E}(\alpha v)}{d\alpha^4}\right]_{\alpha=0} \tag{4.39}$$

and we obtain

$$D\mathcal{E}(0; v) = 0 \quad D^2\mathcal{E}(0; v, v) = 2!\left[\frac{EI}{2}\int_0^L v''^2\, dz - \frac{N}{2}\int_0^L v'^2\, dz\right] \tag{4.40}$$

$$D^3\mathcal{E}(0; v, v, v) = 0 \quad D^4\mathcal{E}(0; v, v, v, v) = 4!\left[\frac{EI}{2}\int_0^L v''^2 v'^2\, dz - \frac{N}{8}\int_0^L v'^4\, dz\right] \tag{4.41}$$

Thus, according to the Taylor theorem (2.99), we get

$$\mathcal{E}(v) = \mathcal{E}_1(v) + \mathcal{E}_2(v) + \mathcal{E}_3(v) + \mathcal{E}_4(v) + 0(|||v|||^4) \tag{4.42}$$

where

$$\mathcal{E}_1(v) = D\mathcal{E}(0; v) \quad \mathcal{E}_2(v) = \frac{1}{2!} D^2\mathcal{E}(0; v) \tag{4.43}$$

$$\mathcal{E}_3(v) = \frac{1}{3!} D^3\mathcal{E}(0; v) \quad \mathcal{E}_4(v) = \frac{1}{4!} D^4\mathcal{E}(0; v) \tag{4.44}$$

and finally, we have

$$\mathcal{E}_1(v) = 0; \quad \mathcal{E}_2(v) = \frac{1}{2}EI\int_0^L v''^2\, dz - \frac{N}{2}\int_0^L v'^2\, dz \tag{4.45}$$

$$\mathcal{E}_3(v) = 0;\ \mathcal{E}_4(v) = \frac{1}{2} EI \int_0^L v'^2\, v''^2\, dz - \frac{N}{8}\int_0^L v'^4\, dz \tag{4.46}$$

The strong differentiability of the functional $\mathcal{E}(v)$ is specified by the following theorem (Como and Grimaldi[2]): The potential energy functional $\mathcal{E}(v)$ is at least 4 times Fréchet differentiable at the origin of the space H_A. First we give the proof for second differentiability. According to the Lagrange mean value theorem, we have

$$\Delta\mathcal{E} = \mathcal{E}(0+v) - \mathcal{E}(0) = \frac{1}{2}\, D^2\mathcal{E}(0;\, v) + \frac{1}{2}\left[D^2\mathcal{E}(\alpha v;\, v) - D^2\mathcal{E}(0;\, v)\right]\quad 0 < \alpha < 1 \tag{4.47}$$

On the other hand, from (4.40) and (4.20) we get

$$\left|D^2\mathcal{E}(0;\, v)\right| \le \left(1 + N\, k_2^2\right)|||v|||^2 \tag{4.48}$$

where k_2 is the constant of (4.29).The functional $D^2\mathcal{E}(0;\, v)$ is quadratic and continuous in the displacement function v. Further,

$$r_2(v) = \frac{1}{2}\left|D^2\mathcal{E}(\alpha v;\, v) - D^2\mathcal{E}(0;\, v)\right| = \frac{EI}{2}\int_0^L \left(\frac{1}{1-\alpha^2 v'^2} - 1\right) v''^2 + \frac{5\alpha^2 v'^2\, v''^2}{\left(1-\alpha^2 v'^2\right)^2}$$

$$+ \frac{4\alpha^4 v'^4\, v''^2}{\left(1-\alpha^2 v'^2\right)^3} - \frac{N}{EI}\left[\left(\frac{1}{\sqrt{1-\alpha^2 v'^2}} - 1\right) v'^2 + \frac{\alpha^2 v'^4}{\left(1-\alpha^2 v'^2\right)^{3/2}}\right] dz \tag{4.49}$$

Thus we get

$$r_2(v) = \frac{1}{2}\left|D^2\mathcal{E}(\alpha v;\, v) - D^2\mathcal{E}(0;\, v)\right|$$

$$\le \frac{|||v|||^2}{2}\left[\left(\frac{1}{1-m^2} - 1\right) + \frac{5m^2}{1-m^2} + \frac{4m^4}{\left(1-m^2\right)^3} + N\left(\frac{1}{\left(1-m^2\right)^{1/2}} - 1 + \frac{m^2\, k_2^2}{\left(1-m^2\right)^{3/2}}\right)\right] \tag{4.50}$$

where

$$m = \max_{z\in[0,L]} \left|v'(z)\right| \tag{4.51}$$

Taking in account that $v'^2(z) \le m^2$ and inequalities (4.20) and (4.21), we get

$$\lim_{|||v|||\to 0} \frac{\left|D^2\mathcal{E}(\alpha v;\, v) - D^2\mathcal{E}(0;\, v)\right|}{|||v|||^2} = 0 \tag{4.52}$$

and the functional $\mathcal{E}(v)$ is two times Fréchet differentiable at $v = 0$. The fourth-order strong differentiability of $\mathcal{E}(v)$ can be proven with the same technique. The increment $\Delta\mathcal{E}$ at the rectilinear configuration, i.e., at $v = 0$, of the rod can in fact be written as

$$\Delta\mathcal{E} = \mathcal{E}(0+v) - \mathcal{E}(0) = \frac{1}{2} D^2\mathcal{E}(0;\, v) + \frac{1}{4} D_4\mathcal{E}(0;\, v) + r_4(v) \tag{4.47'}$$

and the remainder $r_4(v)$ can be expressed as

$$r_4(v) = \frac{1}{4!}\left[D_4\mathcal{E}(\alpha v; v) - D_4\mathcal{E}(0; v)\right] \tag{4.49'}$$

The evaluation of the fourth-order differential of $\mathcal{E}(v)$ from (4.46) yields the following inequality

$$|r_4(v)| \leq |||v|||^4 f(m) \tag{4.50'}$$

where

$$\lim_{m \to 0} f(m) = 0 \tag{4.50''}$$

and $\mathcal{E}(v)$ is at least four times strong differentiable. In the expressions of the second and fourth differentials of $\mathcal{E}(v)$, see for instance (4.36), the second derivatives v" of the displacements appear only with the square power; powers higher than 2 involve only the first derivative of v(z). On the other hand, the energy norm is proportional to the mean square value of the second derivative v"(z). The first derivative v'(z), according to (4.21), is uniformly bounded by $|||v|||$. From these properties follow the inequalities (4.50) and (4.50′) involving the remainders r_2 and r_4.

4.2.1.4 The Buckling Load; Stability at the Critical State

We have

$$\mathcal{L}_2(v) = \frac{N}{2}\int_0^L v'^2\, dz \tag{4.53}$$

and the following inequalities hold

$$|||v||| \geq k_1 \|v\|_{H^{(2)}} \qquad \mathcal{L}_2(v) \leq k_2 \int_0^L v'^2\, dz \leq k_2 \|v\|_{H^{(1)}} \tag{4.54}$$

where $\|v\|_{H^{(2)}}$, $\|v\|_{H^{(1)}}$ denote the norms in the Sobolev spaces $H^{(2)}$ and $H^{(1)}$. Any set of displacement fields bounded in H_A [0,l] is also bounded in $H^{(2)}$[0,l]. The embedding of $H^{(2)}$ in $H^{(1)}$ is compact; i.e., any set bounded in $H^{(2)}$ is compact in $H^{(1)}$ and the assumption of the theorem stated in Section 3.4.2.2 is satisfied. Conditions (3.74) and (3.73) hold. Consequently, the variational formulation of the second differential energy criterion can be applied to inextensible rods. From (3.89) we get

$$\int_0^L (EI\, v''\delta v'' - Nv'\,\delta v')\, dz = 0 \quad \forall\, \delta v \in H_A \tag{4.55}$$

Integrating by part with the more restrictive assumption that v, $\delta v \in C^{(4)}[0,L]$ yields

$$\int_0^L \left(EI\, v^{IV} - Nv''\right) \delta v\, dz = 0 \quad \forall\, \delta v \in C^{(4)}\,[0,L] \tag{4.56}$$

From this condition, by means (2.42), we obtain the Eulerian equation

$$EI\, v^{IV} - Nv'' = 0 \tag{4.57}$$

coupled to the boundary conditions (1.20'). Equation (4.57) is a self-adjoint eigenvalue equation of the type

$$Au - \lambda Bu = 0 \tag{4.58}$$

with A, B linear, positive-definite differential operators with constant coefficents, whose domains D_A, D_B are such that $D_B \supset D_A$. Let H_A, H_B be the energy spaces corresponding to the operators A and B. According to the assumptions, every set bounded in H_A is compact in H_B: for instance, in the case of equation (4.57) $H_A = H^{(2)}[0,L]$, $H_B = H^{(1)}[0,L]$. Thus equation (4.58) has a countable set of eigenvalues, where infinity is the only condensation point, and a complete set of eigenvectors in H_A. Equation (4.58) corresponds to the strong formulation of the variational equation (3.89). Thus the problem arises: to which extent can we say that solutions of the minimum problem for the functional $\mathcal{E}_2(\mathbf{u}^*)$ do satisfy equation (4.58)? As a rule we can say that, according to Mikhlin,[30] that the function $\mathbf{u}_c \in H_A$ that achieves the minimum (3.88) of $\mathcal{E}_2(\mathbf{u}^*)$ has derivative of all orders and satisfies the equation (4.58). Thus the smallest eigenvalue and the corresponding eigenvector of (4.57) give the minimum of the functional $\mathcal{E}_2(\mathbf{u}^*)$ in H_A. For instance, in the case of the rod with hinged ends of Fig. 4.3, equilibrium is Liaupounov stable[2,18] for $N < N_c$ where

$$N_c = \pi^2 \frac{EI}{L^2} \tag{4.59}$$

is the Euler buckling load. The equilibrium at the critical state is stable.[2] In fact, according to (3.101) and taking into account that, because of (4.46), $\mathcal{E}_{21}(u_c^*,\eta) = 0$ for any η, we have $\mathbf{\Phi}_1 = \mathbf{0}$. Thus, the stability condition at the critical state in this case gives

$$\mathcal{E}_4(u_c^*) > 0 \tag{4.60}$$

This inequality is satisfied. The critical displacement in fact is

$$v_c = A\sin\frac{\pi z}{L} \tag{4.61}$$

and, according to (4.46), we get, with some calculations, as first shown by Koiter,[3]

$$\mathcal{E}_4\left(u_c^*\right) = A^4 \frac{\pi^6}{64}\frac{EI}{L^5} > 0 \tag{4.62}$$

4.2.2 THE INFLUENCE OF SHEAR

In the previous analysis we have assumed that the transversal beam deflections are produced only by bending. No consideration was given to the fact that shearing forces developed along any element of the beam are accompanied by a distorsion of the element, which produces an additional deflection. (Fig. 4.4). Shear deformations are usually very small compared with those due to bending. On the other hand, in the case of short beams, these shear deformations have to be taken into account for a more accurate description of the elastic deformation of the beams.

For the usual slender columns, the influence of shearing deformations on the buckling loads can be neglected. For columns consisting of struts connected by lacing bars or batten plates, the shear effect, on the contrary, may become of practical importance (Fig. 4.5) (see for instance Timoshenko and Gere[4]).These shear deformations are usually taken into account by assuming a distorsion angle, representative of the mean shearing strain, occurring between the cross section and the normal to the tangent line to the deformed axis of the rod. (Fig. 4.6). With reference to Fig. 4.7, let

1. α be the rotation angle of the cross section
2. β be the angle of the tangent line to the deformed axis of the rod

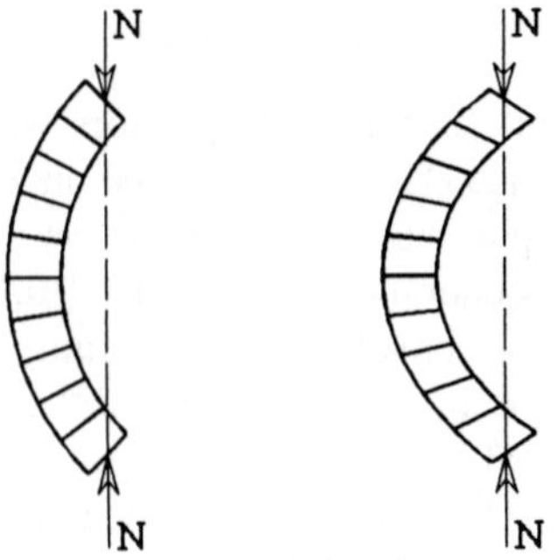

FIGURE 4.4

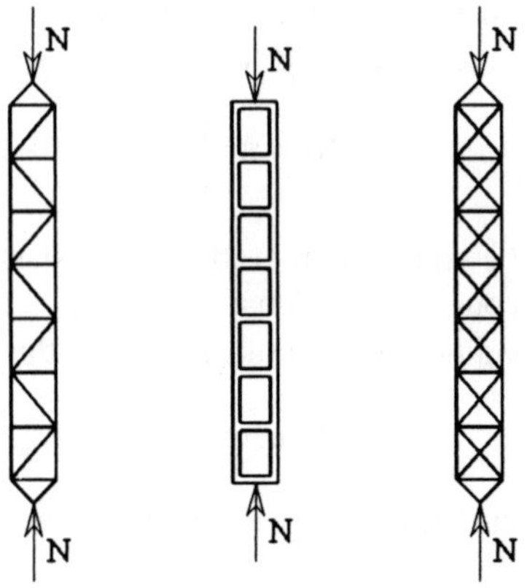

FIGURE 4.5

Thus we have

$$\frac{dv}{ds} = \sin\beta \tag{4.63}$$

or

$$\frac{dv}{dz}\frac{dz}{ds} = \sin\beta \tag{4.64}$$

The cross section of the beam rotates at an angle $\alpha < \beta$. A measure of the average shear deformation γ of the cross section is thus the angle $\beta - \alpha$. Hence we write

$$\gamma = \beta - \alpha \tag{4.65}$$

Also in this model of beam we assume inextensibility of the beam axis. Then we get, according to (4.5) and (4.6)

$$\frac{ds}{dz} = 1 \quad w' = \sqrt{1 - v'^2} - 1 \tag{4.66}$$

Taking into account (4.64) and (4.66), we have

$$\beta = \arcsin v' \quad (\ )' = \frac{d(\ )}{dz} \tag{4.67}$$

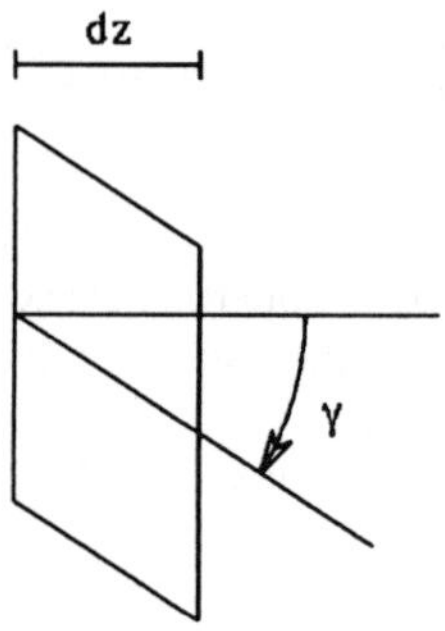

FIGURE 4.6

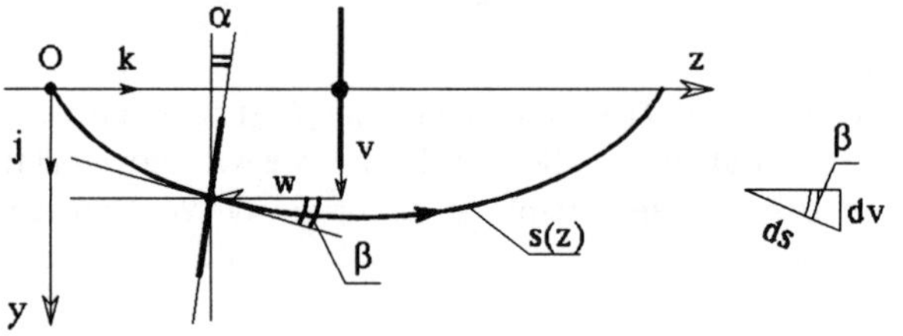

FIGURE 4.7

The rotation function α is given by

$$\alpha = \arcsin v' - \gamma \tag{4.65'}$$

and, consequently, the curvature of the rod axis, according to the assumption of Section 4.2.1, is

$$\frac{1}{\rho} = -\alpha' = -\frac{v''}{\sqrt{1-v'^2}} + \gamma' \tag{4.68}$$

defined for

$$v'^2(z) \le 1 \quad z \in [0, L] \tag{4.69}$$

The strain energy of the beam is

$$\mathcal{V}(v) = \frac{1}{2}\, EI \int_0^L \left[-\frac{v''}{\sqrt{1-v'^2}} + \gamma' \right]^2 dz + \frac{1}{2}\frac{GA}{\chi}\int_0^L \gamma^2\, dz \tag{4.70}$$

where G is the shear modulus, A is the area of the cross section, and χ the shear correction factor. When the shear strain γ vanishes, the expression (4.70) finds again (4.27). With reference to the case of loading sketched in Fig.4.3, we observe that the potential energy of the load N is still given by (4.28). At conclusion, in place of (4.31), we get

$$\mathcal{E}(v) = \frac{EI}{2}\int_0^L \left[-\frac{v''}{\sqrt{1-v'^2}} + \gamma' \right]^2 dz + \frac{GA}{2\chi}\int_0^L \gamma^2\, dz + \frac{N}{2}\int_0^L \left(\sqrt{1-v'^2} - 1\right) dz \tag{4.71}$$

The energy space H_A is composed by all displacement vector functions **u**(z) of components [v(z), γ(z)] satisfying linear homogeneous boundary conditions excluding any rigid displacement, and for which is defined the norm

$$|||\mathbf{u}|||^2 = EI \int_0^L (-v'' + \gamma')^2 \, dz + \frac{GA}{\chi} \int_0^L \gamma^2 \, dz \tag{4.72}$$

According to the Sobolev theorems discussed in Section 2.9 the vector functions $\mathbf{u}(z)$, describing the deformation of the inextensible rod in flexure and shear, have components $v(z) \in C^1[0,L]$ and $\gamma(z) \in C[0,L]$ satisfying the following inequalities

$$\exists\, \gamma_1 > 0 : \max_{z \in [0,L]} |v(z)| \le \gamma_1 |||\mathbf{u}||| \tag{4.73}$$

$$\exists\, \gamma_2 > 0 : \max_{z \in [0,L]} |v'(z)| \le \gamma_2 |||\mathbf{u}||| \tag{4.73′}$$

$$\exists\, \gamma_3 > 0 : \max_{z \in [0,L]} |\gamma(z)| \le \gamma_3 |||\mathbf{u}||| \tag{4.73″}$$

The potential energy functional $\mathcal{E}(\mathbf{u})$ is defined in a spherical neighborhood of the origin O of H_A. With the same procedure used at Section 4.2.1.3, we will show that the functional (4.71) is at least four times strong differentiable at the origin O. We consider the strain energy functional (4.70). We get

$$D^2\mathcal{V}(\alpha\mathbf{u}; \mathbf{u}) = EI \int_0^L \left(\left[-\frac{v''}{\sqrt{1-\alpha^2 v'^2}} \right]^2 - 5\alpha^2 \left[-\frac{v''}{\sqrt{1-\alpha^2 v'^2}} \right] \frac{v'^2 v''}{\left(1-\gamma^2 v'^2\right)^{3/2}} \right) dz$$

$$+ EI \int_0^L \frac{\alpha^4 v'^4}{\left(1-\alpha^2 v'^2\right)^3} \left[v''^2 - 3\left(-\frac{v''}{\sqrt{1-\alpha^2 v'^2}} + \gamma' \right) v'' \sqrt{1-\alpha^2 v'^2} \right] dz + \frac{GA}{\chi} \int_0^L \gamma^2 dz \tag{4.74}$$

and

$$D^2\mathcal{V}(0; \mathbf{u}) = EI \int_0^L (-v'' + \gamma')^2 \, dz + \frac{GA}{\chi} \int_0^L \gamma^2 \, dz \tag{4.75}$$

Then, taking into account (4.73), (4.73′) and (4.73″), we obtain

$$\lim_{|||\mathbf{u}||| \to 0} \frac{\left| D^2\mathcal{V}(\alpha\mathbf{u}; \mathbf{u}) - D^2\mathcal{V}(0; \mathbf{u}) \right|}{|||\mathbf{u}|||^2} = 0 \tag{4.76}$$

With the same technique it is possible to show the Fréchet differentiability of $\mathcal{V}(\mathbf{u})$ of the fourth order. The differentials of $\mathcal{E}(\mathbf{u})$ at the origin are

$$\mathcal{E}_2(\mathbf{u}) = \frac{EI}{2} \int_0^L (-v'' + \gamma')^2 \, dz + \frac{GA}{2\chi} \int_0^L \gamma^2 \, dz - \frac{N}{2} \int_0^L v'^2 \, dz \tag{4.77}$$

$$\mathcal{E}_3(\mathbf{u}) = 0 \tag{4.78}$$

$$\mathcal{E}_4(\mathbf{u}) = \frac{EI}{2} \int_0^L (v'' - \gamma') \, v'^2 \, v'' \, dz - \frac{N}{8} \int_0^L v'^4 \, dz \tag{4.79}$$

Let us evaluate the buckling load of the strut.

According to (3.38) and taking into account (4.77), with the same procedure used in the previous section we can say that the minimizing function $\mathbf{u}_c^*$ of $\mathcal{E}_2(\mathbf{u}^*)$ solves the following linear differential equations and boundary conditions

$$\mathrm{EI}\left(\mathrm{v}^{\mathrm{IV}}-\gamma'''\right)+\mathrm{Nv}''=0 \qquad \mathrm{EI}(\mathrm{v}'''-\gamma'')+\frac{\mathrm{GA}}{\chi}\gamma=0 \tag{4.80}$$

$$\left|\mathrm{EI}(\mathrm{v}''-\gamma')\,(\delta\mathrm{v}'-\delta\gamma)\right|_0^L=0 \quad \left|\left[\mathrm{EI}(\mathrm{v}'''-\gamma'')+\mathrm{Nv}'\right]\delta\mathrm{v}\right|_0^L=0 \tag{4.81}$$

For the case of hinged ends, solutions of the above boundary value problem are the functions

$$\mathrm{v(z)}=\mathrm{A}_1\sin\frac{\pi z}{L} \qquad \gamma(z)=\mathrm{A}_2\cos\frac{\pi z}{L} \tag{4.82}$$

Substitution of (4.82) into (4.80) gives

$$\begin{vmatrix} \mathrm{N_c}-\mathrm{N} & \mathrm{N_c} \\ \mathrm{N_c} & \mathrm{N_c}\left(1+\dfrac{1}{\nu}\right) \end{vmatrix}=0 \tag{4.83}$$

where

$$\mathrm{N_c}=\frac{\mathrm{EI}\pi^2}{L^2} \qquad \nu=\frac{\mathrm{N_c}\chi}{\mathrm{GA}} \tag{4.84}$$

and the buckling load is

$$\mathrm{N_c}=\frac{\mathrm{N_c}}{1+\nu} \tag{4.85}$$

The equilibrium of the rod is stable at buckling. We have in fact, from (4.78), (4.79) and (4.85)

$$\mathcal{E}_4=\frac{A_1^4}{64}\,\frac{\mathrm{EI}\pi^6}{(1+\nu)L^5}>0 \tag{4.86}$$

4.2.3 THE EXTENSIBLE ROD

Let us study the finite deformation of rods in bending and extension. In the framework of the model of the "Elastica", assume assumptions 1, 2 and 3 are still valid. Thus, in this case, too, displacements and strains of each point in the beam will depend only on the displacements of the rod axis. With reference to Fig. 4.8 the displacement of the generic point P is given by

$$\mathrm{v(y,z)}=\mathrm{v(z)}-\mathrm{y}(1-\cos\alpha) \quad \mathrm{w(y,z)}=\mathrm{w(z)}-\mathrm{y}\sin\alpha \tag{4.87}$$

Here v(z) and w(z) are the displacement components of a point on the beam axis and α the rotation angle of the cross section that, before deformation, occupies the position z. The angle α, on the other hand, is determined by the orthogonality condition between the cross section and the deformed axis of the rod, i.e., by assumption 2. Thus

$$\sin\alpha=\frac{dv}{ds} \tag{4.88}$$

To the point P, of coordinates (x,y,z), before deformation, corresponds, after deformation, the point P,' of coordinates (x',y',z') given, according to (4.87), by

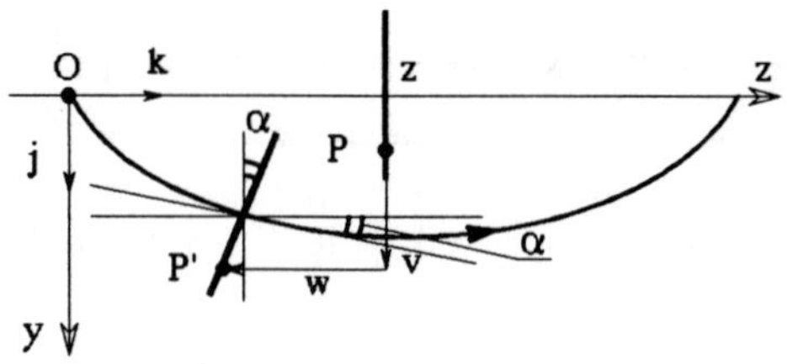

FIGURE 4.8

$$x' = x$$

$$y' = y + v(y,z) = v(z) + y\cos\alpha$$

$$z' = z + w(y,z) = z + w(z) - y\sin\alpha \tag{4.89}$$

Points on the axis of the rod have coordinates

$$x' = 0\ ;\ y' = v(z);\ z' = z + w(z) \tag{4.90}$$

and

$$ds^2 = dy'^2 + dz'^2 = [\,v'^2 + (1 + w')^2]\,dz^2 \tag{4.91}$$

Thus, the orthogonality condition (4.88) gives

$$\sin\alpha = \frac{v'}{\left[v'^2 + (1 + w'^2)^2\right]^{1/2}} \tag{4.88'}$$

Equation (4.89) defines the change of configuration of the rod, and the local deformation, defined by the gradient tensor of transformation (4.89), is

$$\left[F_{ij}\right] = \left[\frac{\partial x'_i}{\partial x_j}\right] = \begin{bmatrix} 1 & 0 & 0 \\ 0 & \cos\alpha & v' - (y\sin\alpha)\alpha' \\ 0 & -\sin\alpha & 1 + w' - (y\cos\alpha)\alpha' \end{bmatrix} \tag{4.92}$$

Evaluate the components of the finite strains and the corresponding strain energy. In the gradient of the transformation (4.92) are included both the local strains and the rigid rotations. If we indicate with ε, the strain tensor, we have

$$(1 + \varepsilon)^2 = \mathbf{F}^T\mathbf{F} \tag{4.93}$$

where the tensor product $\mathbf{F}^T\mathbf{F}$ is given by

$$\left[\mathbf{F}^T\,\mathbf{F}\right]_{ij} = \begin{bmatrix} 1 & 0 & 0 \\ 0 & 1 & \cos\alpha(v' - \alpha' y\sin\alpha) - \sin\alpha(1 + w' - \alpha' y\cos\alpha) \\ 0 & \cos\alpha(v' - y\alpha' y\sin\alpha) - \sin\alpha(1 + w' - \alpha' y\cos\alpha) & (v' - y\alpha' y\sin\alpha)^2 + (1 + w' - \alpha' y\cos\alpha)^2 \end{bmatrix} \tag{4.94}$$

From the orthogonality between cross sections and the deformed axis of the rod, we get

$$\cos\alpha(v' - y\alpha' \sin\alpha) - \sin\alpha(1 + w' - y\alpha' \cos\alpha) = v' \cos\alpha - \sin\alpha(1 + w') \tag{4.95}$$

We have also

$$(v' - \alpha' y \sin\alpha)^2 + [(1 + w') - \alpha' y \cos\alpha]^2$$

$$= v'^2 + (1 + w')^2 + y^2 \alpha'^2 - 2\alpha' y \, [v' \sin\alpha + (1 + w') \cos\alpha] \tag{4.95′}$$

On the other hand

$$\cos\alpha = (1 + w') \frac{dz}{ds} \qquad \sin\alpha = v' \frac{dz}{ds} \tag{4.96}$$

and we have

$$v' \sin\alpha + (1 + w') \cos\alpha = [v'^2 + (1 + w')^2] \frac{dz}{ds} = \frac{ds}{dz} \tag{4.97}$$

Thus, (4.95′) becomes

$$(v' - \alpha' y \sin\alpha)^2 + [(1 + w') - \alpha' y \cos\alpha]^2 = (\frac{ds}{dz} - \alpha' y)^2 \tag{4.98}$$

and the matrix $[\mathbf{F}^T\mathbf{F}]_{ij}$ is

$$[\mathbf{F}^T\mathbf{F}]_{ij} = \begin{bmatrix} 1 & 0 & 0 \\ 0 & 1 & 0 \\ 0 & 0 & \left(\frac{ds}{dz} - y\alpha\right)^2 \end{bmatrix} \tag{4.99}$$

Thus, we get

$$\sqrt{[\mathbf{F}^T\mathbf{F}]_{ij}} = (1 + \varepsilon)_{ij} = \begin{bmatrix} 1 & 0 & 0 \\ 0 & 1 & 0 \\ 0 & 0 & \frac{ds}{dz} - y\alpha' \end{bmatrix} \tag{4.100}$$

Therefore the pure strain components are

$$\varepsilon_{ij} = \begin{bmatrix} 0 & 0 & 0 \\ 0 & 0 & 0 \\ 0 & 0 & \left(\frac{ds}{dz} - 1\right) - y\alpha' \end{bmatrix} \tag{4.101}$$

These components are all zero except ε_{zz}, which is composed of extensional and flexural fractions due to the elongation and bending of the rod axis. We have in fact

$$\varepsilon_{zz} = e + y \frac{1}{\rho} \tag{4.102}$$

where

$$e = \frac{ds}{dz} - 1 \qquad \frac{1}{\rho} = -\alpha' \tag{4.103}$$

which, respectively, represent the extension and the flexural curvature of the rod. The assumed kinematics of the deformation of the rod implies that the material is rigid as far as deformations along x and y and shears between x, y, and z are concerned.Therefore, the rod deformability is only represented by elongations along the longitudinal fibers. The elasticity equations are assumed to be represented by a simple linear relation between the longitudinal stress σ_{zz} and strain ε_{zz} components

$$\sigma_{zz} = E\, \varepsilon_{zz} \tag{4.104}$$

where E indicates the Young modulus. The strain energy of the rod is therefore a quadratic function of the components of the pure strains.

$$\mathcal{V} = \frac{E}{2}\int_V \varepsilon_{zz}^2\, dV \tag{4.105}$$

and, according to (4.101),

$$\mathcal{V} = \frac{EA}{2}\int_0^L \left(\frac{ds}{dz} - 1\right)^2 dz + \frac{EI}{2}\int_0^L \alpha'^2\, dz \tag{4.106}$$

Define the corresponding energy space H_A for the assumed model of the rod. With the assumption of infinitesimal strains, the linearized expressions of quantities (4.103) are

$$e = w' \qquad \frac{1}{\rho} = -v'' \tag{4.107}$$

and the corresponding strain energy takes the form

$$\mathcal{V}_2 = \frac{1}{2} EA\int_0^L w'^2\, dz + \frac{1}{2}\, EI\int_0^L v''^2\, dz \tag{4.108}$$

The space H_A is thus composed of all the displacement functions $\mathbf{u}(z)$ of components $[v(z),w(z)]$ for which the quantity

$$|||\mathbf{u}|||^2 = EA\int_0^L w'^2\, dz + EI\int_0^L v''^2\, dz \tag{4.109}$$

is defined. Every $\mathbf{u} \in H_A$ has components $v \in C^1[0,L]$ and $w(z) \in C[0,L]$. It is not possible to evaluate the finite strain components defined by (4.103) corresponding to a generic displacement function $[v(z),w(z)]$ of H_A.We get, in fact,

$$e = [v'^2 + (1 + w')^2]^{1/2} - 1; \quad \frac{1}{\rho} = -\frac{v''(1 + w') - w''v'}{v'^2 + (1 + w')^2} \tag{4.110}$$

We notice the presence of the second derivative w" that is not defined. At the same time, the strain energy functional $\mathcal{V}$, which, according to (4.106), takes the form

$$\mathcal{V} = \frac{1}{2} EA\int_0^L \left(\left[v'^2 + (1+w')^2\right]^{1/2} - 1\right)^2 dz + \frac{1}{2} EI\int_0^L \left[\frac{v''(1+w') - w''v'}{v'^2 + (1+w')^2}\right]^2 dz \tag{4.111}$$

does not satisfy the smoothness properties, as far as its Frechét differentiability is concerned. In fact the functional (4.111) is not Frechét differentiable with respect to the energy norm (4.109).

The stability criteria previously stated and based on the property of strong differentiability of the energy functional thus cannot be applied in this case.

4.3 TORSIONAL-FLEXURAL BUCKLING OF INEXTENSIBLE RODS

4.3.1 FINITE FLEXURAL-TORSIONAL DEFORMATION OF SLENDER RODS

In this section we will analyze the flexural-torsional deformation of beams on the basis of the assumptions already considered for the model of the "Elastica":

1. The cross sections of the beam are not deformable in their planes.
2. The cross sections remain orthogonal to the beam axis after the deformation.
3. The beam axis is inextensible.

Within this context, which corresponds essentially to the classical analysis given by Love[1], a model of bars in finite flexure and torsion is given. The flexural and torsional curvatures and the corresponding bending and twisting moments represent the basic forces and deformation variables. The relationships between moments and curvatures are taken to be linear, according to the large displacements-small strains assumption.

According to Fig. 4.9, let O123 be the reference coordinate system: the axis 1(z) is directed along the beam axis, while 2(x) and 3(y) are parallel to the principal directions of the cross sections at their centroids in the unstressed state. Let G be the centroid of the generic cross section S, and let G1,2,3 be the local reference system, parallel to the fixed system O123. Let $\mathbf{D}_1$, $\mathbf{D}_2$, and $\mathbf{D}_3$ be three unitary vectors issued from G in the directions of the axes z, x, and y. When the rod gets deformed, cross section S takes position S′, local axes G1,2,3 transform into G′1′,2′,3′ and the vector system $G\mathbf{D}_1\mathbf{D}_2\mathbf{D}_3$ into $G'\mathbf{d}_1\mathbf{d}_2\mathbf{d}_3$. Here the unitary vector $\mathbf{d}_1$, issued from G′, is tangent to the deformed axis of the beam, and $\mathbf{d}_2$ and $\mathbf{d}_3$ are unit vectors issued from G′ and acting along the directions 2′ and 3′, the principal axes of the section in its deformed position S′ (Fig. 4.9). The rigid displacement of the vector system from $G\mathbf{D}_1\mathbf{D}_2\mathbf{D}_3$ to $G'\mathbf{d}_1\mathbf{d}_2\mathbf{d}_3$ can be obtained by a rigid translation and a sequence of three rotations (Fig. 4.10)

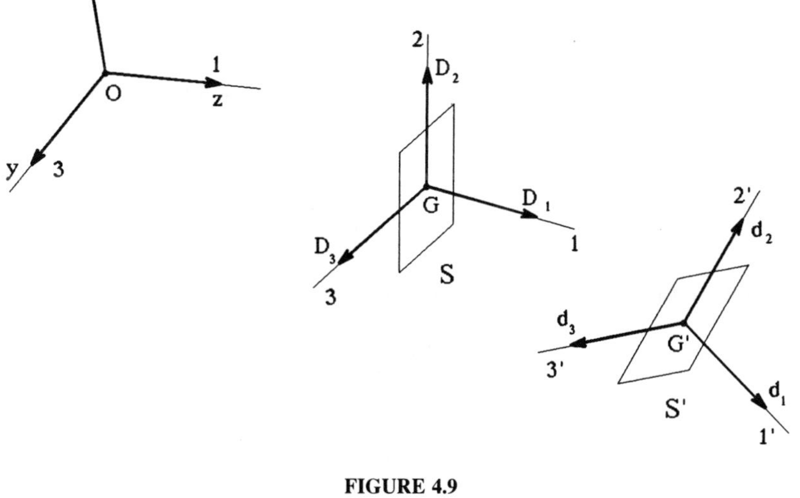

FIGURE 4.9

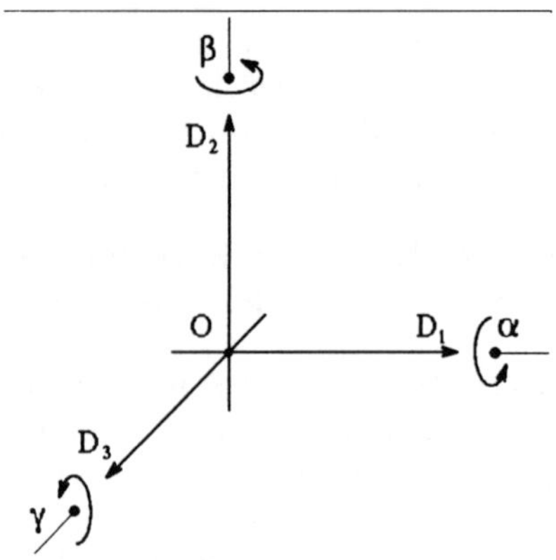

FIGURE 4.10

$$\alpha, \beta, \gamma \tag{4.112}$$

respectively, around axes 1, 2, and 3 of the fixed system O123. The values of these three angles depend on the order with which the three rotations are applied. We will consider the sequence (4.112) of rotations around the axes 1, 2, and 3. The matrix of this rotation transformation is immediately obtained. According to the established sequence, let us first apply the rotation α around the axis 1; a vector X is thus transformed into the vector X′ given by

$$X' = AX \tag{4.113}$$

where

$$A = \begin{bmatrix} 1 & 0 & 0 \\ 0 & \cos\alpha & -\sin\alpha \\ 0 & \sin\alpha & \cos\alpha \end{bmatrix} \tag{4.114}$$

Then, the rotation β around the axis 2 will occur. The matrix B of this last transformation is

$$B = \begin{bmatrix} \cos\beta & 0 & \sin\beta \\ 0 & 1 & 0 \\ -\sin\beta & 0 & \cos\beta \end{bmatrix} \tag{4.115}$$

Finally, the corresponding matrix C of the last rotation γ around the axis 3 is

$$C = \begin{bmatrix} \cos\gamma & -\sin\gamma & 0 \\ \sin\gamma & \cos\gamma & 0 \\ 0 & 0 & 1 \end{bmatrix} \tag{4.116}$$

Summing up these three rotation according to the above defined sequence, we get

$$X' = \mathcal{R}X = CBAX \tag{4.117}$$

where

$$\mathcal{R} = CBA = \begin{bmatrix} d_{11}\, d_{12}\, d_{13} \\ d_{21}\, d_{22}\, d_{23} \\ d_{31}\, d_{32}\, d_{33} \end{bmatrix} \tag{4.118}$$

is the rotation matrix, corresponding to the sequence (4.112). Components d_{ij} are

$$d_{11} = \cos\gamma\cos\beta;\ d_{12} = \cos\gamma\sin\beta\sin\alpha - \sin\gamma\cos\alpha;\ d_{13} = \cos\gamma\sin\beta\cos\alpha + \sin\gamma\sin\alpha$$
$$d_{21} = \sin\gamma\cos\beta;\ d_{22} = \sin\gamma\sin\beta\sin\alpha + \cos\gamma\cos\alpha;\ d_{23} = \sin\gamma\sin\beta\cos\alpha - \cos\gamma\sin\alpha$$
$$d_{31} = -\sin\beta;\quad d_{32} = \cos\beta\sin\alpha;\quad d_{33} = \cos\beta\cos\alpha \tag{4.119}$$

Columns of the rotation matrix (4.118) give the components of the transformed vectors $\mathbf{d}_1$, $\mathbf{d}_2$, and $\mathbf{d}_3$ of the unit vectors $\mathbf{D}_1$, $\mathbf{D}_2$, and $\mathbf{D}_3$, evaluated with respect to the fixed axis O123. So, for instance, we have

$$\mathbf{d}_1 = \cos\gamma\cos\beta\mathbf{D}_1 + \sin\gamma\cos\beta\,\mathbf{D}_2 - \sin\beta\mathbf{D}_3 \tag{4.120}$$

Of course, the rotation matrix (4.118) changes if, in the passage from $G\mathbf{D}_1\mathbf{D}_2\mathbf{D}_3$ to $G'\mathbf{d}_1\mathbf{d}_2\mathbf{d}_3$, we assume a different order of application of the three rotations α, β, and γ around the fixed axes 1, 2, and 3. Let us assume in fact a different sequence of the transformations: for instance, the sequence 1-3-2. In this case we have, in place of (4.118),

$$\overline{\mathcal{R}} = BCA = \begin{bmatrix} \bar{d}_{11} & \bar{d}_{12} & \bar{d}_{13} \\ \bar{d}_{21} & \bar{d}_{22} & \bar{d}_{23} \\ \bar{d}_{31} & \bar{d}_{32} & \bar{d}_{33} \end{bmatrix} \tag{4.118$'$}$$

where

$$\bar{d}_{11} = \cos\gamma\cos\beta;\ \bar{d}_{12} = -\sin\gamma\cos\beta\cos\alpha + \sin\beta\sin\alpha;\ \bar{d}_{13} = \sin\gamma\cos\beta\sin\alpha + \sin\beta\cos\alpha$$
$$\bar{d}_{21} = \sin\gamma;\quad \bar{d}_{22} = \cos\gamma\cos\alpha;\quad \bar{d}_{23} = -\cos\gamma\sin\alpha$$
$$\bar{d}_{31} = -\sin\beta\cos\gamma;\ \bar{d}_{32} = \sin\beta\sin\gamma\cos\alpha + \cos\beta\sin\alpha;\ \bar{d}_{33} = -\sin\beta\sin\alpha\sin\gamma + \cos\beta\cos\alpha \tag{4.119$'$}$$

Let us consider now the cross section of the beam that in the unstressed state occupies the position z, $z \in [0,L]$. For every z we can define the three rotations (4.112) in the previously defined sequence

$$\alpha(z),\ \beta(z),\ \gamma(z) \qquad z \in [0,L] \tag{4.121}$$

These three functions are assumed continuous together with their first derivatives in the whole interval [0,L]. A continuous sequence of deformed reference systems running along the deformed beam axis is thus defined. (Fig. 4.11). The origin G′ of the vector frame $G'\mathbf{d}_1\mathbf{d}_2\mathbf{d}_3$ moves along the deformed central line of the rod. Let us define the vector $\omega(z)$ that represents the rate of change of the frame rotation. This vector has components directed along the axes $G'\mathbf{d}_1\mathbf{d}_2\mathbf{d}_3$ and can be expressed as

$$\omega(z) = \omega_1(z)\mathbf{d}_1 + \omega_2(z)\mathbf{d}_2 + \omega_3(z)\mathbf{d}_3 \tag{4.122}$$

where

$$\omega_1(z) = \tau;\ \omega_2(z) = k_2;\ \omega_3(z) = k_3 \tag{4.123}$$

Thus τ is the twist of the rod at G′ and k_2 and k_3 the curvatures of the projections of the deformed central line of the beam on the planes $G\mathbf{d}_1\mathbf{d}_3$ and $G\mathbf{d}_1\mathbf{d}_2$, i.e., the curvature components of the deformed central line at G′. (Fig. 4.12).

The resultant of $\omega_2(z)\mathbf{d}_2$ and $\omega_3(z)\mathbf{d}_3$ is a vector directed along the binormal of the strained central line and its modulus equals the curvature $1/\rho$ of this curve. The rate of change of the

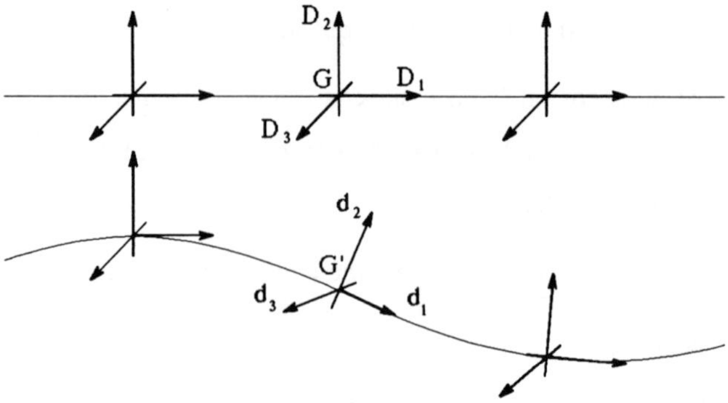

FIGURE 4.11

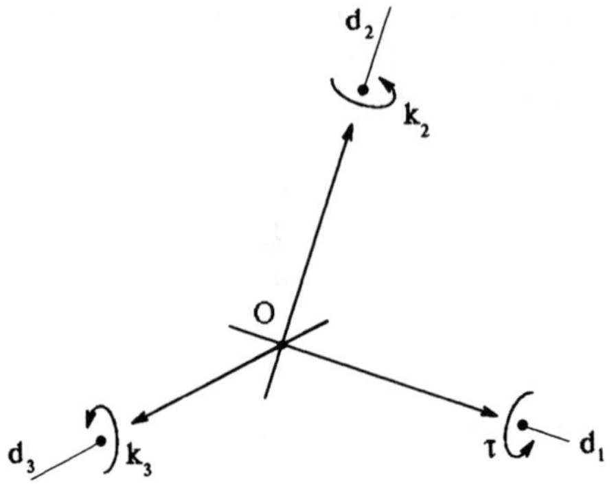

FIGURE 4.12

rotation frame ω(z) can be expressed by means the rates of change $\dot{\mathbf{d}}_1$, $\dot{\mathbf{d}}_2$, and $\dot{\mathbf{d}}_3$ of the unitary vectors $\mathbf{d}_1$, $\mathbf{d}_2$, and $\mathbf{d}_3$, when the solid frame $G'\mathbf{d}_1\mathbf{d}_2\mathbf{d}_3$ moves along the deformed central line of the rod on which the curvilinear abscissa s(z) is defined. (Fig. 4.13). We have

$$\dot{\mathbf{d}}_1 = \omega \wedge \mathbf{d}_1;\ \dot{\mathbf{d}}_2 = \omega \wedge \mathbf{d}_2;\ \dot{\mathbf{d}}_3 = \omega \wedge \mathbf{d}_3,\ (\dot{\ }) = \frac{d(\ )}{ds} \tag{4.124}$$

and, consequently,

$$\dot{\mathbf{d}}_1 \cdot \mathbf{d}_2 = \omega \wedge \mathbf{d}_1 \cdot \mathbf{d}_2;\ \dot{\mathbf{d}}_1 \cdot \mathbf{d}_3 = \omega \wedge \mathbf{d}_1 \cdot \mathbf{d}_3;\ \dot{\mathbf{d}}_2 \cdot \mathbf{d}_3 = \omega \wedge \mathbf{d}_2 \cdot \mathbf{d}_3 \tag{4.125}$$

On the other hand

$$\omega \wedge \mathbf{d}_1 \cdot \mathbf{d}_2 = k_3;\ \omega \wedge \mathbf{d}_1 \cdot \mathbf{d}_3 = -\ k_2;\ \omega \wedge \mathbf{d}_2 \cdot \mathbf{d}_3 = \tau \tag{4.126}$$

Thus

$$\begin{aligned} &k_3 = \dot{d}_{11}d_{21} + \dot{d}_{12}d_{22} + \dot{d}_{13}d_{23} - k_2 = \dot{d}_{11}d_{31} + \dot{d}_{12}d_{32} + \dot{d}_{13}d_{33} \\ &\tau = \dot{d}_{21}d_{31} + \dot{d}_{22}d_{32} + \dot{d}_{23}d_{33} \end{aligned} \tag{4.127}$$

Thus, according to (4.119), i.e., according to the assumed order (4.112) of the rotations sequence, we get

$$k_2 = \dot{\beta}\cos\alpha + \dot{\gamma}\sin\alpha\cos\beta\ \ k_3 = -\dot{\beta}\sin\alpha + \dot{\gamma}\cos\alpha\ \sin\beta\ \ \tau = \dot{\alpha} - \dot{\gamma}\sin\beta \tag{4.128}$$

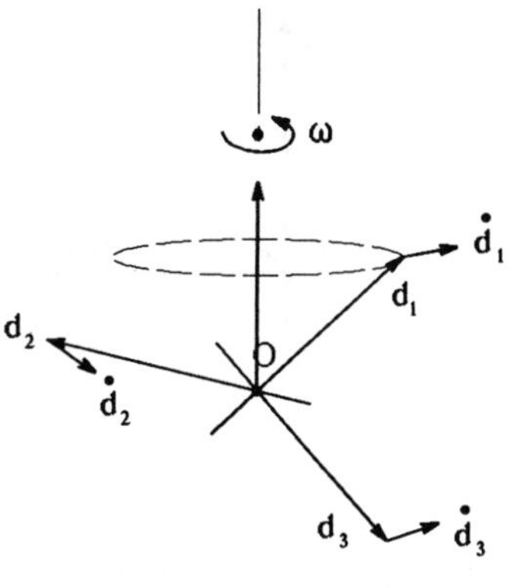

FIGURE 4.13

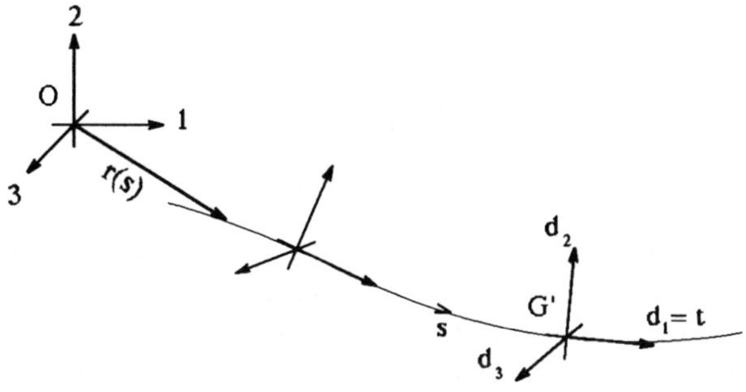

FIGURE 4.14

We can express the flexural curvatures and the twist of the beam in terms of displacement components of the beam axis u(z), v(z), and w(z) and of the angle of twist α(z). (Fig. 4.14). The position vector **r**(z) of the displaced centroid G′ is defined on the fixed axes system O123, as

$$\mathbf{r}(z) = (z + w)\mathbf{D}_1 + u\mathbf{D}_2 + v\mathbf{D}_3 \tag{4.129}$$

This position vector **r**(s) runs along the deformed central line of the rod (Fig. 4.15) where the curvilinear abscissa s = s(z) is defined. Thus, the unitary vector **t** tangent to this curve is given by

$$\mathbf{t} = \frac{d\mathbf{r}}{ds} = (1 + \frac{dw}{ds})\mathbf{D}_1 + \frac{du}{ds}\mathbf{D}_2 + \frac{dv}{ds}\mathbf{D}_3 \tag{4.130}$$

or by

$$\mathbf{t} = (1 + \frac{dw}{dz}\frac{dz}{ds})\mathbf{D}_1 + \frac{du}{dz}\frac{dz}{ds}\mathbf{D}_2 + \frac{dv}{dz}\frac{dz}{ds}\mathbf{D}_3 \tag{4.131}$$

By taking into account the inextensibility of the central axis of the beam, i.e., that

$$\frac{dz}{ds} = 1 \tag{4.132}$$

we have

$$\mathbf{t} = (1 + w')\mathbf{D}_1 + u'\,\mathbf{D}_2 + v'\,\mathbf{D}_3 \tag{4.133}$$

where

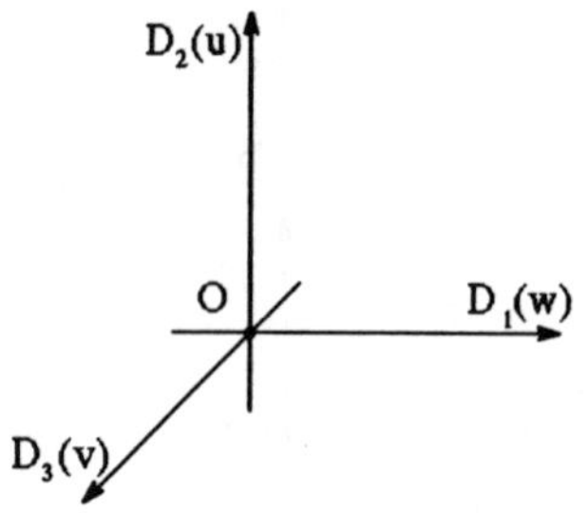

FIGURE 4.15

$$ds^2 = du^2 + dv^2 + (dz + dw)^2 \tag{4.134}$$

Thus, with (4.132), we get the condition

$$(1 + w')^2 = 1 - u'^2 - v'^2 \tag{4.135}$$

where we have used the common notation

$$(\ )' = d(\ )/dz \tag{4.136}$$

On the other hand the position vector $\mathbf{r}(z)$ cannot define the whole deformation of the rod because it does not define the position of the principal axes of the cross sections of the deformed beam. To define this deformation, let us consider the local orthogonal axes issued from G′, the centroid of the cross section S′ of the deformed rod. This system is represented by the tangent vector $\mathbf{t}$ to the deformed central line and by the two principal axes, solid to the rotated cross section.

Every position of this displaced system can be reached by the corresponding system $\mathbf{GD_1D_2D_3}$ of the unstressed state by means a rigid translation plus an established sequence of rotations $\alpha(z)$, $\beta(z)$, $\gamma(z)$. Then, to obtain the corresponding values of these rotation angles according to the assumed order of the sequence, we have to take into account that

$$\mathbf{t} = \mathbf{d}_1 \tag{4.137}$$

Thus, comparing (4.120) and (4.133) gives

$$\sqrt{1 - u'^2 - v'^2} = \cos\gamma\cos\beta \qquad u' = \sin\gamma\cos\beta \qquad v' = -\sin\beta \tag{4.138}$$

and we obtain

$$\dot{\gamma}\sin\beta = \left[-u''v' - \frac{u'v'^2v''}{(1-v'^2)}\right]\frac{1}{\sqrt{1-u'^2-v'^2}}$$

$$\dot{\gamma}\cos\beta = \left[u'' + \frac{u'v'v''}{(1-v'^2)}\right]\frac{\sqrt{1-v'^2}}{\sqrt{1-u'^2-v'^2}}$$

$$\dot{\beta} = -\frac{v''}{\sqrt{1-v'^2}} \tag{4.139}$$

Now we can express the flexural curvatures and the twist of the rod as appropriate functions of the twist angle α and of the transversal displacements u(z) and v(z) of the beam axis

$$k_2 = -\frac{v''}{\sqrt{1-v'^2}}\cos\alpha + \left(u'' + \frac{u'v'v''}{1-v'^2}\right)\sqrt{\frac{1-v'^2}{1-u'^2-v'^2}}\sin\alpha$$

$$k_3 = -\frac{v''}{\sqrt{1-v'^2}}\sin\alpha + \left(u'' + \frac{u'v'v''}{1-v'^2}\right)\sqrt{\frac{1-v'^2}{1-u'^2-v'^2}}\cos\alpha$$

$$\tau = \alpha' + \left(u''v' + \frac{u'v'^2v''}{1-v'^2}\right)\frac{1}{\sqrt{1-u'^2-v'^2}} \tag{4.140}$$

With a different sequence of the three rotations — for instance, the sequence 1-3-2 — by using the transformation matrix (4.118′), we will have, on the other hand, in place of (4.140),

$$\bar{k}_2 = \frac{u''}{\sqrt{1-u'^2}}\sin\alpha - \left(v'' + \frac{v'u'u''}{1-u'^2}\right)\sqrt{\frac{1-u'^2}{1-u'^2-v'^2}}\cos\alpha$$

$$\bar{k}_3 = \frac{u''}{\sqrt{1-u'^2}}\cos\alpha + \left(v'' + \frac{v'u'u''}{1-u'^2}\right)\sqrt{\frac{1-u'^2}{1-u'^2-v'^2}}\sin\alpha$$

$$\bar{\tau} = \alpha' - \left(v''u' + \frac{v'u'^2u''}{1-u'^2}\right)\frac{1}{\sqrt{1-u'^2-v'^2}} \tag{4.141}$$

The position of the inextensible axis of the rod is defined by the functions u(z) and v(z), while the deformation of the rod by the curvature functions $k_2(z)$, $k_3(z)$, and $\tau(z)$ — or, equivalently, by the functions u(z), v(z), and $\alpha(z)$ — is connected with a given order of the sequence (4.121).

With a different order of the rotation sequence, but taking as fixed the functions u(z), v(z), and $\alpha(z)$, the axis of the beam, will move, in fact, in the same position. On the contrary, if we want the beam to occupy the same previous configuration, while the functions u(z) and v(z) remain unchanged, suitable changes have to be given to the twist angle function $\alpha(z)$. Only in this case, in fact, the curvatures $k_2(z)$, $k_3(z)$, and $\tau(z)$ will remain the same.

We conclude the analysis of the bending-torsion of the elastic rod by giving some further expressions of the torsional and flexural curvatures of the bar. These curvatures, given by (4.140) and expanded up to terms of the fourth order in the displacements, take the form

$$\tau = \tau^{(1)} + \tau^{(2)} + \tau^{(3)} + \tau^{(4)} + \dots \tag{4.142a}$$

$$k_2 = k_2^{(1)} + k_2^{(2)} + k_2^{(3)} + k_2^{(4)} + \dots \tag{4.142b}$$

$$k_3 = k_3^{(1)} + k_3^{(2)} + k_3^{(3)} + k_3^{(4)} + \dots \tag{4.142c}$$

where,

$$\tau^{(1)} = \alpha';\quad \tau^{(2)} = u''v';\quad \tau^{(3)} = 0;\quad \tau^{(4)} = u'v'^2v'' + \frac{1}{2}u''v'\left(u'^2 + v'^2\right) \tag{4.143a}$$

$$k_2^{(1)} = -v'';\quad k_2^{(2)} = u''\alpha;\quad k_2^{(3)} = \frac{1}{2}v''\left(\alpha^2 - v'^2\right);\quad k_2^{(4)} = \alpha\left(\frac{u''u'^2}{2} + u'v'v'' - \frac{u''\alpha^2}{6}\right) \tag{4.143b}$$

$$k_3^{(1)} = u'';\quad k_3^{(2)} = v''\alpha;\quad k_3^{(3)} = \frac{u''}{2}\left(u'^2 - \alpha^2\right) + u'v'v'';\quad k_3^{(4)} = \alpha v''\left(\frac{v'^2}{2} - \frac{\alpha^2}{6}\right) \tag{4.143c}$$

4.3.1.1 The Strain Energy Functional of the Rod with Torsional-Flexural Finite Deformations

To obtain a useful expression of the potential energy of the beam, assume that the strains are small quantities with respect to unity, while displacements, on the contrary, can be relatively large with respect to the main dimensions of the rod. This last statement is justified by the assumption of the high stiffness of the material, such as steel, of which the beam is composed.

For the unidimensional model of the rod, the assumption of small strains implies small values of the flexure and torsion curvatures; in this case the energy functional can be approximated by the quadratic form

$$\mathcal{V} = \frac{1}{2} C \int_0^L \tau^2 \, dz + \frac{1}{2} B_2 \int_0^L k_2^2 \, dz + \frac{1}{2} B_3 \int_0^L k_3^2 \, dz \tag{4.144}$$

where C, B_2, and B_3 are respectively the torsional and flexural stiffnesses of the beam. We observe that, due to the assumption of a solid cross section, the warping effects have been neglected; further, the nonlinear torsional response has not been taken into account.

4.3.1.2 The Energy Space and the Strong Differentiability of the Strain Energy Functional

The set of admissible configurations of the beam, and the corresponding energy space H_A, correspond to the set of vectorial functions **u**(z) which meet the kinematical boundary conditions and such that the quantity

$$|||u|||^2 = C \int_0^L \alpha'^2 \, dz + B_2 \int_0^L u''^2 \, dz + B_3 \int_0^L v''^2 \, dz \tag{4.145}$$

is finite. By using the Sobolev theorems recalled above, it is easy to show that, if the displacement vector functions **u**(z) belong to H_A, their displacement components u(z) and v(z) are continuous with their first derivatives u'(z) and v'(z), while the torsional component $\alpha(z)$ is only continuous in the interval [0,L]. Further, the following inequalities hold:

$$\exists\, k_1, k_2, k_3, k_4, k_5 > 0: \max_{z\in[0,L]} |u'(z)| \le k_1 |||\mathbf{u}|||; \ \max_{z\in[0,L]} |v'(z)| \le k_2 |||\mathbf{u}||| \tag{4.146}$$

$$\max_{z\in[0,L]} |\alpha(z)| \le k_3 |||\mathbf{u}|||; \ ||u'|| = \left(\int_0^1 u'^2 \, dz \right)^{1/2} \le k_4 |||\mathbf{u}|||; \ ||v'|| = \left(\int_0^1 v'^2 \, dz \right)^{1/2} \le k_5 |||\mathbf{u}||| \tag{4.147}$$

The strain energy functional (4.144) is defined for any function **u**(z) such that

$$|u'(z)| \le c_1 < 1; \ |v'(z)| \le c_2 < 1; \ z \in [0,L] \tag{4.148}$$

i.e., taking account of (4.148), $\mathcal{V}(\mathbf{u})$ is defined in the set

$$S_\rho = \{\mathbf{u} \in H_A : |||\mathbf{u}||| < \rho\} \tag{4.149}$$

According to Ascione and Grimaldi[5] the potential energy functional, with the strain energy (4.144), is at least four times strong differentiable at the equilibrium state C_0. We will give a detailed proof of this statement only for the strain energy functional. The strong differentiability of the external load potential, for the several examples that will be examined in the next sections, can be proved with similar arguments. We will now prove the second strong differentiability, i.e., we will show that

$$\mathcal{V}(\mathbf{u}_0+\mathbf{u})-\mathcal{V}(\mathbf{u}_0)=\mathrm{D}\mathcal{V}(\mathbf{u}_0;\mathbf{u})\frac{1}{2}\mathrm{D}^2\mathcal{V}(\mathbf{u}_0;\mathbf{u},\mathbf{u})+\mathrm{r}_2(\mathbf{u}) \tag{4.150}$$

where $\mathrm{D}^2\mathcal{V}(\mathbf{u}_0;\ \mathbf{u},\ \mathbf{u}) \leq \mathrm{k}|||\mathbf{u}|||^2$ and

$$\lim_{|||\mathbf{u}|||\to 0}\frac{|\mathrm{r}_2(\mathbf{u})|}{|||\mathbf{u}|||^2}=0 \tag{4.151}$$

A more convenient expression of the remainder $\mathrm{r}_2(\mathbf{u})$ can be obtained by writing the increment

$$\begin{aligned}\mathcal{V}(\mathbf{u}_0;\mathbf{u})-\mathcal{V}(\mathbf{u}_0)&=\mathrm{D}\mathcal{V}(\mathbf{u}_0;\mathbf{u})+\frac{1}{2}\mathrm{D}^2\mathcal{V}(\mathbf{u}_0;\mathbf{u},\mathbf{u})\\&+\frac{1}{2}\left[\mathrm{D}^2\mathcal{V}(\mathbf{u}_0+\varepsilon\mathbf{u};\mathbf{u},\mathbf{u})-\mathrm{D}^2\mathcal{V}(\mathbf{u}_0;\mathbf{u},\mathbf{u})\right]\end{aligned} \tag{4.152}$$

with $0 \leq \varepsilon \leq 1$. Therefore, the remainder takes the form

$$\mathrm{r}_2(\mathbf{u})=\left[\frac{1}{2}\mathrm{D}^2\mathcal{V}(\mathbf{u}_0+\varepsilon\mathbf{u};\mathbf{u},\mathbf{u})-\frac{1}{2}\mathrm{D}^2\mathcal{V}(\mathbf{u}_0;\mathbf{u},\mathbf{u})\right] \tag{4.153}$$

To get a simpler form of these differentials let us introduce the nondimensional quantities

$$\zeta = z/L \quad d(\ )/d\zeta = (\ )' \quad U = u/L \quad V = v/L \tag{4.154}$$

$$\chi_1 = B_3/B_2 \quad \chi_2 = C/B_2 \tag{4.155}$$

Thus, by using (4.144), the strain energy functional $\mathcal{V}(\mathbf{u})$ can be expressed in the form

$$\mathcal{V}(\mathbf{u})=\frac{1}{2}\frac{B_2}{L}\int_0^1[\chi_2\,\alpha'^2+U''^2g+V''^2g_1+U''\alpha'g_2+V''\alpha'g_3+U''V''g_4]\,d\zeta \tag{4.156}$$

where g, g_1, g_2, g_3, and g_4 are continuously differentiable functions of U', V', and α, and linear functions of χ_1, and χ_2. For instance,

$$g=\chi_2\frac{V'^2}{1-U'^2-V'^2}+\frac{1-V'^2}{1-U'^2-V'^2}\sin^2\alpha+\chi_1\frac{1-V'^2}{1-U'^2-V'^2}\cos^2\alpha \tag{4.157}$$

It is obvious that the first term in bracket of (4.156) is twice strong differentiable with respect to the energy norm. All the other terms will behave in the same way. Let us analyze, for instance, the differentiability of the functional

$$F(\mathbf{u})=\int_0^1 U''^2\,g(U',V',\alpha)\,d\zeta \tag{4.158}$$

The second weak differential of $F(\mathbf{u})$ at the point $\mathbf{u}$ and in the direction $\mathbf{u}_1$ is

$$D^2F(\mathbf{u};\mathbf{u}_1) = \int_0^1 (2\,U_1''^2 + 2U''U_1''^2\,d^1g + U''d^2g)d\zeta \tag{4.159}$$

where, if we denote $y_1 = U_1'$, $y_2 = U_2'$, $y_3 = \alpha_1$,

$$d^1g = \frac{\partial g}{\partial U'}U_1' + \frac{\partial g}{\partial V'}V_1' + \frac{\partial g}{\partial \alpha}\alpha_1 = \frac{\partial g}{\partial y_i}y_i \tag{4.160}$$

$$d^2g = \frac{1}{2}\left[\frac{\partial}{\partial U'}(d^1g)\,U_1' + \frac{\partial}{\partial V'}(d^1g)\,V_1' + \frac{\partial g}{\partial \alpha}(d^1g)\,\alpha_1\right] = \frac{1}{2}\frac{\partial^2 g}{\partial y_i \partial y_j}y_i y_j\ (i,j=1,2,3) \tag{4.161}$$

The remainder $r_2(\mathbf{u})$, given by (4.153), can now be analyzed. We have

$$\begin{aligned}
|r_2(u_1)| &= \frac{1}{2}\left|D^2F(\mathbf{u}+\varepsilon\mathbf{u}_1;\mathbf{u}_1,\mathbf{u}_1) - D^2F(\mathbf{u};\mathbf{u}_1,\mathbf{u}_1)\right| \\
&= \left|\int_0^1 \left[U_1''^2[g(\mathbf{u}+\varepsilon\mathbf{u}_1) - g(\mathbf{u})] + 2(U''+\varepsilon U_1'')U_1''d^1g(\mathbf{u}+\varepsilon\mathbf{u}_1)\right.\right. \\
&\quad \left.\left. -2U''U_1''d^1g(\mathbf{u}) + (U''+\varepsilon U_1'')^2 d^2g(\mathbf{u}+\varepsilon\mathbf{u}_1) - U''^2 d^2g(\mathbf{u})\right]d\zeta\right| \\
&= \left|\int_0^1 \left\{U_1''^2[g(\mathbf{u}+\varepsilon\mathbf{u}_1) - g(\mathbf{u})] + 2U''U_1''[d^1g(\mathbf{u}+\varepsilon\mathbf{u}_1) - d^1g(\mathbf{u})] + 2\varepsilon U_1''^2 d^1g(\mathbf{u}+\varepsilon\mathbf{u}_1)\right.\right. \\
&\quad + U''^2[d^2g(\mathbf{u}+\varepsilon\mathbf{u}_1) - d^2g(\mathbf{u})] + 2\varepsilon U''U_1''d^2g(\mathbf{u}+\varepsilon\mathbf{u}_1) \\
&\quad \left.\left. +\varepsilon^2 U_1''^2 d^2g(\mathbf{u}+\varepsilon\mathbf{u}_1)\right\}d\zeta\right| \\
&\le \max_{\zeta\in[0,1]}\left|g(\mathbf{u}+\varepsilon\mathbf{u}_1) - g(\mathbf{u})\right|\int_0^1 U_1''^2 d\zeta \\
&\quad +2\max_{\zeta\in[0,1]}\left|d^1g(\mathbf{u}+\varepsilon\mathbf{u}_1) - d^1g(\mathbf{u})\right|\left(\int_0^1 U''^2 d\zeta\right)^{1/2}\left(\int_0^1 U_1''^2 d\zeta\right)^{1/2} \\
&\quad + \max_{\zeta\in[0,1]}\left|d^1g(\mathbf{u}+\varepsilon\mathbf{u}_1)\right|\int_0^1 U_1''^2 d\zeta + \max_{\zeta\in[0,1]}\left|d^2g(\mathbf{u}+\varepsilon\mathbf{u}_1) - d^2g(\mathbf{u})\right|\int_0^1 U''^2 d\zeta \\
&\quad +2\max_{\zeta\in[0,1]}\left|d^2g(\mathbf{u}+\varepsilon\mathbf{u}_1)\right|\left(\int_0^1 U''^2 d\zeta\right)^{1/2}\left(\int_0^1 U_1''^2 d\zeta\right)^{1/2} \\
&\quad + \max_{\zeta\in[0,1]}\left|d^2g(\mathbf{u}+\varepsilon\mathbf{u}_1)\right|\left(\int_0^1 U_1''^2 d\zeta\right)
\end{aligned} \tag{4.162}$$

It is useful to recall that the displacement fields $\mathbf{u}(\zeta) \in S \subset H_A$ satisfy (4.146) and (4.147). Hence, $|||\mathbf{u}||| \to 0$ yields $\max_{\zeta\in[0,1]} |U_1'| \to 0$, $\max_{\zeta\in[0,1]} |V_1'| \to 0$, and $\max_{\zeta\in[0,1]} |\alpha_1| \to 0$. Thus, taking into account the continuity of g, $\frac{\partial g}{\partial y_i}, \frac{\partial^2 g}{\partial y_i \partial y_j}$, for $u \in S$ it easy to verify that the following conditions hold:

$$\lim_{|||\mathbf{u}_1|||\to 0} \max\left|g(\mathbf{u}+\varepsilon\mathbf{u}_1) - g(\mathbf{u})\right| = 0 \quad \lim_{|||\mathbf{u}_1|||\to 0} \frac{\max\left|d^1 g(\mathbf{u}+\varepsilon\mathbf{u}_1) - d^1 g(\mathbf{u})\right|}{|||\mathbf{u}_1|||} = 0 \tag{4.163}$$

$$\lim_{|||\mathbf{u}_1|||\to 0} \max\left|d^1 g(\mathbf{u}+\varepsilon\mathbf{u}_1) - g(\mathbf{u})\right| = 0 \quad \lim_{|||\mathbf{u}_1|||\to 0} \frac{\max\left|d^2 g(\mathbf{u}+\varepsilon\mathbf{u}_1) - d^2 g(\mathbf{u})\right|}{|||\mathbf{u}_1|||^2} = 0 \tag{4.164}$$

$$\lim_{|||\mathbf{u}_1|||\to 0} \frac{\max\left|d^2 g(\mathbf{u}+\varepsilon\mathbf{u}_1)\right|}{|||\mathbf{u}_1|||} = 0 \tag{4.165}$$

From these conditions and the previous inequality (4.162) we can obtain the final result

$$\lim_{|||\mathbf{u}_1|||\to 0} \frac{|r_2(\mathbf{u})|}{|||\mathbf{u}_1|||^2} = 0 \tag{4.166}$$

In conclusion, the functional F(**u**) is twice Frechét differentiable. The same result can be obtained for the other terms contained in the expression (4.156) of the strain energy functional and of the potential energy of the external loads. With the same technique it is possible to prove fourth-order strong differentiability with respect to the energy norm.

4.3.2 STABILITY IN THE TRIDIMENSIONAL SPACE OF THE AXIALLY LOADED INEXTENSIBLE ROD

In the previous analysis of buckling of axially loaded rods, it was assumed that the column would buckle by bending in a plane of symmetry of the cross sections. By using the bending-torsional analysis of rods given in the previous section we will analyze the problem of the buckling of the rod in the tridimensional space, i.e., buckling for double flexure and torsion.

Let us consider the rectilinear inextensible compressed rod of Fig. 4.16. Let C, B_2, and B_3 be the torsional and flexural stiffnesses of the cross section of the beam. During the application of the axial load N, the beam will remain in its undeformed rectilinear configuration $C_O = C_i$. If the rod is then displaced from this configuration with a double flexure and torsion, the additional strain energy of the beam will be given by (4.144) because we have

$$(u)_{C_0} = (v)_{C_0} = (\alpha)_{C_0} = 0 \tag{4.167}$$

Consequently the additional strain energy of the beam is expressed as

$$\mathcal{V}(\mathbf{v}) = \mathcal{V}_1(\mathbf{v}) + \mathcal{V}_2(\mathbf{v}) + \mathcal{V}_3(\mathbf{v}) + \mathcal{V}_4(\mathbf{v}) + o(|||\mathbf{v}|||^4) \tag{4.168}$$

where

$$\mathcal{V}_1(\mathbf{v}) = 0$$

$$\mathcal{V}_2(\mathbf{v}) = \frac{1}{2} C \int_0^L \alpha'^2\, dz + \frac{1}{2} B_2 \int_0^L v''^2\, dz + \frac{1}{2} B_3 \int_0^L u''^2\, dz \tag{4.169}$$

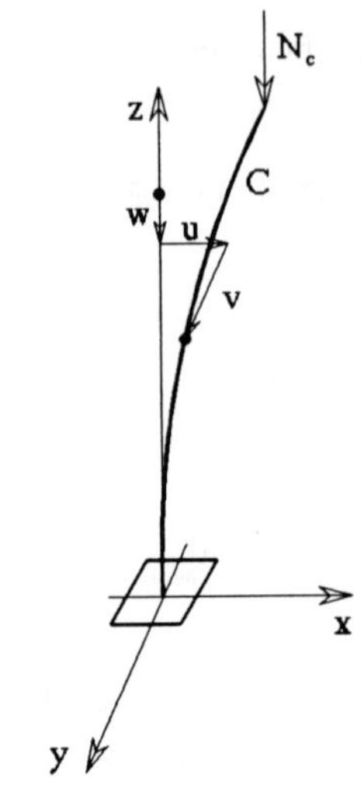

FIGURE 4.16

$$\mathcal{V}_3(\mathbf{v}) = C\int_0^L \alpha' u''v' \, dz - B_2\int_0^L \alpha u'' v'' \, dz + B_3\int_0^L \alpha u''v'' \, dz \tag{4.170}$$

$$\mathcal{V}_4(\mathbf{v}) = \frac{1}{2} C\int_0^L v''^2 v'^2 \, dz + \frac{1}{2} B_2\int_0^L (\alpha^2 u''^2 + v''^2 v'^2 - v''^2 \alpha^2) \, dz$$

$$+ \frac{1}{2} B_3\int_0^L (\alpha^2 v''^2 + 2u'u''v'v'' + u'^2 u''^2 - u''^2 \alpha^2) \, dz \tag{4.171}$$

Let us evaluate the potential energy of the external axial force. We have

$$\mathcal{U}(\mathbf{v}) = N\, w(L) = N\int_0^L w' \, dz = N\int_0^L \left[(1 - u'^2 - v'^2)^{1/2} - 1\right] dz \tag{4.172}$$

or

$$\mathcal{U}(\mathbf{v}) = \mathcal{U}_1(\mathbf{v}) + \mathcal{U}_2(\mathbf{v}) + \mathcal{U}_3(\mathbf{v}) + \mathcal{U}_4(\mathbf{v}) + o(|||\mathbf{v}|||^4) \tag{4.173}$$

where

$$\mathcal{U}_1(\mathrm{v}) = 0; \; \mathcal{U}_2(\mathbf{v}) = -\frac{1}{2} N \int_0^L (u'^2 + v'^2) \, dz \tag{4.174}$$

$$\mathcal{U}_3(\mathbf{v}) = 0; \; \mathcal{U}_4(\mathbf{v}) = -\frac{1}{8} N\int_0^L (u'^4 + v'^4 + 2u'^2 v'^2) \, dz \tag{4.175}$$

Thus, according to (3.89), the critical state is given by

$$C\int_0^L \alpha'\delta\alpha' dz + B_2\int_0^L v''\delta v'' dz + B_3\int_0^L u''\delta u'' dz - N\int_0^L (u'\ \delta u' + v'\ \delta v')dz = 0 \tag{4.176}$$

and we get the following differential equations

$$C\ \alpha'' = 0;\ B_2\ v^{IV} + Nv'' = 0;\ \ B_3\ u^{IV} + Nu'' = 0 \tag{4.177}$$

and boundary conditions

$$[C\ \alpha'\ \delta\alpha]_0^L = 0 \tag{4.178}$$

$$[B_2\ v''\ \delta v']_0^L = 0;\ \ [(-B_2\ v''' - Nv')\ \ \delta v]_0^L = 0 \tag{4.179}$$

$$[B_3\ u''\ \delta u']_0^L = 0;\ \ [(-B_3\ u''' - Nu')\ \ \delta u]_0^L = 0 \tag{4.180}$$

With equal constraints at the end sections, the buckling of the bar occurs in the plane orthogonal to the plane of minimum moment of inertia of the section. The buckling mode is represented by a simple plane flexural displacement, which, with $B_2 < B_3$, is

$$\mathbf{u}_c = (0,\ v_c(z),\ 0) \tag{4.181}$$

where $v_c(z)$ is obtained by solving (4.177) and (4.179). Let us analyze stability at buckling. We have

$$\mathcal{E}_3(0, v_c(z), 0) = 0 \tag{4.182}$$

The differential $\mathcal{E}_3(\mathbf{u})$ is not identically zero. Thus, according to (3.99), we have to calculate the functional $\mathcal{E}_{21}(\mathbf{u}_c^*, \Phi)$. Taking into account (4.170) and taking the two vector functions

$$\mathbf{u}_1 = [u_1(z);\ v_1(z);\ \alpha_1(z)];\ \mathbf{u}_2 = [u_2(z);\ v_2(z);\ \alpha_2(z)] \tag{4.183}$$

we get

$$\mathcal{E}_{21}(\mathbf{u}_1, \mathbf{u}_2) = C\int_0^L (\alpha_1' u_1'' v_2' + \alpha_1' v_1' u_2'' + u_1'' v_1' \alpha_2')dz$$

$$+ (B_3 - B_2)\int_0^L (\alpha_1' u_1'' v_2'' + \alpha_1 v_1'' u_2'' + u_1'' v_1'' \alpha_2)\ dz \tag{4.184}$$

Hence, according to (4.181), $\mathcal{E}_{21}(\mathbf{u}_c^*, \Phi) = 0$ and (3.101) we obtain

$$\Phi = 0 \tag{4.185}$$

$\mathcal{E}_4(\mathbf{u}_c)$ is still given by (4.62). Thus, at buckling, the equilibrium of the bar is stable.

4.3.3 TORSIONAL BUCKLING AND STABILITY OF THE AXIALLY LOADED THIN CRUCIFORM SECTION

There are cases in which a thin-walled bar, subjected to uniform axial compression, buckles torsionally while its longitudinal axis remains straight. Such a torsional buckling failure, first

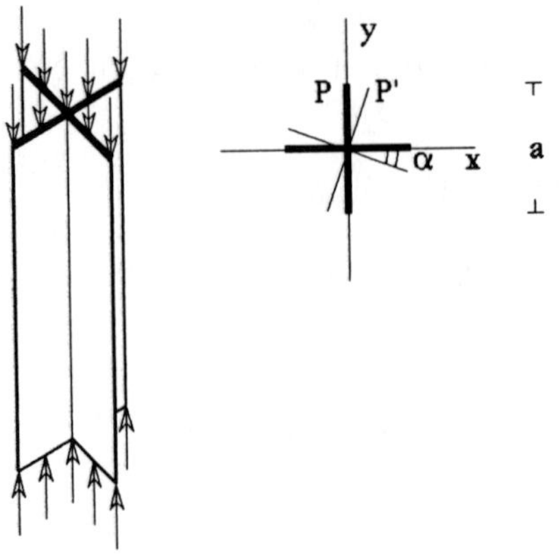

FIGURE 4.17

studied by Wagner,[6] occurs when the torsional stiffness of the section is very low. We will show that this torsional buckling of the thin-walled beam is unstable. Let us consider the double symmetric bar of cruciform section of length L, with four identical flanges of width a and thickness t. When the torsional buckling occurs, the axis of the bar remains straight while each flange buckles, rotating about the z axis. To determine the torsional buckling load, examine the flange deflection that occurs at buckling. The displacement components of the point P on the flange, with reference to Fig. 4.17, are

$$u = y \sin \alpha \quad v = -y(1 - \cos \alpha) \tag{4.186}$$

Thus, by assuming inextensibility of the longitudinal fibers of the flanges, we have

$$w' = \sqrt{1 - u'^2 - v'^2} - 1 = \sqrt{1 - y^2 \alpha'^2} - 1 \tag{4.187}$$

The potential energy $\mathcal{U}(\mathbf{u})$ of the external axial forces, represented by a constant compression p applied at the top section and having resultant N, is given by

$$\mathcal{U} = 2 \int_{A_F} p \, dA \int_0^L \left[\sqrt{1 - \alpha'^2 y^2} - 1\right] dz = 2 \frac{N}{A} \int_{A_F} \int_0^L \left[\sqrt{1 - \alpha'^2 y^2} - 1\right] dz \, dA \tag{4.188}$$

where A_F is the single flange area. Thus, we have

$$\mathcal{U}_2 = -\frac{N}{A} \frac{ta^3}{12} \int_0^L \alpha'^2 dz \, ; \quad \mathcal{U}_3 = 0 \, ; \quad \mathcal{U}_4 = -\frac{N}{A} \frac{ta^5}{320} \int_0^L \alpha'^4 dz \tag{4.189}$$

Buckling of the cruciform section involves both torsion and bending of the flange fibers. Consequently, the strain energy is given by

$$\mathcal{V} = \mathcal{V}_t + \mathcal{V}_b \tag{4.190}$$

where, according to (4.144) and (4.143a)

$$\mathcal{V}_t = \frac{C}{2} \int_0^L \alpha'^2 \, dz \tag{4.191}$$

Bending of the flanges does not involve transversal displacements of the beam axis. The flexural strain energy $\mathcal{V}_b$, depending on the flanges bending, according to (4.27) and to (4.186), is

$$\mathcal{V}_f = 4 \int_{A_F/2} \frac{Et^3}{24} \int_0^L \frac{u''^2}{1 - u'^2} dz\, dA \tag{4.192}$$

where $A_F/2$ is the upper semi-flange area and u = u(y,z) is the displacement component (4.186). Thus, substitution of (4.186) into (4.191) gives

$$\mathcal{V}_f = \frac{Et^3}{6} \int_0^{a/2} y^2 \int_0^L (\alpha'' \cos\alpha - \alpha'^2 \sin\alpha)^2 (1 + y^2\alpha'^2 \cos^2\alpha \ldots) dy\, dz$$

$$= \frac{Et^3}{6}\frac{a^3}{24}\int_0^L \alpha''^2 dz + \frac{Et^3}{6}\left(\frac{a^3}{24}\int_0^L (-\alpha^2\alpha''^2 - 2\alpha\alpha''\alpha'^2) dz + \frac{a^5}{160}\int_0^L \alpha''^2 \alpha'^2 dz + \ldots\right) \tag{4.193}$$

The second differential $\mathcal{E}_2(\mathbf{u})$ of the total potential energy $\mathcal{E}(\mathbf{u})$ is given by

$$\mathcal{E}_2(\mathbf{u}) = \frac{E}{2}\frac{a^3t^3}{3\cdot 4\cdot 6}\int_0^L \alpha''^2 dz - \frac{1}{2}\left(\frac{NI_0}{A} - C\right)\int_0^L \alpha'^2 dz \tag{4.194}$$

Thus, the torsional buckling mode of the cruciform section is represented by

$$\alpha_c = A \sin\frac{\pi z}{L} \tag{4.195}$$

and the buckling load is

$$N_c = \frac{CA}{I_0}\left(1 + \frac{E}{G}\frac{\pi^2}{3\cdot 6\cdot 8}\frac{a^2}{L^2}\right) \tag{4.196}$$

that, when the ratio $a/L \ll 1$, gives the Wagner critical load $N_c = CA/I_0$.

The load N_c (4.196) is the torsional buckling load of the cruciform profile. In (4.196) $I_0 = ta^3/6$ is the polar moment of inertia of the cross section about the centroid and C (= GJ) is the torsional stiffness of the section, where G is the shearing modulus of elasticity and J (= $\frac{1}{3}$ 2a t^3) is the torsion constant.

We can immediately recognize that the torsional buckling state of the thin-walled section is unstable. In fact

$$\mathcal{E}_3(\mathbf{u}_c) = 0 \quad \mathcal{E}_4(\mathbf{u}_c) = \frac{Et^3}{6}\left[\frac{a^3}{24}\int_0^L (-\alpha_c^2\alpha_c''^2 - 2\alpha_c\alpha_c''\alpha_c'^2) dz + \frac{a^5}{160}\int_0^L \alpha_c''^2\alpha_c'^2 dz\right]$$

$$- \frac{N_c}{A}\frac{t a^5}{320}\int_0^L \alpha_c'^4 dz = -A^4 \frac{\pi^4}{L^4}\frac{L}{8}\frac{Et^3a^3}{6}\left[\frac{1}{24}\left(1 - \frac{3\pi^2}{20}\frac{a^2}{L^2}\right) + \frac{27}{40}\frac{G}{E}\left(1 + \frac{E}{G}\frac{\pi^2}{3\cdot 6\cdot 8}\frac{a^2}{L^2}\right)\right] \tag{4.197}$$

Assuming $(a/L)^2 < 1$, the critical equilibrium is unstable.

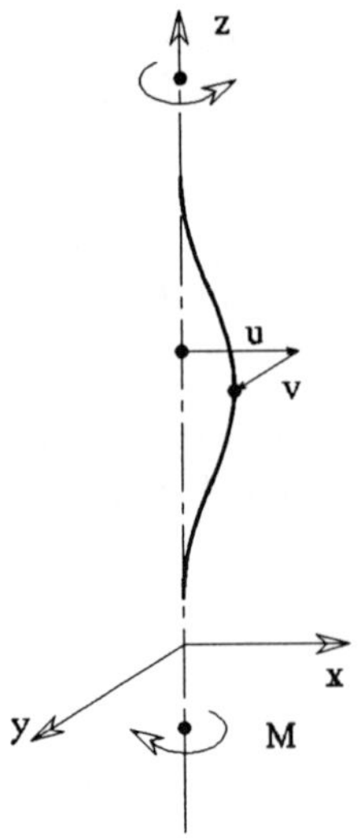

FIGURE 4.18

4.3.4 FLEXURAL BUCKLING AND STABILITY OF FLEXIBLE SHAFTS IN TORSION

Let us consider the weightless rectilinear shaft with clamped end sections loaded by torque M. The shaft has circular cross section with moment of inertia J, length L, and flexural and torsional stiffnesses B and C (Fig. 4.18). Under increasing torque M the shaft can buckle in a double flexure, as first shown by Greenhill[7] and Nicolai.[8]

We will analyze the buckling of the shaft, inquiring whether or not this critical state is stable. The principal deformation $\mathbf{u}_P$ of the shaft is represented by a pure torsion, while the additional deformation $\mathbf{u}$ is represented by torsion and double flexure. Thus we have

$$\mathbf{u}_P = \begin{bmatrix} \alpha_P \\ 0 \\ 0 \end{bmatrix} \qquad \mathbf{u} = \begin{bmatrix} \alpha \\ u \\ v \end{bmatrix} \tag{4.198}$$

Therefore the additional potential energy of the bar is

$$\mathcal{E}(\mathbf{u}) = \mathcal{E}(\mathbf{u}_P + \mathbf{u}) - \mathcal{E}(\mathbf{u}_P) = \mathcal{V}(\mathbf{u}_P + \mathbf{u}) - \mathcal{V}(\mathbf{u}_P) - M|\alpha|_0^L \tag{4.199}$$

where $\mathcal{V}(\mathbf{u}_P + \mathbf{u}) - \mathcal{V}(\mathbf{u}_P)$ indicates the additional strain energy corresponding to the occurrence of the additional displacement $\mathbf{u}$.Thus we have

$$\mathcal{E}_1(\mathbf{u}) = C\int_0^L \alpha'_p\alpha'\,dz - M|\alpha|_0^L = M\int_0^L \alpha'\,dz - M|\alpha|_0^L = 0 \tag{4.200}$$

$$\mathcal{E}_2(\mathbf{u}) = \frac{C}{2}\int_0^L \alpha'^2\,dz + M\int_0^L u''v'\,dz + \frac{B}{2}\int_0^L (u''^2 + v''^2)\,dz \tag{4.201}$$

$$\mathcal{E}_3(\mathbf{u}) = C\int_0^L \alpha'u''v'\,dz \tag{4.202}$$

$$\mathcal{E}_4(\mathbf{u}) = \frac{C}{2}\int_0^L u''^2 v'^2 dz + M\int_0^L [u'v'^2v'' + \frac{u''v'}{2}(u'^2+v'^2)]\, dz$$

$$+\frac{B}{2}\int_0^L (v''^2v'^2 + u''^2u'^2 + 2u'u''v'v'')\, dz \tag{4.203}$$

Condition (3.89), that defines the critical state, yields the following system of differentials equations

$$Bv^{IV} - Mu''' = 0;\ Bu^{IV} + Mv''' = 0 \tag{4.204}$$

Both buckling components u(z) and v(z) will satisfy the clamping conditions v = v' = 0. It is convenient to introduce, according to Nicolai,[8] a complex deflection of the shaft

$$S = u + iv \tag{4.205}$$

together with the notation

$$\beta = \frac{M}{B} \tag{4.206}$$

The differential equations (4.204) are then equivalent to the single complex equation

$$S'''' - i\beta S''' = 0 \tag{4.207}$$

and the boundary conditions become

$$S(0) = S'(0) = S(L) = S'(L) = 0 \tag{4.208}$$

The general solution of (4.207) is

$$S(z) = Ae^{i\beta z} + Bz^2 + Cz + D \tag{4.209}$$

where A, B, C, and D are complex constants. The boundary conditions (4.208) require

$$\begin{array}{lllll} A & & & +D & =0 \\ Ai\beta & & +C & & =0 \\ Ae^{i\beta L} & +BL^2 & +CL & +D & =0 \\ Ai\beta e^{i\beta L} & +2BL & +C & & =0 \end{array} \tag{4.210}$$

Thus, the characteristic equation is

$$(1 - i\lambda)e^{i\lambda} = (1 + i\lambda)e^{-i\lambda} \tag{4.211}$$

or, equivalently,

$$\mathrm{tg}\lambda = \lambda \tag{4.211'}$$

with

$$\lambda = \frac{\beta L}{2} \tag{4.212}$$

The smallest solution of this equation is

$$\lambda \cong 4.4934 \tag{4.213}$$

and the buckling torque is

$$M_c = 8.9868\,\frac{B}{L} \tag{4.214}$$

Evaluations of the complex constants by using (4.211) yields the real components u_c, and v_c of the complex buckling deflection. Hence we have

$$u_c(z) = A[(\cos 2\lambda\zeta - 1) + \lambda\zeta^2 \sin 2\lambda] \tag{4.215}$$

$$v_c(z) = A[(\sin 2\lambda\zeta - 2\lambda\zeta) + \lambda\zeta^2(1 - \cos 2\lambda)] \tag{4.216}$$

where

$$\zeta = \frac{z}{L} \tag{4.217}$$

is the nondimensional abscissa. The bar thus buckles with a double flexure. We now analyze whether the buckling state is stable or unstable. According to (4.202)

$$\mathcal{E}_3(\mathbf{u}_c) = 0 \tag{4.218}$$

and

$$\mathcal{E}_4(\mathbf{u}_c) = \frac{C}{2}\int_0^L u_c''^2 v_c'^2 dz + M\int_0^L [u_c' v_c'^2 v_c'' + \frac{u_c'' v_c'}{2}(u_c'^2 + v_c'^2)]dz$$

$$+ \frac{B}{2}\int_0^L (v_c''^2 v_c'^2 + u_c''^2 u_c'^2 + 2u_c' u_c'' v_c' v_c'')\, dz \tag{4.219}$$

Let us evaluate the function Φ taking into account that this unknown vectorial function is constrained by the orthogonality condition

$$[u_c, \Phi] = 0 \tag{4.220}$$

From (4.202) we get, for a generic couple of vectorial functions $\mathbf{u}_1$, and $\mathbf{u}_2$

$$\mathcal{E}_{21}(\mathbf{u}_1, \mathbf{u}_2) = C\int_0^L (\alpha_1' u_1'' v_2' + \alpha_1' v_1' u_2'' + u_1'' v_1' \alpha_2')\, dz \tag{4.221}$$

But the buckling mode $\mathbf{u}_c$ and the additional postbuckling displacement vector Φ_1 are

$$\mathbf{u}_c = \begin{bmatrix} 0 \\ u_c \\ v_c \end{bmatrix} \qquad \Phi_1 = \begin{bmatrix} \alpha_1 \\ u_1 \\ v_1 \end{bmatrix} \tag{4.222}$$

Consider, according (3.101), the functional $\mathcal{E}_{21}(\mathbf{u}_c^*, \delta\mathbf{v})$ where $\delta\mathbf{v}$ has components $\delta\alpha$, δu, δv. We get, taking into account (4.221) and (4.222)

$$\mathcal{E}_{21}(\mathbf{u}_c, \delta\mathbf{v}) = C\int_0^L u_c'' v_c' \delta\alpha' \, dz \tag{4.223}$$

Let us consider the functional $\mathcal{E}_{11}(\Phi_1,\delta\mathbf{v})$. From (4.201) we get

$$\mathcal{E}_{11}(\Phi_1,\delta\mathbf{v}) = C\int_0^L \alpha_1' \delta\alpha' \, dz + M_c\int_0^L (u_1''\delta v' + v_1'\delta u'')dz + B\int_0^L (u_1''\delta u'' + v_1''\delta v'')dz \tag{4.224}$$

Condition (3.101), then, gives

$$C\int_0^L (\alpha_1' + u_c'' v_c')\, \delta\alpha' \, dz = 0 \tag{4.225}$$

$$M_c\int_0^L (u_1''\delta v' + v_1'\delta u'')dz + B\int_0^L (u_1''\delta u'' + v_1''\delta v'')dz = 0 \tag{4.226}$$

for any choice of $\delta\alpha$, δu, and δv. Thus we have

$$u_1 = v_1 = 0 \quad \alpha_1' = -\, u_c'' v_c' \tag{4.227}$$

The additional vector Φ_1 is purely torsional. Evaluate the value of $\mathcal{E}_2(\mathbf{u})$ at $\mathbf{u} = \Phi_1$. We have

$$\mathcal{E}_2(\Phi_1) = \frac{C}{2}\int_0^L u_c''^2 v_c'^2 \, dz \tag{4.228}$$

Taking into account (4.215), (4.216), and (4.227) we get

$$\mathcal{E}_4(\mathbf{u}_c) - \mathcal{E}_2(\Phi_1) = M_c\int_0^L [u_c' v_c'^2 v_c'' + \frac{u_c'' v_c'}{2}(u_c'^2 + v_c'^2)]\, dz$$

$$+ \frac{B}{2}\int_0^L (v_c''^2 v_c'^2 + u_c''^2 u_c'^2 + 2u_c' u_c'' v_c' v_c'')\, dz \tag{4.229}$$

With some calculations we obtain

$$\mathcal{E}_4(\mathbf{u}_c) - \mathcal{E}_2(\Phi_1) = 6907.1\,\frac{A^4 B}{L^5} \tag{4.230}$$

and the equilibrium of the shaft at the critical state is stable.

4.3.5 LATERAL BUCKLING OF DEEP BEAMS

4.3.5.1 The Potential Energy of the External Loads and the Potential Energy Functional

Let us consider the beam of rectangular cross section of Fig. 4.19 loaded at the ends by a distribution of loads, increasing with the load parameter λ

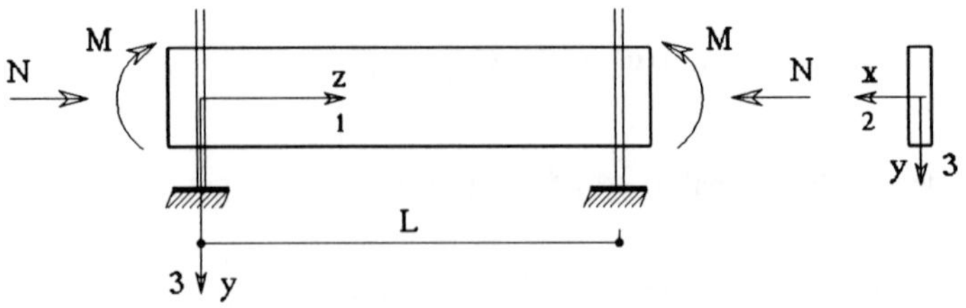

FIGURE 4.19

$$q(x,y) = q_1(x,y)\mathbf{D}_1 \tag{4.231}$$

This load distribution has resultant moment and force

$$\int_\Omega q_1 X_2\, d\Omega = 0 \quad \int_\Omega q_1 X_3\, d\Omega = M \quad \int_\Omega q_1\, d\Omega = -N \tag{4.232}$$

Thus, these loads are equivalent to an eccentric thrust applied on a point of the vertical axis 3 passing through the centroid of the end section of the beam. To evaluate the potential energy of these loads let us examine the end section at z = L. Let $\mathbf{u}(x,y,L,\lambda)$ be a displacement process with $\mathbf{u}(x,y,L,0) = 0$.(Fig. 4.19). The work done by the load $\mathbf{q}$ at $\lambda = \lambda_0$ is

$$\mathcal{L}_q = \int_0^{\lambda_0} d\lambda \int_\Omega \mathbf{q}.\frac{d\mathbf{u}}{d\lambda}\, d\Omega \tag{4.233}$$

Taking into account the non-deformability of the cross section, the displacement $\mathbf{u}(x,y,L,\lambda)$ can be defined by the displacement function $\mathbf{u}_G(0,0,L,\lambda)$ of the section centroid and by the rotation matrix $\mathcal{R}$ (4.118). Thus the field $du/d\lambda$ can be expressed as

$$\mathbf{u} = (\mathcal{R} - \mathcal{I})\mathbf{X} + \mathbf{u}_G \tag{4.234}$$

Hence

$$\frac{d\mathbf{u}}{d\lambda} = \frac{dR}{d\lambda}\mathbf{X} + \frac{d\mathbf{u}_G}{d\lambda} \tag{4.235}$$

and the work $\mathcal{L}_q$ is given by

$$\mathcal{L}_q = \int_0^{\lambda_0} d\lambda \int_\Omega \mathbf{q}\cdot\frac{d\mathcal{R}}{d\lambda}\mathbf{X} + \mathbf{q}\cdot\frac{d\mathbf{u}_G}{d\lambda}\, d\Omega \tag{4.236}$$

or

$$\mathcal{L}_q = \int_0^{\lambda_0}\int_\Omega q_i \frac{d\mathcal{R}_{ij}}{d\lambda}X_j + q_i\frac{du_{Gi}}{d\lambda}\, dSd\lambda = \left[\frac{d\mathcal{R}_{ij}}{d\lambda}\int_\Omega q_i X_j\, d\Omega + \int_\Omega q_i \frac{du_{Gi}}{d\lambda}\, d\Omega\right] d\lambda \tag{4.237}$$

Thus, taking into account that N is a compression load,

$$\mathcal{L}_q = \int_0^{\lambda_0}\left[\frac{d\mathcal{R}_{13}}{d\lambda}M + N\frac{dw}{d\lambda}\right] d\lambda = M\,\mathcal{R}_{13}(\lambda_0) + Nw(L,\lambda_0) \tag{4.238}$$

Substitution of w and $\mathcal{R}_{13}$ as functions of the parameters (u,v,α) gives finally, from (4.119) and (4.138)

$$\mathcal{U}_q = -M\left|\cos\gamma\sin\beta\cos\alpha + \sin\gamma\sin\alpha\right|_0^L + N\int_0^L w'\,dz$$

$$= -M\left|-\cos\alpha\frac{\sqrt{1-u'^2-v'^2}}{\sqrt{1-v'^2}}v' + \sin\alpha\frac{u'}{\sqrt{1-v'^2}}\right|_0^L + \int_0^L\left[N\sqrt{1-u'^2-v'^2}-1\right]dz \quad (4.239)$$

The total potential energy of the beam for the loading condition shown in Fig. 4.19 is thus

$$\mathcal{E}(\mathbf{u}) = \mathcal{V} + \mathcal{U}_q = \frac{1}{2}C\int_0^L \tau^2\,dz + \frac{1}{2}B_2\int_0^L k_2^2\,dz + \frac{1}{2}B_3\int_0^L k_3^2\,dz$$

$$-M\left|-\cos\alpha\frac{\sqrt{1-u'^2-v'^2}}{\sqrt{1-v'^2}}v' + \sin\alpha\frac{u'}{\sqrt{1-v'^2}}\right|_0^L + \int_0^L\left[N\sqrt{1-u'^2-v'^2}-1\right]dz \quad (4.240)$$

4.3.5.2 Differentials of the Potential Energy

Analyze the principal equilibrium configurations of the beam, defined by the displacement $\mathbf{u}_P(\lambda)$, corresponding to the applied loads, proportionally increasing with the loading parameter λ. Taking into account that x and y are the principal axes of the cross section, the principal equilibrium path $\mathbf{u}_P(\lambda)$ is characterized only by flexural displacements occurring in the plane (y,z)

$$\mathbf{u}_P(\lambda) = (0, v_P(\lambda), 0) \quad (4.241)$$

Due to the high flexural stiffness B_2 of the cross sections in the plane (y,z), the equilibrium path (4.241) can be represented by the linear elastic solution, i.e., by $\mathbf{u}_P(\lambda) = \lambda\mathbf{u}_P$. With increasing λ, the beam can suddenly buckle with combined flexural-torsional deformations.

To analyze the buckling and the stability at buckling of the beam, the second, third and fourth differentials of the potential energy at the principal equilibrium state $\mathbf{u}_P(\lambda)$ have to be explicitly evaluated. We will use the nondimensional quantities (4.154) and (4.155).

Every principal equilibrium state is characterized by the vertical displacements $V_P(\zeta)$

$$V_P(\zeta) = \lambda\mu g(\zeta) \quad (4.242)$$

where

$$\mu = \sqrt{\chi_1\chi_2} \quad (4.243)$$

In expression (4.242), $g(\zeta)$ is a nondimensional function of ζ depending on the applied load and on the boundary conditions. Flexural and torsional stiffnesses B_3 and C will, as usual, be very small compared to the vertical stiffness B_2. Thus we will have

$$\chi_1 << 1 \quad \chi_2 << 1 \quad (4.244)$$

Finally let η be the ratio

$$\eta = \frac{NL}{M} \quad (4.243')$$

that defines the eccentricity of the axial force N. Inequalities (4.244) imply that the prebuckling displacements are very small with respect to the main dimensions of the beam. Therefore, it is consistent to retain only linear terms in χ_1, χ_2.

After these preliminaries, we can give the expressions of the differentials of the potential energy functional obtained from (4.240). We have

$$D^2\mathcal{E}(V_P,\lambda;\mathbf{u},\mathbf{u})=\frac{B_2}{L}\left|2\lambda\mu g''U'\alpha\right|_0^1$$

$$+\frac{B_2}{L}\int_0^1\left[V''^2+\chi_1 U''^2+\chi_2\alpha'^2-2\lambda\mu g''U''\alpha-\lambda\eta\mu\left(U'^2+V'^2\right)\right]d\zeta \qquad (4.245)$$

$$D^3\mathcal{E}(V_P,\lambda;\mathbf{u},\mathbf{u},\mathbf{u})=+6\frac{B_2}{L}\left|\frac{1}{2}\lambda\mu g''\left(U'^2V'+V'\alpha^2\right)\right|_0^1$$

$$+6\frac{B_2}{L}\int_0^1\left[-U'V''\alpha+\chi_1U''V'\alpha+\chi_2U''V''\alpha+\lambda\mu g'V'V''^2+\lambda_c\mu g''\left(V'^2V''-V''^2\right)\right]d\zeta \qquad (4.246)$$

$$D^4\mathcal{E}(V_P,\lambda;\mathbf{u},\mathbf{u},\mathbf{u},\mathbf{u})=12\frac{B_2}{L}\int_0^1\left\{V'^2V''^2-V''^2\alpha^2+U''^2\right.$$

$$+\chi_1\left(V''^2\alpha^2-U''^2\alpha^2+U''^2U'^2+2U'U''V'V''\right)$$

$$\chi_2U''^2V'^2-2\lambda\mu g''\left(-\frac{2}{3}U''\alpha^3+\frac{1}{2}U'^2U''\alpha+\frac{1}{2}U'V'V''\alpha\right)$$

$$\left.+\lambda\mu g'\left(-2U'V''^2\alpha-2U''V'V''\alpha\right)-\lambda\eta\mu\frac{1}{4}\left(U'^2+V'^2\right)^2\right\}d\zeta$$

$$+12\frac{B_2}{L}\left|\lambda\mu g''\left(U'V'^2\alpha-\frac{1}{3}U'\alpha^3+6V'^2\alpha^2+3V'^4\right)\right|_0^1 \qquad (4.247)$$

4.4.5.3 The Energy Space and the Strong Differentiability of the Energy Functional

The set of the admissible configurations of the beam, i.e., the corresponding energy space H_A, is given by the vectorial functions $\mathbf{u}(z)$ that satisfy the kinematical boundary conditions and such that the quantity (4.145) is finite. By using the above recalled Sobolev theorems, it is easy to show that, if the displacement vector functions $\mathbf{u}(z)$ belong to H_A, the components u(z) and v(z) are continuous with their first derivatives u'(z) and v'(z), while the other component $\alpha(z)$ is only continuous in the interval [0,L]. Further, inequalities (4.146) and (4.147) hold. The potential energy functional $\mathcal{E}(\mathbf{u})$, given by (4.240), is defined in the set (4.149). Thus it is possible to show that the so defined functional (4.240) is, at least, four times strong differentiable at the equilibrium state C_0.[5]

4.3.5.4. Critical State and Stability

This state is governed by the variational condition

$$D^2\mathcal{E}\left[\mathbf{u}_P(\lambda_c),\lambda_c;\,\mathbf{u}_1,\delta\mathbf{u}\right]=0\quad\forall\,\delta\mathbf{u} \qquad (4.248)$$

that defines the critical load λ_c and the buckling mode

$$\mathbf{u}_1=\mathbf{u}_c \qquad (4.249)$$

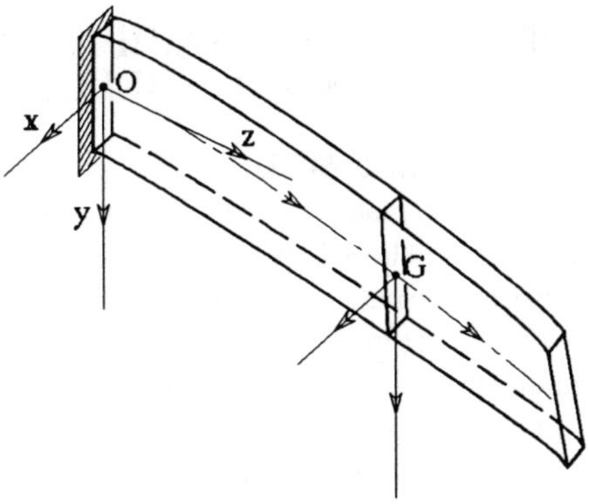

FIGURE 4.20

Taking into account expression (4.245), we get the explicit form of the condition (4.248)

$$\int_0^1 \left[V_c'' \delta V'' + \chi_1 U_c'' \delta U'' + \chi_2 \alpha_c' \delta \alpha' - \lambda_c \mu g'' \left(U_c'' \delta \alpha + \alpha_c \delta U'' \right) \right.$$

$$\left. - \lambda_c \mu \eta \left(U_c' \delta U' + \chi_1 V_c' \delta V' \right) \right] d\zeta + \left| \lambda_c \mu g'' \left(U_c' \delta \alpha + \alpha_c \delta U' \right) \right|_0^1 = 0 \qquad (4.248')$$

Thus, from (4.248) we obtain the following system of differential equations and boundary conditions involving the buckling horizontal displacement U_c and the torsional rotation α_c:

$$\left| \lambda_c \mu g'' \left(U_c' \delta \alpha + \alpha_c \delta U' \right) \right|_0^1 = 0 \qquad (4.250)$$

$$-\chi_2 \alpha_c'' - \lambda_c \mu g'' U_c^{II} = 0 \qquad (4.251)$$

$$\chi_1 U_c^{IV} - \lambda_c \mu (g'' \alpha_c)'' + \lambda_c \eta \mu U_c^{II} = 0 \qquad (4.252)$$

$$\left| (\chi_1 U_c^{II}) \delta U' \right|_0^1 = 0; \quad \left| \left[-\chi_1 U_c^{III} + \lambda_c \mu (g'' \alpha_c)' - \lambda_c \mu \eta U_c' \right] \delta U \right|_0^1 = 0 \qquad (4.253)$$

$$\left| \left(\chi_2 \alpha_c' + \lambda_c \mu g'' U_c' \right) \delta \alpha \right|_0^1 = 0 \qquad (4.254)$$

The buckling mode does not include vertical displacements, i.e., $V_c = 0$ (Fig. 4.20). Thus we have, from (4.246)

$$D^3E(U_P, \lambda_c; u_c^3) = 0 \qquad (4.255)$$

and the equilibrium bifurcation at the critical state is symmetric. The postbuckling behavior of the deep beams will be examined in detail in Chapter 6 and there it will be shown that the critical state is stable.

4.4 BASIC PROBLEMS IN STABILITY ANALYSIS OF PLATES AND SHELLS

4.4.1 THE BIDIMENSIONAL SHELL MODEL

The nonlinear elastic behavior of plates and shells is usually studied according to the shallow shell theory, defined by Donnell,[9] which accounts only for the nonlinear strains of the middle surface of the shell. A first formulation of the stability analysis of shallow shells was given by Koiter[11] by means of a direct inspection of the energy functional. Later it was also shown by Como and Grimaldi[10] that the energy functional of the shallow shell model meets the regularity conditions, previously shown, required to the applicability of the energy method. We will analyse these conditions in detail.

Let $\mathbf{u}(x^\alpha)$ be the displacement field, as from the reference configuration C_i, of the middle surface of the shell, with tangential components u_α and normal component w and where x^α denotes a pair of Gaussian surface coordinates. According to the linearized elastic theory, the tensor $\theta_{\alpha\beta}$ of the middle surface strain, the tensor $\rho_{\alpha\beta}$ of the changes of curvature and the tensor of the elastic moduli $E^{\alpha\beta\lambda\mu}$ of the isotropic material, are so defined:

$$\theta_{\alpha\beta} = \frac{1}{2}(u_{\alpha|\beta} + u_{\beta|\alpha}) - wb_{\alpha\beta};\ \rho_{\alpha\beta} = w_{|\alpha\beta} \tag{4.256}$$

$$E^{\alpha\beta\lambda\mu} = G\left[a^{\alpha\lambda}a^{\beta\mu} + a^{\alpha\mu}a^{\beta\lambda} + \frac{2\nu}{1-\nu}\,a^{\alpha\beta}a^{\lambda\mu}\right] \tag{4.257}$$

In (4.256) and (4.257) the vertical stroke denotes covariant differentiation, $a^{\alpha\beta}$ and $b_{\alpha\beta}$ the first and the second fundamental tensors of the middle surface in the reference configuration C_i, G the shear modulus, and ν Poisson's ratio. The strain energy corresponding to the displacement field **u** is:

$$\mathcal{V}(\mathbf{u}) = \int_S \left[\frac{1}{2}hE^{\alpha\beta\lambda\mu}\,\theta_{\alpha\beta}\theta_{\lambda\mu} + \frac{1}{24}h^3E^{\alpha\beta\lambda\mu}\,\rho_{\alpha\beta}\rho_{\lambda\mu}\right]dS \tag{4.258}$$

where h indicates the shell thickness. The energy product and the energy norm are defined according to (4.258); for instance:

$$|||\mathbf{u}|||^2 = 2\mathcal{V}(\mathbf{u}) \tag{4.259}$$

Under the assumption that the shell is properly supported, i.e., that rigid body displacements are excluded, the following inequalities hold:

$$||w||^{1/2} = \left[\int_S w^2\,dS\right]^{1/2} \le c|||\mathbf{u}||| \tag{4.260}$$

$$\left[\int_S \left(a^{\alpha\beta}\,w_{,\alpha}w_{,\beta}\right)^{q/2} dS\right]^{1/q} \le c_1\,|||\mathbf{u}||| \tag{4.261}$$

where $w_{,\alpha}$ is the derivative of w with respect to x^α and q is any positive number. Inequalities (4.260) and (4.261) can be derived from the Sobolev embedding theorems and the condition of properly supported shell; an explicit proof, obtained by direct analysis, is essentially given in Koiter's paper.[11]

Let us now consider the nonlinear theory of the elastic shallow shell. The strain tensor $\gamma_{\alpha\beta}$ of the middle surface, defined as:

$$\gamma_{\alpha\beta} = \theta_{\alpha\beta} + \frac{1}{2} w_{,\alpha} w_{,\beta} \tag{4.262}$$

is the only nonlinear quantity present in the displacement **u**. Inequality (4.261) yields

$$\left[\int_S \gamma^{\alpha\beta}\gamma_{\alpha\beta}\, dS\right]^{1/2} \leq c_2 |||\mathbf{u}||| \tag{4.263}$$

Thus energy convergence implies convergence, in the mean square sense, of the nonlinear strain fields γ_{ab}. The same property holds for the stress field defined as $\sigma^{\alpha\beta} = E^{\alpha\beta\lambda\mu}(\gamma_{\lambda\mu} + z\rho_{\lambda\mu})$ if z is directed across the thickness of the shell and for the stress resultants $N^{\alpha\beta}$ and the stress couples $M^{\alpha\beta}$

$$N^{\alpha\beta} = hE^{\alpha\beta\lambda\mu}\gamma_{\lambda\mu} \quad ; \quad M^{\alpha\beta} = \frac{h^3}{12} E^{\alpha\beta\lambda\mu} \rho_{\lambda\mu} \tag{4.264}$$

4.4.2 THE ADDITIONAL POTENTIAL ENERGY OF SHALLOW SHELLS

Examine directly the increment of potential energy $\mathcal{E}(\mathbf{u})$ as from an equilibrium configuration C_0. We shall restrict our attention to cases in which the deformations at the equilibrium state C_0 may be ignored. The potential energy increment, according to Koiter ([11] eq. 3.6), is specified by

$$\mathcal{E}(\mathbf{u}) = \int_S \left[\frac{1}{2} N^{\alpha\beta} w_{,\alpha} w_{,\beta} + \frac{1}{2} hE^{\alpha\beta\lambda\mu}\gamma_{\alpha\beta}\gamma_{\lambda\mu} + \frac{1}{24} h^3 E^{\alpha\beta\lambda\mu} \rho_{\alpha\beta}\rho_{\lambda\mu}\right] dS \tag{4.265}$$

where $N^{\alpha\beta}$ denotes the tensor of the initial stress resultant in the equilibrium state C_0. The energy functional (4.265) can be defined in the energy space H_A. Let us now evaluate the weak differentials of the potential energy functional (4.265). The shell is in the equilibrium configuration C_0, coincident with the undeformed configuration $C_i = 0$ but characterized by the presence of the inplane forces $N^{\alpha\beta}$. Thus, we get:

$$D^{(1)}\mathcal{E}(0;\mathbf{u}) = 0 \tag{4.266}$$

$$D^{(2)}\mathcal{E}(0;\mathbf{u}) = \int_S \left[\frac{1}{2} N^{\alpha\beta} w,_{\alpha} w,_{\beta} + \frac{1}{2} hE^{\alpha\beta\lambda\mu}\theta_{\alpha\beta}\theta_{\lambda\mu} + \frac{h^3}{24} E^{\alpha\beta\lambda\mu} \rho_{\alpha\beta}\rho_{\lambda\mu}\right] dS \tag{4.267}$$

$$D^{(3)}\mathcal{E}(0;\mathbf{u}) = \int_S \frac{h}{2} E^{\alpha\beta\lambda\mu}\left(\frac{1}{2}\theta_{\alpha\beta} w,_{\lambda} w,_{\mu} + \frac{1}{2}\theta_{\lambda\mu} w,_{\alpha} w,_{\beta}\right) dS \tag{4.268}$$

$$D^{(4)}\mathcal{E}(0;\mathbf{u}) = \int_S \frac{h}{2} E^{\alpha\beta\lambda\mu}\left(\frac{1}{4} w,_{\alpha} w,_{\beta} w,_{\lambda} w,_{\mu}\right) dS \tag{4.269}$$

We have also

$$D^{(n)}\mathcal{E}(0;\mathbf{u}) = 0 \quad \forall\, n > 4 \tag{4.270}$$

We can state the following regularity theorem[10] of the energy functional $\mathcal{E}(\mathbf{u})$: the energy functional (4.265) is indefinitely Fréchet differentiable at the origin **0**, with respect to the "energy norm". In fact the potential energy increment $\mathcal{E}$ can be written in the form

$$\Delta \mathcal{E} = \frac{1}{2!} D^{(2)}\mathcal{E}(0;\mathbf{u}) + \frac{1}{3!} D^{(3)}\mathcal{E}(0;\mathbf{u}) + \frac{1}{4!} D^{(4)}\mathcal{E}(0;\mathbf{u}) \quad (4.271)$$

Taking into account inequalities (4.260) and (4.261) we easily get the conditions

$$\left|D^{(2)}\mathcal{E}(0;\mathbf{u})\right| \leq k_2 |||\mathbf{u}|||^2; \; \left|D^{(3)}\mathcal{E}(0;\mathbf{u})\right| \leq k_3 |||\mathbf{u}|||^3; \; \left|D^{(4)}\mathcal{E}(0;\mathbf{u})\right| \leq k_4 |||\mathbf{u}|||^4 \quad (4.272)$$

with k_2, k_3, and k_4 positive constants. Therefore the functionals $D^{(2)}\mathcal{E}(0;\mathbf{u})$, $D^{(3)}\mathcal{E}(0;\mathbf{u})$, $D^{(4)}\mathcal{E}(0;\mathbf{u})$, are continuous with respect to **u**. Further, we observe that the condition of Fréchet differentiability of $\mathcal{E}(0;\mathbf{u})$ is satisfied for every integer n.

4.5 TRIDIMENSIONAL ELASTICITY

4.5.1 THE "ENERGY" HILBERT SPACE H_A

We recall now the definition of energy norm and of energy space, with particular reference to the case of the tridimensional elasticity. Let us consider an elastic body T and let Ω be the region of the space occupied by T at the natural (stress-free) initial configuration C_i of T.The unstressed configuration C_i is assumed to be the reference configuration of T. Let

$$\mathbf{u}(P)$$

be a vector function, with domain Ω, that defines the displacement field that moves T to C. We will suppose that **u**(P), continuous with all its derivatives, will satisfy suitable boundary conditions that we assume linear, homogeneous, and able to prevent any rigid displacement of T. Let **X** and **x** be the vectors that define the position of a generic particle P in the configurations C_i and C; we therefore have

$$\mathbf{u}(P) = \mathbf{x} - \mathbf{X} \quad (4.273)$$

Components x_i and X_i of vectors **x** and **X** are respectively called spatial and material coordinates. The set of all the vector function **u**(P), satisfying the mentioned conditions above, constitutes a vector space that we will indicate by the symbol M. The space M is a subspace of $L_2(\Omega)$ if, according to the definition given in Chapter 2, we define over M the inner product

$$< \mathbf{u},\mathbf{v} > = \int_\Omega \mathbf{u} \cdot \mathbf{v} \, d\Omega = \int_\Omega u_i v_i \, d\Omega \quad (4.274)$$

with the usual notation. The completion of the space M, by adding all the limit elements of the fundamental sequences in M, transforms the vector space M into the Hilbert space $L_2(\Omega)$. Of course, M is dense in $L_2(\Omega)$. In this space two configurations of the body T, defined by the displacement fields $\mathbf{u}_1$ and $\mathbf{u}_2$, have distance given by the norm

$$d(\mathbf{u}_1,\mathbf{u}_2) = ||\mathbf{u}_1 - \mathbf{u}_2||_{L_2} = ||\mathbf{u}_1 - \mathbf{u}_2|| = \left[\int_\Omega (\mathbf{u}_1 - \mathbf{u}_2).(\mathbf{u}_1 - \mathbf{u}_2) \, d\Omega\right]^{1/2} \quad (4.275)$$

and angle

$$\alpha(\mathbf{u}_1,\mathbf{u}_2) = \cos^{-1}\frac{< \mathbf{u}_1,\mathbf{u}_2 >}{||\mathbf{u}_1|| \, ||\mathbf{u}_2||} \quad (4.276)$$

However, the scalar product (4.274) and the ordinary metric (4.275) don't fit for evaluating "mechanical" closeness among configurations of continuous bodies. For instance, configuration

C_1 of T cannot be considered close to configuration C_2 solely because the norm $\|\mathbf{u}_1 - \mathbf{u}_2\|$ is small. In fact, as far as strain and stress fields are concerned, C_1 and C_2 can be very distant. On the contrary, the "energy" space H_A answers the purpose. Let A be the "stiffness" operator of the linear elastic equilibrium of T that transforms each displacement vector **u** into the vector **f** that defines the load producing, if applied to T at C_i, the same displacement **u** (see, for instance, Mikhlin[13]):

$$\mathbf{Au} = \mathbf{f} \tag{4.277}$$

The operator A is linear, symmetric, and positive definite, and therefore satisfies the three conditions (2.10), (2.47), and (2.49). Construct, now, a new Hilbert space, as follows. The elements of the new space include all the elements of the set M, and we define on them a new scalar product

$$[\mathbf{u},\mathbf{v}] = \langle \mathbf{Au},\mathbf{v} \rangle \tag{4.278}$$

satisfying the three axioms of symmetry, linearity, and positivity. The new scalar product generates for the tridimensional body the "energy norm"

$$|||\mathbf{u}||| = ([\mathbf{u},\mathbf{u}])^{1/2} \tag{4.279}$$

If $\mathbf{u} \in$ M, then

$$|||\mathbf{u}||| = \langle \mathbf{Au},\mathbf{u} \rangle \tag{4.280}$$

and, by the positive-definitiveness inequality, we have

$$|||\mathbf{u}||| \geq \mu^{1/2}\|\mathbf{u}\| \quad \mu \in R^+ \tag{4.281}$$

Indeed, the square of the "energy" norm is twice the internal strain "energy" of T, evaluated according to the linear elastic theory. Let

$$\varepsilon_{ij}^{(1)} = \frac{1}{2}\left(u_{i,j} + u_{j,i}\right) \tag{4.282}$$

be the components of the first order strain tensor, where

$$(\nabla\mathbf{u})_{ij} = u_{i,j} = \frac{\partial u_i}{\partial X_j} \tag{4.283}$$

is the displacement gradient tensor and X_i the coordinates of the generic point P in the configuration C_i. The corresponding stress tensor, according to the generalized Hooke's law, is

$$\sigma_{ij}^{(1)} = C_{ijpq}\,\varepsilon_{pq}^{(1)} \tag{4.284}$$

where $[C_{ijpq}]$ is the elastic tensor. The strain energy is therefore given by

$$\mathcal{V}(\mathbf{u}) = \frac{1}{2}\int_\Omega \sigma_{ij}^{(1)}\varepsilon_{ij}^{(1)}\,d\Omega = \frac{1}{2}\,|||\mathbf{u}|||^2 \tag{4.285}$$

The space M, with the energy product (4.278), will be designated by M_A; of course, the space M_A, as a rule, is not complete. The completion of M_A is made by adding to it elements of the space $L_2(\Omega)$. According to inequality (4.281) we associate to any fundamental sequence in the space M_A, a sequence in the space $L_2(\Omega)$ that converges to an element of $L_2(\Omega)$: this element will be the limit of the fundamental sequence in M_A. The "energy" product and the "energy" norm of the limit elements **u** and **v** will be obtained by continuity

$$\lim_{n \to \infty} |||\mathbf{u}_n||| = |||\mathbf{u}|||, \quad \lim_{n \to \infty} [\mathbf{u}_n, \mathbf{v}_n] = [\mathbf{u}, \mathbf{v}] \tag{4.286}$$

The space so completed will be designated by H_A and constitutes a separable Hilbert space.

4.5.2 CONVERGENCE IN THE ENERGY SPACE OF FINITE DEFORMATION FIELDS

According to the previous definitions, let us introduce some useful notation. Let **S** and **T** be two second order tensor fields defined in the region Ω. The local scalar product, i.e., the scalar product at the generic point P in Ω, is

$$\mathbf{S} \cdot \mathbf{T} = S_{ij}\, T_{ij} = \mathrm{tr}(\mathbf{S}^T\mathbf{T}) \tag{4.287}$$

The modulus of **S**, i.e., the local norm of **S**, is

$$|\mathbf{S}| = (\mathbf{S} \cdot \mathbf{S})^{1/2} = (\, S_{ij}\, S_{ij})^{1/2} = [\mathrm{tr}(\mathbf{S}^T\mathbf{S})]^{1/2} \tag{4.288}$$

Then the following inequalities hold

$$|\mathbf{S} + \mathbf{T}| \leq |\mathbf{S}| + |\mathbf{T}| \tag{4.289}$$

$$|\mathbf{S} - \mathbf{T}| \geq \big|\, |\mathbf{S}| - |\mathbf{T}| \,\big| \tag{4.290}$$

$$|\mathbf{S} \cdot \mathbf{T}| \leq |\mathbf{S}|\, |\mathbf{T}| \tag{4.291}$$

and

$$|\mathbf{S}\,\mathbf{T}| \leq |\mathbf{S}|\, |\mathbf{T}| \tag{4.292}$$

The L_p norm ($p > 0$) of S will be denoted by

$$||\mathbf{S}||_{L_p} = \left[\int_\Omega |\mathbf{S}|^p d\Omega \right]^{1/p} \tag{4.293}$$

In particular the $L_2(\Omega)$ and $L_1(\Omega)$ norms are respectively

$$||\mathbf{S}||_{L_2} = \left(\int_\Omega |\mathbf{S}|^2 d\Omega \right)^{1/2} \tag{4.294}$$

$$||\mathbf{S}||_{L_1} = \int_\Omega |\mathbf{S}| d\Omega \tag{4.295}$$

We have also

$$|\langle \mathbf{S},\mathbf{T} \rangle| \leq (||\mathbf{S}||_{L_2}\, ||\mathbf{T}||_{L_2}) \tag{4.296}$$

where with the symbol <**S**,**T**> we, indicate the product

$$< \mathbf{S},\mathbf{T} > = \int_\Omega \mathbf{S} \cdot \mathbf{T}\, d\Omega = \int_\Omega \mathrm{tr}(\mathbf{S}^T\mathbf{T})\, d\Omega = \int_\Omega S_{ij}\, T_{ij}\, d\Omega \tag{4.297}$$

while

$$\int_\Omega [\mathbf{ST}]_{ik}\, d\Omega = \int_\Omega S_{ij}\, T_{jk}\, d\Omega \tag{4.298}$$

After these preliminary definitions, let us examine the mechanical meaning in tridimensional elasticity of the "energy convergence". Thus, with $n \in N$, let $\mathbf{u}_n$ be a sequence of displacement fields of $\mathbf{H}_A$ converging to the element $\mathbf{u}_0 \in \mathbf{H}_A$ of "energy" norm $|||\mathbf{u}_0|||$. Thus we have

$$\lim_{n \to \infty} |||\mathbf{u}_n - \mathbf{u}_0||| = 0 \tag{4.299}$$

To examine if also the other fundamental quantities that define the finite deformation of T behave like the displacement fields $\mathbf{u}_n$, with $n \in N$, it is of great importance now to recall the Korn inequality,[13,14] implying the positive definitiveness of the linear elastic operator **A**,

$$|||\mathbf{u}||| \geq k\, ||\nabla\mathbf{u}|| \quad k \in R^+ \tag{4.300}$$

where $\nabla\mathbf{u}$ indicates the material displacement gradient tensor (4.283), produced by partial differentiation of the vector displacement **u** with respect to the material coordinates X_i.[15,16] The norm of $\nabla\mathbf{u}$ is given by

$$||\nabla\mathbf{u}||^2 = \int_\Omega \mathrm{tr}(\nabla\mathbf{u}^T\nabla\mathbf{u})d\Omega = \int_\Omega u_{i,j}u_{i,j}\, d\Omega \tag{4.301}$$

From inequality (4.299) and condition (4.298) it follows that

$$\lim_{n \to \infty} ||\nabla\mathbf{u}_n - \nabla\mathbf{u}_0|| = 0 \tag{4.302}$$

Likewise, it is possible to show that, if $\varepsilon^{(1)}$ and $\omega^{(1)}$ indicate respectively the first-order strain and rotation tensors,

$$\lim_{n\to\infty} ||\varepsilon_n^{(1)} - \varepsilon_0^{(1)}|| = \lim_{n\to\infty} ||\omega_n^{(1)} - \omega_0^{(1)}|| = 0 \tag{4.303}$$

Therefore, if a sequence of displacement fields $\{\mathbf{u}_n\}_{n\in N}$ converges in the "energy" norm to the displacement field $\mathbf{u}_0$, the corresponding sequences of the "infinitesimal" strain and rotation tensors $\varepsilon_n^{(1)}$ and $\omega_n^{(1)}$ converge in the ordinary norm to $\varepsilon_0^{(1)}$ and $\omega_0^{(1)}$, the infinitesimal strain and rotation tensors corresponding to $\mathbf{u}_0$. Can we say that the same results remain valid for finite deformations? An answer to this question is given by the following theorem[17]: Let $\{\mathbf{u}_n\}_{n\in N}$ be a sequence of displacement fields converging in the "energy" norm to $\mathbf{u}_0$. Then the corresponding sequences of the deformation gradient tensor and, the finite pure strain and finite rotation tensors converge in the L_2 norm; the Green deformation tensor sequence converges in the L_1 norm. In fact, according to the assumption

$$\exists \left\{\mathbf{u}_n \in H_A\right\}_{n\in N} : \mathbf{u}_0 \in H_A,\ \lim_{n\to\infty} |||\mathbf{u}_n - \mathbf{u}_0||| = 0 \tag{4.304}$$

The deformation gradient tensor **F**, produced by partial differentiation of the spatial coordinates x_i with respect to the material coordinates X_j, has components

$$F_{ij} = x_{i,j} \tag{4.305}$$

and L_2 norm

$$\|\mathbf{F}\|^2 = \int_\Omega \mathrm{tr}(\mathbf{F}^T\mathbf{F})d\Omega = \int_\Omega F_{ij}F_{ij}\, d\Omega \tag{4.306}$$

Now

$$\mathbf{F} = \mathbf{I} + \nabla\mathbf{u} \tag{4.307}$$

where **I** is the identity tensor. Thus

$$\lim_{n\to\infty} |||\mathbf{u}_n - \mathbf{u}_0||| = 0 \Rightarrow \lim_{n\to\infty} \|\nabla\mathbf{u}_n - \nabla\mathbf{u}_0\| = \lim_{n\to\infty} \|\mathbf{F}_n - \mathbf{F}_0\| = 0 \tag{4.308}$$

Moreover, the right Cauchy deformation tensor is given by

$$\mathbf{C} = \mathbf{F}^T\mathbf{F} \tag{4.309}$$

Then, using the Schwarz inequality, we obtain

$$\begin{aligned}\int_\Omega |\mathbf{C}_n - \mathbf{C}_o| d\Omega &= \int_\Omega \left|\mathbf{F}_n^T(\mathbf{F}_n - \mathbf{F}_o) + (\mathbf{F}_n^T - \mathbf{F}_o^T)\mathbf{F}_o\right| d\Omega \\ &\le \int_\Omega \left|\mathbf{F}_n^T(\mathbf{F}_n - \mathbf{F}_o)\right| d\Omega + \int_\Omega \left|(\mathbf{F}_n^T - \mathbf{F}_o^T)\mathbf{F}_o\right| d\Omega \\ &\le \int_\Omega |\mathbf{F}_n|\,|\mathbf{F}_n - \mathbf{F}_o| d\Omega + \int_\Omega |\mathbf{F}_n - \mathbf{F}_o|\,|\mathbf{F}_o| d\Omega \le (\|\mathbf{F}_n\| + \|\mathbf{F}_o\|)\,\|\mathbf{F}_n - \|\mathbf{F}_o\|\end{aligned} \tag{4.310}$$

Hence, because of (4.308), we have

$$\lim_{\substack{\text{en} \\ u_n \to u_o}} \int_\Omega |\mathbf{C}_n - \mathbf{C}_o|\, d\Omega = \lim_{\substack{\text{en} \\ u_n \to u_o}} \|\mathbf{C}_n - \mathbf{C}_o\|_{L_1} = 0 \tag{4.311}$$

The analogous result for the limit $\|\mathbf{C}_n - \mathbf{C}_o\|_{L_2}$ cannot be established. In fact, terms with power four of $\nabla\mathbf{u}$ are present in the norm $\|\mathbf{C}_n - \mathbf{C}_o\|_{L_2}$ and convergence of terms with power two of $\nabla\mathbf{u}$ does not guarantee convergence of higher power terms of $\nabla\mathbf{u}$. The Green tensor $\boldsymbol{\gamma}$ is given by

$$\boldsymbol{\gamma} = \frac{1}{2}(\mathbf{C} - \mathbf{I}) \tag{4.312}$$

and

$$\int_\Omega |\boldsymbol{\gamma}_n - \boldsymbol{\gamma}_o|\, d\Omega = \|\boldsymbol{\gamma}_n - \boldsymbol{\gamma}_o\|_{L_1} \to 0 \tag{4.313}$$

Consider the pure strain tensor $\boldsymbol{\varepsilon}$, connected to the Cauchy tensor by means the relation

$$\mathbf{C} = (\mathbf{I} + \boldsymbol{\varepsilon})^2 \tag{4.314}$$

It is also possible to show[17] that the pure strain tensors converges in the L_2 norm, i.e.,

$$\lim_{n\to\infty}\left(\int_{\Omega}|\boldsymbol{\varepsilon}_n - \boldsymbol{\varepsilon}_o|^2\, d\Omega\right)^{\frac{1}{2}} = \lim_{n\to\infty}\|\boldsymbol{\varepsilon}_n - \boldsymbol{\varepsilon}_o\|_{L_2} = 0 \tag{4.315}$$

To prove this statement it is convenient to use the convergence in measure. Then, with an arbitrary quantity $\sigma > 0$, let us consider the following partition of the set Ω

$$\Omega = \Omega_n' \cup \Omega_n'' \tag{4.316}$$

where

$$\Omega_n' = \{P \in \Omega : |\mathbf{C}_n - \mathbf{C}_0| \leq \sigma\} \tag{4.317}$$

$$\Omega_n'' = \{P \in \Omega : |\mathbf{C}_n - \mathbf{C}_0| \geq \sigma\} \tag{4.318}$$

When, according to (4.311), the tensor fields $\mathbf{C}_n$ converge in L_1 to $\mathbf{C}_0$, the measures of the sets Ω_n' approach the measure of Ω while the measures of the sets Ω_n'' approach zero. Taking into account that

$$\int_{\Omega_n''}\left|\mathbf{C}_n - \mathbf{C}_o\right| d\Omega \geq \sigma\mu\left(\Omega_n''\right) \tag{4.319}$$

thus we get

$$\lim_{n\to\infty}\mu\left(\Omega_n''\right) = 0 \tag{4.320}$$

On the other hand

$$\mu(\Omega) = \mu\left(\Omega_n'\right) + \mu\left(\Omega_n''\right) \tag{4.321}$$

$$\lim_{n\to\infty}\mu\left(\Omega_n'\right) = \mu(\Omega) \tag{4.322}$$

Thus, we can write

$$\|\boldsymbol{\varepsilon}_n - \boldsymbol{\varepsilon}_o\|^2 = \int_{\Omega_n'}\left|\boldsymbol{\varepsilon}_n - \boldsymbol{\varepsilon}_o\right|^2 d\Omega + \int_{\Omega_n''}\left|\boldsymbol{\varepsilon}_n - \boldsymbol{\varepsilon}_o\right|^2 d\Omega \tag{4.323}$$

Let us consider now the first integral at the second side of (4.323), that is evaluated over the region Ω_n', where the modulus of the difference $\mathbf{C}_n - \mathbf{C}_0$ is less than the arbitrarily chosen positive number σ. The local tensorial function

$$\boldsymbol{\varepsilon} = \boldsymbol{\varepsilon}(\mathbf{C}) = \mathbf{C}^{1/2} - \mathbf{I} \tag{4.324}$$

is continuous and, consequently, no matter how small we chose δ there is a positive number σ such that

$$|\mathbf{C}_n - \mathbf{C}_0| < \sigma \Rightarrow |\boldsymbol{\varepsilon}_n - \boldsymbol{\varepsilon}_0| < \delta \tag{4.325}$$

Thus, for every $\delta > 0$

$$\int_{\Omega_n'} \left|\boldsymbol{\varepsilon}_n - \boldsymbol{\varepsilon}_o\right|^2 d\Omega < \delta^2 \mu\left(\Omega_n'\right) \tag{4.326}$$

and

$$\lim_{n\to\infty} \int_{\Omega_n'} \left|\boldsymbol{\varepsilon}_n - \boldsymbol{\varepsilon}_o\right|^2 d\Omega = 0 \tag{4.327}$$

Consider the second integral at the right side of (4.323). We get

$$\begin{aligned}
|\boldsymbol{\varepsilon}_n - \boldsymbol{\varepsilon}_o|^2 &= (\boldsymbol{\varepsilon}_n - \boldsymbol{\varepsilon}_o)\cdot(\boldsymbol{\varepsilon}_n - \boldsymbol{\varepsilon}_o) = [(\boldsymbol{\varepsilon}_n + \mathbf{I}) - (\boldsymbol{\varepsilon}_o + \mathbf{I})]\cdot[(\boldsymbol{\varepsilon}_n + \mathbf{I}) - (\boldsymbol{\varepsilon}_o + \mathbf{I})] \\
&= \left|(\boldsymbol{\varepsilon}_n + \mathbf{I})\cdot(\boldsymbol{\varepsilon}_n + \mathbf{I}) + (\boldsymbol{\varepsilon}_o + \mathbf{I})\cdot(\boldsymbol{\varepsilon}_o + \mathbf{I}) - 2(\boldsymbol{\varepsilon}_n + \mathbf{I})\cdot(\boldsymbol{\varepsilon}_o + \mathbf{I})\right| \\
&\le |\mathrm{tr}\mathbf{C}_n| + |\mathrm{tr}\mathbf{C}_o| + 2\left|\mathrm{tr}(\boldsymbol{\varepsilon}_n + \mathbf{I})^T(\boldsymbol{\varepsilon}_n + \mathbf{I})\right|^{1/2}\left(\left|\mathrm{tr}(\boldsymbol{\varepsilon}_o + \mathbf{I})^T(\boldsymbol{\varepsilon}_o + \mathbf{I})\right|\right)^{1/2} \\
&= |\mathrm{tr}\mathbf{C}_n| + |\mathrm{tr}\mathbf{C}_o| + 2(|\mathrm{tr}\mathbf{C}_n|)^{1/2}(|\mathrm{tr}\mathbf{C}_o|)^{1/2}
\end{aligned} \tag{4.328}$$

Hence

$$\begin{aligned}
\int_{\Omega_n''}\left|\boldsymbol{\varepsilon}_n - \boldsymbol{\varepsilon}_o\right|^2 d\Omega &\le \int_{\Omega_n''}|\mathrm{tr}\mathbf{C}_n|d\Omega + \int_{\Omega_n''}|\mathrm{tr}\mathbf{C}_o|d\Omega + 2\int_{\Omega_n''}|\mathrm{tr}\mathbf{C}_n|^{1/2}\left(|\mathrm{tr}\mathbf{C}_o|\right)^{1/2} d\Omega \\
&\le \int_{\Omega_n''}|\mathrm{tr}\mathbf{C}_n|d\Omega + \int_{\Omega_n''}|\mathrm{tr}\mathbf{C}_o|d\Omega + 2\left(\int_{\Omega_n''}|\mathrm{tr}\mathbf{C}_n|d\Omega\right)^{1/2}\left(\int_{\Omega_n''}|\mathrm{tr}\mathbf{C}_o|d\Omega\right)^{1/2} \\
&= \left\{\left(\int_{\Omega_n''}|\mathrm{tr}\mathbf{C}_n|d\Omega\right)^{1/2} + \left(\int_{\Omega_n''}|\mathrm{tr}\mathbf{C}_o|d\Omega\right)^{1/2}\right\}^2
\end{aligned} \tag{4.329}$$

Since the measure of Ω_n'' converges to zero, these integrals will also approach to zero. In fact, since

$$\int_{\Omega_n''}\left|\mathbf{C}_n - \mathbf{C}_o\right|d\Omega \ge \int_{\Omega_n''}\left(|\mathbf{C}_n| - |\mathbf{C}_o|\right)d\Omega \tag{4.330}$$

from (4.311) we get

$$\lim_{n\to\infty}\int_{\Omega_n''}|\mathbf{C}_n|d\Omega = \lim_{n\to\infty}\int_{\Omega_n''}|\mathbf{C}_o|d\Omega \tag{4.331}$$

Since the measure of the set Ω is not zero and the integral $\int_\Omega |\mathbf{C}_o|\, d\Omega$ exists, we have

$$\lim_{n\to\infty}\int_{\Omega_n''}|\mathbf{C}_o|d\Omega = \lim_{n\to\infty}\int_{\Omega_n''}|\mathbf{C}_n|d\Omega = 0 \tag{4.332}$$

and consequently

$$\lim_{n\to\infty}\int_{\Omega_n''}\left|\boldsymbol{\varepsilon}_n - \boldsymbol{\varepsilon}_o\right|^2 d\Omega = 0 \tag{4.333}$$

From (4.327) and (4.333) we conclude that for every choice of the positive number δ, we can find a $\sigma > 0$, such that

$$\lim_{n\to\infty} \|\boldsymbol{\varepsilon}_n - \boldsymbol{\varepsilon}_o\|^2 = \lim_{n\to\infty} \int_{\Omega_n'} |\boldsymbol{\varepsilon}_n - \boldsymbol{\varepsilon}_o|^2 \, d\Omega + \lim_{n\to\infty} \int_{\Omega_n'} |\boldsymbol{\varepsilon}_n - \boldsymbol{\varepsilon}_o|^2 \, d\Omega \le \delta^2 \mu(\Omega) \qquad (4.333')$$

Consequently, $\|\boldsymbol{\varepsilon}_n - \boldsymbol{\varepsilon}_o\|^2 \to 0$ when $\mathbf{u}_n$ converges in "energy" to $\mathbf{u}_0$. Likewise, the same result holds for the finite rotation tensor **R**, defined according to the classical polar decomposition theorem[16]

$$\mathbf{F} = \mathbf{R}\,\mathbf{V} \qquad (4.334)$$

Hence we have

$$\lim_{n\to\infty} \int_{\Omega} |\mathbf{R}_n - \mathbf{R}_o|^2 \, d\Omega = 0 \qquad (4.335)$$

4.5.3 THE POTENTIAL ENERGY FUNCTIONAL

4.5.3.1 Definitions

Let $\mathcal{E}(\mathbf{u})$ be the additional potential energy functional, with domain belonging to M_A. Thus, we have

$$\mathcal{E}(\mathbf{u}) = \mathcal{V}(\mathbf{u}) + \mathcal{U}(\mathbf{u}) \qquad (4.336)$$

where $\mathcal{U}(\mathbf{u})$ is the potential energy of the external loads and $\mathcal{V}(\mathbf{u})$ the strain energy. With the assumption of hyperelasticity, this last functional $\mathcal{V}(\mathbf{u})$ can be evaluated along any path connecting C_i and C in M_A — for instance, along the linear path defined by the displacement set

$$\alpha\mathbf{u}, \ \forall\alpha : 0 \le \alpha \le 1 \qquad (4.337)$$

Let $\boldsymbol{\sigma}$ indicate the first Piola-Kirchhoff stress tensor

$$\boldsymbol{\sigma} = \boldsymbol{\sigma}(\nabla\mathbf{u}) \qquad (4.338)$$

which we will suppose to be a continuous function of $\nabla\mathbf{u}$. Thus we can write

$$\frac{d\mathcal{V}}{d\alpha} = \int_{\Omega} \sigma_{ij}(\nabla\alpha\mathbf{u}) \frac{d}{d\alpha}\left(\alpha u_{i,j}\right) d\Omega = \langle \sigma(\nabla\alpha\mathbf{u}), \nabla\mathbf{u} \rangle \qquad (4.339)$$

if $\sigma(\alpha\mathbf{u})$ represents the stress at the configuration C', defined by the displacement $\alpha\mathbf{u}$. According to the assumption of hyperelasticity, the differential form is integrable.Thus,

$$\mathcal{V}(\mathbf{u}) = \int_0^1 d\alpha \int_{\Omega} \sigma_{ij}(\Delta\alpha\mathbf{u}) \frac{d}{d\alpha}\left(\alpha u_{i,j}\right) d\Omega = \int_0^1 \langle \boldsymbol{\sigma}(\Delta\alpha\mathbf{u}), \Delta\mathbf{u} \rangle \, d\alpha \qquad (4.340)$$

Likewise, for the potential of the external loads, represented by body forces **q**,

$$\mathcal{U}(\mathbf{u}) \int_0^1 d\alpha \int_{\Omega} q_i(\alpha\mathbf{u}) \frac{d}{d\alpha}\left(\alpha u_i\right) d\Omega \qquad (4.341)$$

4.5.3.2 Continuity and Differentiability of $\mathcal{E}$(u) in M_A.

The differential of $\mathcal{E}(\mathbf{u})$, at the point $\mathbf{u} = \mathbf{u}_1$ and in the direction **v**, is given by

$$D\mathcal{E}\left(\mathbf{u}_1;\mathbf{v}\right) = \left[\frac{\partial\mathcal{E}\left(\mathbf{u}_1 + \beta\mathbf{v}\right)}{\partial\beta}\right]_{\beta=0} \qquad (4.342)$$

according to the definitions given in Section 2.8.1 For instance, the weak differential $D\mathcal{V}(\mathbf{u};\mathbf{v})$ of $\mathcal{V}(\mathbf{u})$ is

$$D\mathcal{V}(\mathbf{u};\mathbf{v}) = <\sigma(\Delta\mathbf{u}), \nabla\mathbf{v}> = \int_\Omega \sigma_{ij}(\nabla\mathbf{u})\, v_{i,j}\, d\Omega \tag{4.343}$$

With a different notation we can also write

$$D\mathcal{V}(\mathbf{u};\ \mathbf{v}) = \nabla\mathcal{V}(\mathbf{u})\mathbf{v} \tag{4.344}$$

if ∇ is the gradient operator. Likewise, for the most cases of external loads, the weak differential $D\mathcal{U}(\mathbf{u};\ \mathbf{v})$ of $\mathcal{U}(\mathbf{u})$ exists and is linear and continuous. Hence,[17] if the constitutive equation (4.338) satisfies in mean the following condition :

$$\exists\, \chi_1 > 0:\ \|\sigma(\nabla\mathbf{u}_1) - \sigma(\nabla\mathbf{u}_2)\| \le \chi_1\, \|\nabla\mathbf{u}_1 - \nabla\mathbf{u}_2\| \quad \forall\ \mathbf{u}_1,\, \mathbf{u}_2 \in M_A \tag{4.345}$$

the strain energy is once Fréchet differentiable at $u \in M_A$. According to the Lagrange theorem we have, in fact,

$$\mathcal{V}(\mathbf{u} + \mathbf{v}) - \mathcal{V}(\mathbf{u}) = D\mathcal{V}(\mathbf{u};\ \mathbf{v}) + [D\mathcal{V}(\mathbf{u} + \theta\mathbf{v};\ \mathbf{v}) - D\mathcal{V}(\mathbf{u};\ \mathbf{v})] \tag{4.346}$$

and

$$|D\mathcal{V}(\mathbf{u};\mathbf{v})| = |<\sigma(\nabla\mathbf{u}), \nabla\mathbf{v}>| \le \|\sigma(\nabla\mathbf{u})\|\ \ \|\nabla\mathbf{v}\| \le \frac{1}{k}\|\sigma(\nabla\mathbf{u})\|\ \ |||\mathbf{v}||| \tag{4.347}$$

Thus, the differential $D\mathcal{V}(\mathbf{u};\ \mathbf{v})$ is linear and continuous with respect to $\mathbf{v}$. On the other hand, under assumption (4.345), we have also

$$\begin{aligned} &|D\mathcal{V}(\mathbf{u}+\theta\mathbf{v};\mathbf{v}) - D\mathcal{V}(\mathbf{u};\mathbf{v})| = |<\sigma(\nabla(\mathbf{u}+\theta\mathbf{v})) - \sigma(\nabla\mathbf{u});\ \nabla\mathbf{v}>| \\ &\le \|\sigma(\nabla(\mathbf{u}+\theta\mathbf{v})) - \sigma(\nabla\mathbf{u})\|\ \ \|\nabla\mathbf{v}\| \le \chi_1\theta\, \|\nabla\mathbf{v}\|^2 \le \chi_1\ \theta\, k^2\ |||\mathbf{v}|||^2 \end{aligned} \tag{4.348}$$

Hence, from (4.346) we have

$$\mathcal{V}(\mathbf{u} + \mathbf{v}) - \mathcal{V}(\mathbf{u}) = D\mathcal{V}(\mathbf{u};\ \mathbf{v}) + o(|||\mathbf{v}|||) \tag{4.349}$$

where

$$\lim_{|||\mathbf{v}|||\to 0} \frac{o(|||\mathbf{v}|||)}{|||\mathbf{v}|||} = \lim_{|||\mathbf{v}|||\to 0} \frac{|D\mathcal{V}(\mathbf{u}+\theta\mathbf{v};\mathbf{v}) - D\mathcal{V}(\mathbf{u};\mathbf{v})|}{|||\mathbf{v}|||} = 0 \tag{4.350}$$

Thus, according to the constitutive assumption in mean (4.345), the strain energy functional $\mathcal{V}(\mathbf{u})$ is once continuously Fréchet differentiable in the space M_A. If we admit that $\mathcal{U}(\mathbf{u})$ is also Fréchet differentiable, the potential energy functional $\mathcal{E}(\mathbf{u})$ of the tridimensional elastic body, with assumption (4.345), is Fréchet differentiable in the space M_A, and, consequently, continuous in M_A. Let us consider now differentials of higher order. The second weak differential of $\mathcal{V}(\mathbf{u})$ is

$$D^2\mathcal{V}(\mathbf{u};\mathbf{v},\mathbf{w}) = \left[\frac{d}{d\beta} < \sigma(\nabla(\mathbf{u}+\beta\mathbf{w});\ \nabla\mathbf{v}) >\right]_{\beta=0} = <\nabla\sigma(\nabla\mathbf{u});\ \nabla\mathbf{v}, \nabla\mathbf{w}> \tag{4.351}$$

where

$$\nabla\boldsymbol{\sigma}(\nabla\mathbf{u}) = \left[\frac{\partial\sigma_{ij}}{\partial u_{p,q}}\right]_{\mathbf{u}} \tag{4.352}$$

We have, of course,

$$D^2\mathcal{V}(\mathbf{u};\ \mathbf{v},\mathbf{w}) = D^2\mathcal{V}(\mathbf{u};\ \mathbf{w},\mathbf{v}) \tag{4.353}$$

and, for simplicity, we will write

$$D^2\mathcal{V}(\mathbf{u};\ \mathbf{v},\mathbf{v}) = D^2\mathcal{V}(\mathbf{u};\ \mathbf{v}^2) \tag{4.354}$$

Examine the continuity of $D^2\mathcal{V}(\mathbf{u};\ \mathbf{v},\mathbf{w})$ with respect to $\mathbf{v}$ and $\mathbf{w}$. We have

$$\left|D^2\mathcal{V}(\mathbf{u};\mathbf{v},\mathbf{w})\right| = \left|\int_\Omega \left[\frac{\partial\sigma_{ij}}{\partial u_{p,q}}\right]_{\mathbf{u}} v_{p,q}w_{i,j}\, d\Omega\right| \le \int_\Omega \left|\left[\frac{\partial\sigma_{ij}}{\partial u_{p,q}}\right]_{\mathbf{u}} v_{p,q}w_{i,j}\right| d\Omega \tag{4.355}$$

On the other hand

$$\left|\left[\frac{\partial\sigma_{ij}}{\partial u_{p,q}}\right]_{\mathbf{u}} v_{p,q}w_{i,j}\right| \le \left[\left(\frac{\partial\sigma_{ij}}{\partial u_{p,q}}\right)_{\mathbf{u}}\left(\frac{\partial\sigma_{ij}}{\partial u_{p,q}}\right)_{\mathbf{u}}\right]^{1/2} \cdot (v_{p,q}v_{p,q})^{1/2}(w_{i,j}w_{i,j})^{1/2} \tag{4.356}$$

Therefore

$$\left|D^2\mathcal{V}(\mathbf{u};\mathbf{v},\mathbf{w})\right| \le \int_\Omega \left|\nabla\boldsymbol{\sigma}\right|_{\mathbf{u}} \left(\left|\nabla\mathbf{v}\right|\,\left|\nabla\mathbf{w}\right|\right) d\Omega \le \max_\Omega \left|\nabla\boldsymbol{\sigma}\right|_{\mathbf{u}} \int_\Omega \left(\left|\nabla\mathbf{v}\right|\,\left|\nabla\mathbf{w}\right|\right) d\Omega$$

$$\le \max_\Omega \left|\nabla\boldsymbol{\sigma}\right|_{\mathbf{u}} \|\nabla\mathbf{v}\|\ \|\nabla\mathbf{w}\| \le \frac{1}{k^2}\max_\Omega \left|\nabla\boldsymbol{\sigma}\right|_{\mathbf{u}} \|\mathbf{v}\|\ \|\mathbf{w}\| \tag{4.357}$$

In conclusion

$$\left|D^2\mathcal{V}(\mathbf{u};\mathbf{v},\mathbf{w})\right| \le \frac{1}{k^2}\max_\Omega \left|\nabla\boldsymbol{\sigma}\right|_{\mathbf{u}} |||\mathbf{v}|||\ |||\mathbf{w}|||\quad \forall\ \mathbf{u},\mathbf{v} \in M_A \tag{4.358}$$

and the bilinear form $D^2\mathcal{V}(\mathbf{u};\ \mathbf{v},\mathbf{w})$ is continuous with respect to $\mathbf{v}$ and $\mathbf{w}$ in the space M_A. Now we analyze the strong differentiability of order two of $\mathcal{V}(\mathbf{u})$. According to the Lagrange theorem we write

$$\mathcal{V}(\mathbf{u} + \mathbf{v}) - \mathcal{V}(\mathbf{u}) = D\mathcal{V}(\mathbf{u};\ \mathbf{v}) + 1/2D^2\mathcal{V}(\mathbf{u};\ \mathbf{v}) + 1/2[D^2\mathcal{V}(\mathbf{u} + \theta\mathbf{v};\ \mathbf{v})$$

$$- D^2\mathcal{V}(\mathbf{u};\ \mathbf{v})]\quad 0 < \theta < 1 \tag{4.359}$$

Let us analyze the remainder

$$r(\mathbf{u};\ \mathbf{v}) = 1/2[D^2\mathcal{V}(\mathbf{u} + \theta\mathbf{v};\ \mathbf{v}) - D^2\mathcal{V}(\mathbf{u};\ \mathbf{v})] \tag{4.360}$$

We can write

$$\left|r(\mathbf{u};\ \mathbf{v})\right| \le \int_\Omega \left|(\nabla\boldsymbol{\sigma})_{u+\theta_v} - (\nabla\boldsymbol{\sigma})_{\mathbf{u}}\right| \left|\nabla\mathbf{v}\right|^2 d\Omega \tag{4.361}$$

For strong differentiability of order two it is also required that

$$\lim_{|||\mathbf{v}||| \to 0} \frac{|r(\mathbf{u};\mathbf{v})|}{|||\mathbf{v}|||^2} = 0 \tag{4.362}$$

Now

$$|r(\mathbf{u};\mathbf{v})| \le \max_{\Omega} |(\nabla\boldsymbol{\sigma})_{\mathbf{u}+\theta\mathbf{v}} - (\nabla\boldsymbol{\sigma})_{\mathbf{u}}| \, ||\nabla\mathbf{v}||^2 \le \frac{1}{k^2} \max_{\Omega} |(\nabla\boldsymbol{\sigma})_{\mathbf{u}+\theta\mathbf{v}} - (\nabla\boldsymbol{\sigma})_{\mathbf{u}}| \, |||\mathbf{v}|||^2 \tag{4.363}$$

Hence, it is required that[17]

$$\lim_{|||\mathbf{v}||| \to 0} \max_{\Omega} |(\nabla\boldsymbol{\sigma})_{\mathbf{u}+\theta\mathbf{v}} - (\nabla\boldsymbol{\sigma})_{\mathbf{u}}| = 0 \tag{4.364}$$

in order for condition (4.362) to hold. On the other hand, when $|||\mathbf{v}||| \to 0$, i.e., $||\nabla\mathbf{v}|| \to 0$, it is not possible to ensure that also $\max_{\Omega} |\nabla\mathbf{v}| \to 0$. The previous condition is satisfied if $\nabla\sigma =$ constant.[17] A complete answer to the previous considerations has been given by a theorem due to Martini,[19] which directly states that strain energy functional $\mathcal{V}(\mathbf{u})$ is twice continuosly strong differentiable if and only if $\nabla\sigma$ = constant i.e, when the strain energy density $\Phi(\nabla\mathbf{u})$ is a quadratic function of $\nabla\mathbf{u}$ (see also[28,25]). In conclusion, $\mathcal{V}(\mathbf{u})$ can be twice Fréchet continuosly differentiable in the space M_A only in the case, physically meaningless in finite elasticity, of a strain energy density quadratic in $\nabla\mathbf{u}$. However, if we restrict suitably the set of the increment fields $\mathbf{v}$, the twice strong differentiablility can be obtained. In fact, from (4.361), we get

$$\begin{aligned} |r(\mathbf{u};\mathbf{v})| &\le \left[\int_{\Omega} |(\nabla\boldsymbol{\sigma})_{\mathbf{u}+\theta\mathbf{v}} - (\nabla\boldsymbol{\sigma})_{\mathbf{u}}|^2 \, d\Omega\right]^{1/2} \left[\int_{\Omega} |\nabla\mathbf{v}|^4 \, d\Omega\right]^{1/2} \\ &= ||(\nabla\boldsymbol{\sigma})_{\mathbf{u}+\theta\mathbf{v}} - (\nabla\boldsymbol{\sigma})_{\mathbf{u}}|| \, ||\nabla\mathbf{v}||_{L_4}^2 \end{aligned} \tag{4.365}$$

where

$$||\nabla\mathbf{v}||_{L_4} = \left[\int_{\Omega} |\nabla\mathbf{v}|^4 \, d\Omega\right]^{1/4} = \left[\int_{\Omega} \left(v_{i,j}v_{i,j}\right)^2 d\Omega\right]^{1/4} \tag{4.366}$$

Consider now, for every positive constant c, the subset D_c of M_A

$$D_c = \left\{u \in M_A : ||\nabla\mathbf{v}||_{L_4} \le c||\nabla\mathbf{v}||_{L_2}, \, c > 0\right\} \tag{4.367}$$

The subset D_c is an homogeneous set in the sense that if the function $\mathbf{u}$ belongs to the set D_c the function $\alpha\mathbf{u}$ will also belong to D_c, $\forall\, \alpha \in R$. D_c is not a subspace of M_A but a cone passing through the origin of M_A. Hence, $\forall\, \mathbf{v} \in D_c$ we have

$$|r(\mathbf{u};\mathbf{v})| \le \left(\frac{c}{k}\right)^2 ||\nabla\boldsymbol{\sigma}(\nabla(\mathbf{u}+\theta\mathbf{v})) - (\nabla\boldsymbol{\sigma}(\nabla(\mathbf{u})))|| \, |||\mathbf{v}|||^2 \tag{4.368}$$

Thus, if

$$\exists\, \chi_2 > 0: ||\nabla\sigma(\nabla(\mathbf{u}_1)) - (\nabla\sigma(\nabla(\mathbf{u}_2))|| \le \chi_2 ||\nabla\mathbf{u}_1 - \nabla\mathbf{u}_2|| \;\; \forall\, \mathbf{u}_1, \mathbf{u}_2 \in M_A \tag{4.369}$$

we get

$$\frac{|r(\mathbf{u};\mathbf{v})|}{|||\mathbf{v}|||^2} \leq \frac{c^2}{k^3}\chi_2 \,|||\mathbf{v}||| \tag{4.370}$$

and finally condition (4.362) holds. In conclusion, with assumption (4.369), the strain energy functional $\mathcal{V}(\mathbf{u})$ is twice continuously Fréchet differentiable on the subset $D_c \subset M_A$, i.e., we can write

$$\mathcal{V}(\mathbf{u}+\mathbf{v}) - \mathcal{V}(\mathbf{u}) = D\mathcal{V}(\mathbf{u};\mathbf{v}) + 1/2D^2\mathcal{V}(\mathbf{u};\mathbf{v}) + o(|||\mathbf{v}|||^2) \quad \forall \mathbf{v} \in D_c \tag{4.371}$$

where

$$\lim_{|||\mathbf{v}||| \underset{D_c}{\to} 0} \frac{|r(\mathbf{u};\mathbf{v})|}{|||\mathbf{v}|||^2} = 0 \tag{4.372}$$

The assumption that **v** belongs to the set D_c is not adequate. It corresponds to setting an *a priori* restriction on the solution of the equation of motion of the elastic body. We don't know, in fact, to which class of initial disturbances disturbed motions of the elastic body having displacement fields belonging to the previously defined set D_c correspond. Other proposals to overcome the difficulty related to the problem of the existence of a potential well for the energy functional in tridimensional finite elasticity have been given by various authors.[21-24,26,29] In the framework of multipolar elasticity, and therefore encompassing classical elasticity theory, Koiter[21,22] assumed a generalized constitutive equation for large values of the displacement gradient. Higher spatial derivatives of the displacement fields are thus contained in the strain energy functional, and it becomes possible to prove the existence of a potential well at the fundamental state by using the Sobolev embedding theorems. Other authors[26,27] include dissipative terms in the equations or adopt a semilinear or finite-dimensional approximation.[24] We mention also a paper by Naghdi and Trapp.[22] These last authors deal with an hyperelastic simple material and do not assume any restriction on the elastic constitutive equation. They proved a stability theorem, similar to condition (3.57), but placed an *a priori* restriction on the solution because they assumed that the initial disturbances are such that the displacement fields of the consequent motions have second derivatives that are uniform bounded.

4.5.3.3 Continuity and Differentiability of $\mathcal{E}$(u) in H_A.

All the previous definitions and results hold in the space M_A but can be extended to H_A. In fact, the first Piola-Kirchhoff stress field $\boldsymbol{\sigma}(\nabla\mathbf{u})$ can also be defined for a displacement field **u** belonging to $H_A - M_A$ if the basic assumption (4.345) of boundness of the stress in M_A is satisfied. Let $\{\mathbf{u}_n\}_{n\in N} \subset M_A$ be any sequence of displacement fields converging to the element $\mathbf{u} \in H_A - M_A$; we can write, according to (4.345),

$$\|\boldsymbol{\sigma}(\nabla\mathbf{u}_n) - \boldsymbol{\sigma}(\nabla\mathbf{u}_m)\| \leq \chi_1 \|\nabla\mathbf{u}_n - \nabla\mathbf{u}_m\| \leq \frac{\chi_1}{k} |||\mathbf{u}_n - \mathbf{u}_m||| \tag{4.373}$$

Thus, the sequence $\{\boldsymbol{\sigma}(\nabla\mathbf{u}_n)\}_{n\in N}$ is fundamental with respect to the mean square norm and therefore converges to a square integrable tensor field that can be defined as

$$\boldsymbol{\sigma}(\nabla\mathbf{u}) = \lim_{\mathbf{u}_n \underset{en}{\to} \mathbf{u}} \boldsymbol{\sigma}(\nabla\mathbf{u}_n), \;\; \forall\, \mathbf{u} \in H_A - M_A \tag{4.374}$$

Further on, the strain energy functional $\mathcal{V}(\mathbf{u})$, defined in M_A, can be extended to a continuous functional defined in the Hilbert space H_A with the same technique. In fact, for any sequence $\{\mathbf{u}_n\}_{n\in N}$ converging to the element $\mathbf{u} \in H_A - M_A$, we have:

$$\left|\mathcal{V}(\mathbf{u}_n)-\mathcal{V}(\mathbf{u}_m)\right| = \int_0^{\alpha}\left[<\boldsymbol{\sigma}(\nabla\alpha\mathbf{u}_n)-\boldsymbol{\sigma}(\nabla\alpha\mathbf{u}_m),\nabla\mathbf{u}_n> + <\boldsymbol{\sigma}(\nabla\alpha\mathbf{u}_m)(\nabla\mathbf{u}_n-\nabla\mathbf{u}_m)>\right]d\alpha$$

$$\leq \int_0^{\alpha}\left[\|\boldsymbol{\sigma}(\nabla\alpha\mathbf{u}_n)-\boldsymbol{\sigma}(\nabla\alpha\mathbf{u}_m)\|\ \|\nabla\mathbf{u}_n\| + \|\boldsymbol{\sigma}(\nabla\alpha\mathbf{u}_m)\|\ \|\nabla\mathbf{u}_n-\nabla\mathbf{u}_m\|\right]d\alpha$$

$$\leq \chi_1\ \|\nabla\mathbf{u}_n-\nabla\mathbf{u}_m\|\left[\|\nabla\mathbf{u}_n\| + \|\nabla\mathbf{u}_m\|\right] \tag{4.375}$$

and the sequence $\{\mathcal{V}(\mathbf{u}_n)\}_{n\in N}$ converges. Therefore the functional $\mathcal{V}(\mathbf{u})$ is extended to a continuous functional definite in H_A with position

$$\mathcal{V}(\mathbf{u}) = \lim_{\substack{\mathbf{u}_n \to \mathbf{u} \\ en}} \mathcal{V}(\mathbf{u}_n),\ \forall\ \mathbf{u} \in H_A - M_A \tag{4.376}$$

Likewise, it is also possible to extend, from the space M_A to the whole space H_A, the differentials of $\mathcal{V}(\mathbf{u})$ together with the property of the strong differentiability of $\mathcal{V}(\mathbf{u})$ or of $\mathcal{E}(\mathbf{u})$. Likewise, the second differential of $\mathcal{V}(\mathbf{u})$ or $\mathcal{E}(\mathbf{u})$ can be defined in H_A.

4.5.3.4 Sufficient Conditions on the Constitutive Equation of the Elastic Material to the First-Order Strong Differentiability of the Strain Energy Functional $\mathcal{V}(\mathbf{u})$ in H_A.

Condition (4.345) implies the first-order strong differentiability of $\mathcal{V}(\mathbf{u})$ in H_A and represents a restriction "in mean" on the stress field. A sufficient condition in the local sense, involving only the elastic stress-strain equation, on the contrary, can be more convenient. If V indicates the space of the second-order tensors, the condition

$$\exists k > 0: |\boldsymbol{\sigma}(\nabla\mathbf{u}) - \boldsymbol{\sigma}(\nabla\mathbf{v})| \leq k\,|\nabla\mathbf{u} - \nabla\mathbf{v}| \quad \forall P \in \Omega\ \forall\ \nabla\mathbf{u}, \nabla\mathbf{v} \in V \tag{4.377}$$

is sufficient to the Fréchet differentiability of $\mathcal{V}(\mathbf{u})$ in the space H_A. Likewise, we can formulate an extension of the previous statement to the Fréchet differentibility of the second order on the subset D_c. Condition (4.377) requires that the first Piola-Kirchoff stress tensor, and eventually its gradients, is upper-bounded by the displacement gradient tensor. In connection to this last question, we observe that the constitutive equations of the elastic structural materials have been formulated only for bounded values of the displacement gradients. The elastic behavior of a structure is, in fact, tacitly restricted to the case of bounded strains, and the elastic strain energy is evaluated for displacement fields satisfying the bounding condition $|\nabla\mathbf{u}| \leq k$. At the same time, the experiments, as a rule, give information on the behavior of the materials tested under bounded pure strains. According to the assumed mathematical model, on the other hand, the admissible displacement fields constitute a linear space to which correspond unavoidable unbounded strains. Therefore, the common constitutive equations have to be extended to the case of very large values of the deformation gradients. This extension is sufficiently arbitrary, and we can operate very simply to satisfy condition (4.377). Let

$$\sigma = \sigma_1(\nabla\mathbf{u}) \tag{4.378}$$

be the constitutive equation, valid in the range of sufficiently small values of the displacement gradient $\nabla\mathbf{u}$, i.e., satisfying the condition $|\nabla\mathbf{u}| \leq k$. We will assume, besides, that (4.378) satisfies (4.377) in the range $|\nabla\mathbf{u}| \leq k$ as, for instance, it occurs if (4.378) is a n times continuously differentiable function of $\nabla\mathbf{u}$. Therefore, the constitutive equations (4.378) can be extended as

$$\sigma = \sigma_1(\nabla\mathbf{u}) + \sigma_2(\nabla\mathbf{u}) \tag{4.379}$$

where $\sigma_2(\nabla \mathbf{u})$ is the indeterminate part of (4.379). We will suppose that $\sigma_2(\nabla \mathbf{u})$ is zero for every displacement field satisying the bounding condition $|\nabla \mathbf{u}| \leq k$, and, when $|\nabla \mathbf{u}| > k$, such that the general constitutive law (4.379) satisfies the inequality (4.377). It possible to show that the elastic constitutive equation can always be modified in order to imply the first-order strong differentiability of $\mathcal{V}(\mathbf{u})$.[17] Because of the modification of the constitutive equation, the energy functional $\mathcal{V}(\mathbf{u})$ remains indeterminate. However, restricting the evaluation of $\mathcal{E}(\mathbf{u})$ to all the really possible configurations of the elastic body, namely in the set L of H_A where $|\nabla \mathbf{u}| \leq k$, we will have $\boldsymbol{\sigma}_2(\nabla \mathbf{u}) = 0$. Hence, the strain energy and its differentials will be completely defined in the subset L.

4.5.3.5 Stability of the Equilibrium State

According to the previous analysis, for a tridimensional body the potential energy functional is not twice strong differentiable in the whole space H_A, but only on the subset D_c. We can then use this property of partial differentiability to obtain only an incomplete result. Let us consider the subset of H_{Am}, denoted with $\overline{D}_c$, of all the initial disturbances

$$\begin{bmatrix} \mathbf{u}^{(0)} \\ \dot{\mathbf{u}}^{(0)} \end{bmatrix} \tag{4.380}$$

to which corresponds a motion of the structure that belongs to the subset $D_c \times L_2$. Thus we give the following definition of a restricted Liapounov stability, that is, of stability with respect to disturbing motions belonging to the subset $\overline{D}_c$ of H_{Am}:

$$\forall \varepsilon > 0 \exists \delta > 0 : \begin{bmatrix} \mathbf{u}(0) \\ \dot{\mathbf{u}}(0) \end{bmatrix} \in \overline{D}_c : |||\mathbf{u}(0)||| + \mathcal{T}[\dot{\mathbf{u}}(0)] < \delta \Rightarrow |||\mathbf{u}(t)||| + \mathcal{T}[\dot{\mathbf{u}}(t)] < \varepsilon \forall t > 0 \tag{4.381}$$

The equilibrium of the elastic body is thus stable at C_0, in the sense of definition (4.381), if condition (3.57) holds. We have obtained only an incomplete and inadequate answer to the question of the validity, in finite elasticity, of the second differential criterion as a sufficient condition of stability. In fact the set $\overline{D}_c$ of initial disturbances, to which stability result applies, remains unknown.

4.5.3.6 Stability of the Unstressed Equilibrium State

In the previous sections we have examined some smoothness properties of the strain energy functional and have shown the main difficulties in stating, in finite elasticity, the second differential stability criterion. In some particular cases of stability analysis, however, it is possible to use the general potential energy theorem directly without using the second differential criterion. Here, we give an example considering the natural configuration C_i of the elastic body, i.e., the equilibrium configuration of the unloaded and unstressed elastic body. We will show that, with some particular restrictions on the constitutive equations, it is possible to satisfy the requirement of stability of the natural configuration, as stated in Chapter 3. More precisely, we will show that, under suitable assumptions on the constitutive equation, the following inequality

$$\mathcal{V}(\mathbf{u}) \geq f(|||\mathbf{u}|||) \tag{4.382}$$

holds in a neighborhood of C_i. At the same time continuity at the origin of $\mathcal{V}(\mathbf{u})$ will also be shown, thus the potential energy $\mathcal{E}(\mathbf{u}) = \mathcal{V}(\mathbf{u})$ has a potential well in C_i.

To obtain this result, analyze a preliminary relation between the Green deformation tensor γ and the displacement gradient $\nabla \mathbf{u}$. We have

$$\boldsymbol{\gamma} = \frac{1}{2}\left(\mathbf{F}^T\mathbf{F} - \mathbf{I}\right) = \frac{1}{2}\left(\nabla\mathbf{u}^T + \nabla\mathbf{u}\right) + \frac{1}{2}\nabla\mathbf{u}^T\nabla\mathbf{u} \tag{4.383}$$

and

$$\text{tr}\,\boldsymbol{\gamma} = \text{tr}\,\nabla\mathbf{u} + \frac{1}{2}\,\text{tr}\left(\nabla\mathbf{u}^T\nabla\mathbf{u}\right) = \text{div}\,\mathbf{u} + \frac{1}{2}\,|\nabla\mathbf{u}|^2 \tag{4.384}$$

Integration of (4.384) over the reference region Ω of C_i gives

$$\int_\Omega \text{tr}\,\boldsymbol{\gamma}\,d\Omega = \int_\Omega \text{div}\,\mathbf{u}\,d\Omega + \frac{1}{2}\int_\Omega |\nabla\mathbf{u}|^2\,d\Omega \tag{4.385}$$

We assume now that the structure is completely constrained on the boundary, in the sense that $\mathbf{u} = \mathbf{0}$ along $\partial\Omega$. By means of the divergence theorem we obtain from (4.385)

$$\int_\Omega \text{tr}\,\boldsymbol{\gamma}\,d\Omega = \int_{\partial\Omega} \mathbf{u}\cdot\mathbf{n}\,dS + \frac{1}{2}\int_\Omega |\nabla\mathbf{u}|^2\,d\Omega = \frac{1}{2}\,\|\nabla\mathbf{u}\|^2 \tag{4.386}$$

Finally, we observe that

$$\int_\Omega \text{tr}\,\boldsymbol{\gamma}\,d\Omega \le \int_\Omega |\text{tr}\,\boldsymbol{\gamma}|\,d\Omega \le \sqrt{3}\int_\Omega |\boldsymbol{\gamma}|\,d\Omega = \sqrt{3}\,\|\boldsymbol{\gamma}\|_{L_1} \tag{4.387}$$

and

$$\|\boldsymbol{\gamma}\|_{L_1} \ge \frac{1}{2\sqrt{3}}\,\|\nabla\mathbf{u}\|^2 \tag{4.388}$$

Inequality (4.388) shows that the displacement gradient is directly bounded by the deformation field with the assumed boundary conditions. Consider now the pure strain tensor $\boldsymbol{\varepsilon}$ related to the Green deformation tensor $\boldsymbol{\gamma}$ by the equation

$$\boldsymbol{\gamma} = \boldsymbol{\varepsilon} + \frac{1}{2}\,\boldsymbol{\varepsilon}^2 \tag{4.389}$$

We have

$$\left|\int_\Omega \text{tr}\,\boldsymbol{\varepsilon}\,d\Omega\right| \le \sqrt{3}\int_\Omega |\boldsymbol{\varepsilon}|\,d\Omega \le \sqrt{3\mu(\Omega)}\,\|\boldsymbol{\varepsilon}\| \tag{4.390}$$

and according to (4.387)

$$\frac{1}{2}\,\|\boldsymbol{\varepsilon}\|^2 + \sqrt{3\mu(\Omega)}\,\|\boldsymbol{\varepsilon}\| \ge \frac{1}{2}\,\|\nabla\mathbf{u}\|^2 \tag{4.391}$$

or

$$\|\boldsymbol{\varepsilon}\| \ge \sqrt{3\mu(\Omega)}\left[\left(1 + \frac{\|\nabla\mathbf{u}\|^2}{3\mu(\Omega)}\right)^{\frac{1}{2}} - 1\right] = f(\|\nabla\mathbf{u}\|) \tag{4.392}$$

where $f(\|\nabla\mathbf{u}\|)$ is a positive and increasing function of $\|\nabla\mathbf{u}\|$. Let us express the strain energy density ϕ as function of the strain field $\boldsymbol{\varepsilon}$

$$\mathcal{V}(\mathbf{u}) = \int_{\Omega} \phi(\boldsymbol{\varepsilon}(\nabla \mathbf{u}))\mathrm{d}\Omega \qquad (4.393)$$

Assume now that the elastic equation satisfies the following condition

$$\mathrm{C}' \, |\boldsymbol{\varepsilon}|^2 \leq \phi(\boldsymbol{\varepsilon}) \leq \mathrm{C}'' \, |\boldsymbol{\varepsilon}|^2 \qquad (4.394)$$

corresponding to the requirement of a strain energy positive and increasing with the strain norm. Thus assumption (4.394) on the constitutive equation, together with (4.392) gives

$$\mathcal{V}(\mathbf{u}) \geq \mathrm{C}' \int_{\Omega} |\varepsilon|^2 \, \mathrm{d}\Omega = \mathrm{C}' \, \|\boldsymbol{\varepsilon}\|^2 \geq \mathrm{C}' \, \mathrm{f}^2(\|\nabla \mathbf{u}\|) \qquad (4.395)$$

Hence the energy functional $\mathcal{V}(\mathbf{u})$ has a potential well at the natural configuration with respect to the $\mathbf{L}_2$ norm of the displacement gradient or with respect to the energy norm. Further we can write

$$\mathcal{V}(\mathbf{u}) \leq \mathrm{C}'' \int_{\Omega} |\boldsymbol{\varepsilon}|^2 \, \mathrm{d}\Omega = \mathrm{C}'' \, \|\varepsilon\|^2 \qquad (4.396)$$

and therefore, according to the result (4.315), the energy functional is continuous at $\mathbf{u} = 0$ with respect to the energy norm.

The unstressed configuration of the assumed elastic body constrained at the boundary is thus stable with respect to the energy norm.

REFERENCES

1. Love, A. E. H., A Treatise on the Mathematical Theory of Elasticity. Dover, New York, 1974.
2. Como, M. and Grimaldi, A., Lyapounov stability of the Euler column, Rend. Acc. Naz. le Lincei, Ser.VIII,Vol. LXI, fasc. I, 2, 1976.
3. Koiter, W. T., Over de stabilitet von het elastisch enerwicht, Thesis, Delft, M. J. Paris, Amsterdam, 1945, English Transl. NASA YY-F-10, 833, 1967.
4. Timoshenko, S. P. and Gere, J. M., Theory of Elastic Stability. McGraw-Hill, New York, 1961.
5. Ascione L. and Grimaldi, A., Stability and Postbuckling behavior of elastic beams, Department of Structures, University of Calabria, Reports no. 13,14,19, 1979.
6. Wagner, H., Festschrift Fund-Zwanzig, Jahre Tecnische Hochschule, Danzig, 1926.
7. Greenhill, A. G., On the strength of shafting when exposed both to torsion and an end thrust, Proc. Inst. Mech. Eng., London, 1883.
8. Nikolai, E. L., Stability of rectilinear form of equilibrium, of a compressed and twisted bar, Works in Mechanics, Gostekhizdat, 1955.
9. Donnell, L. H., A new thory for the buckling of thin cylinders under axial compression and bending, Trans. ASME 56, 1934.
10. Como, M. and Grimaldi, A., Stability, buckling and postbuckling of elastic structures, Meccanica 12, 1977.
11. Koiter, W. T., A sufficient condition for the stability of shallow shells, Proc. Kon. Ned. Ak. Wet. B70, 1967.
12. Koiter, W. T., General Equations of Elastic Stability for Thin Shells, Donnell Anniversary Volume, Houston, 1966.
13. Mikhlin, S. G., Variational Methods in Mathematical Physics. Pergamon Press, Oxford, 1964.
14. Korn, A., Die Eigenschwingungen eines elastischen Korpers mit ruhender Oberflache, Akad. Wiss., Munchen, Math. Phys. Kl. Sitz. 36(13), 1906.
15. Gurtin, M. E., The linear theory of elasticity, Handbuch der Physik, Band VIa/2, Springer-Verlag, Berlin,
16. Truesdell, C., The Elements of Continuum Mechanics, Springer-Verlag, Berlin, 1966.
17. Como, M. and Grimaldi, A., Stability, buckling and postbuckling of elastic structures, Meccanica 4, 1975.
18. Caflish, R. E. and Maddocks, J. H., Nonlinear dynamical theory of the Elastica, Proc. R. Soc. of Edinburgh 99A, 1984.
19. Martini, R., On the Fréchet differentiability of certain energy functionals, Proc. Kon. Ned. Akad. Wetenschap, Amsterdam, 1978.

20. Knops, R. J. and Wilkes, E. W., Theory of elastic stability Handbuch der Physik Band VIa/3, Springer-Verlag, NewYork, 1973.
21. Koiter, W. T., The energy criterion of stability for continuous elastic bodies, I and II, Proc. Kon. Ned. Akad. Wetenschap, B68, 1965.
22. Koiter, W. T., A basic open problem in the theory of elastic stability, IUTAM-IMU Symp. on Applications of Methods of Functional Analysis to Problems of Mechanics, Marseille, 1975.
23. Naghdi, P. M. and Trapp, J. A., On the general theory of stability for elastic bodies, Arch. Rational Mech. Anal. 51, 1973.
24. Marsden, J. E., On global solutions of nonlinear Hamiltonian evolution equations, Communs Math. Phys. 30, 1973.
25. Ball, J. M., Knops, R. J., and Marsden, J. E., Two examples in nonlinear elasticity, Journées d'Analyse Nonlinéaire, Proc. 1977, Ed. by P. Benjamin and J. Robert, Lecture Notes in Mathematics 665, Springer-Verlag, Berlin, 1978.
26. Hughes, T. J. R. and Marsden, J. E., Classical elastodynamics as a symmetric hyperbolic system, J. Elasticity 8, 1978.
27. Holmes, P. and Marsden, J. E., Bifurcation of dynamical systems and nonlinear oscillations in engineering systems, Proc. on Conf. on Nonlinear Partial Differential Equations, Ed. by Chadam, Lecture Notes in Mathematics, Springer-Verlag, Berlin, 1978.
28. Knops, R. J. and Payne, L. E., On potential wells and stability in nonlinear elasticity, Math. Proc. Cambridge Philos. Soc. 84, 1978.
29. Knops, R. J., Instability and the ill-posed Cauchy problem in elasticity, in Mechanics of Solids, ed. by H. G. Hopkins and M. J. Sewell, Pergamon Press, Oxford and New York, 1982.
30. Mikhlin, S. G., The problem of the minimun of a quadeatic functional, Holden-Day, Inc., S. Francisco, 1965.

Chapter 5

BUCKLING OF CONTINUOUS ELASTIC STRUCTURES

5.1 INTRODUCTION

In the previous chapter we showed that the energy functionals of many structural models, present relevant smoothness properties in their energy spaces and in the neighborhood of their fundamental equilibrium states. It has thus been possible to analyze, according to methods and procedures defined in Chapter 3, the stability at buckling of many structural elements. The same approach, but limited to buckling analysis, will be now applied to the "structural continuum", able to represent a generic tridimensional structural element. Special constraint conditions among displacements and their derivatives fit the tridimensional elastic body to the chosen structural model.

Our attention will be particularly focused on the general formulation of the neutral equilibrium equations. The assumption of "small strains with respect to rotations", because of the high rigidity of the structural materials, will allow us to obtain a simple variational formulation of the neutral equilibrium. This approach was developed during earlier decades,[10,11,14] but it is still useful and efficient.

A short survey of buckling problems, particularly the variational formulation of the Donnell equation in the buckling analysis of cylindrical shells, ends the chapter.

5.2 THE NEUTRAL EQUILIBRIUM OF THE STRUCTURAL CONTINUUM

As was already pointed out in Section 1.1, the critical state of equilibrium of a structural system can be analyzed by means of the classical "adjacent equilibrium method". The system attains the "critical" or the "neutral" state C_0 if, under the same loads $\lambda_c\mathbf{p}$, it is in equilibrium at C_0 as well as at all neighboring configurations C, infinitesimally displaced from C_0 along the buckling mode $\mathbf{u}_c$.

We will therefore analyze the response of a structural element, in equilibrium at C_0, to a small additional displacement $\mathbf{u}$. We will determine the local response of the continuum, in a state of initial stress at C_0, passing to an adjacent state C.

Let T be a generic structural element T, as a beam, a plate or a shell, in equilibrium under the action of the loads $\lambda\mathbf{p}$ at a known fundamental configuration C_0. Let Ω be the region occupied by T at C_0. As usual, we will indicate with C_i the unstressed initial configuration of T. The configuration C_0 is assumed to be near to C_i.

We consider a small neighborhood N_{P_0} of a point P_0 of T at C_0 and focus our attention on the initial stress state acting at P_0. In this chapter we will use the technical notation x, y, z for the coordinate system. For simplicity we will first analyze a two-dimensional stress field; let

$$S_{xx}, S_{xy}, S_{yx}, S_{yy} \tag{5.1}$$

be the stress components defining the initial stress at P_0 of coordinates x and y in the plane. The dependence of the initial state of stress on the load multiplier λ is implicitly assumed.

A small displacement field is then superimposed. For instance, the medium T is displaced from C_0 to a neighboring configuration C with the application of additional loads ΔF. A point P_0 of T at C_0 that has coordinates

$$x, y, z \tag{5.2}$$

displaces to the new position P with coordinates

$$\xi = x + u(x,y), \ \eta = y + v(x,y) \tag{5.3}$$

where u, v, continuous with their derivatives, are assumed to be small quantities with respect to the dimensions of T. Of course, special constraint conditions will connect the displacement functions and their derivatives to represent the kinematics of the considered structural element.

When the structural continuum is deformed and the point P_0 is displaced to a point P of coordinates ξ, η, given by (5.3), the small region N_{P_0} around P_0 undergoes a translation, a solid body rotation, and a pure strain. The differential relations

$$d\xi = \left(1 + \frac{\partial u}{\partial x}\right) dx + \frac{\partial u}{\partial y} dy$$

$$d\eta = \frac{\partial v}{\partial x} dx + \left(1 + \frac{\partial v}{\partial y}\right) dy \tag{5.4}$$

represent a linear transformation of the infinitesimal vector of components dx, dy in the vicinity of the point P_0 into an infinitesimal vector of components $d\xi$, $d\eta$ in the vicinity of the point P. The neighborhood N_{P_0} of P_0, according to (5.4), undergoes a small rigid rotation

$$\omega = \frac{1}{2}\left(\frac{\partial v}{\partial x} - \frac{\partial u}{\partial y}\right) \tag{5.5}$$

and the local reference axes, solid with N_{P_0}, become the locally rotated directions 1, 2 of Fig. 5.1.

Further, a pure small deformation of N_{P_0} will occur. The corresponding pure strain components referred to the rotated axes 1, 2 up to the second order in the additional displacement derivatives, are[1-3]

$$\varepsilon_{xx} = e_{xx} + \varepsilon_{xx}^{(2)} \quad \varepsilon_{yy} = e_{yy} + \varepsilon_{yy}^{(2)} \quad \varepsilon_{xy} = e_{xy} + \varepsilon_{xy}^{(2)} \tag{5.6}$$

where

$$e_{xx} = \frac{\partial u}{\partial x} \quad e_{yy} = \frac{\partial v}{\partial y} \quad e_{xy} = \frac{1}{2}\left(\frac{\partial v}{\partial x} + \frac{\partial u}{\partial y}\right) \tag{5.7}$$

and

$$\varepsilon_{xx}^{(2)} = e_{xy}\,\omega + \frac{1}{2}\,\omega^2 \quad \varepsilon_{yy}^{(2)} = e_{xy}\,\omega + \frac{1}{2}\,\omega^2 \quad \varepsilon_{xy}^{(2)} = \frac{1}{2}(e_{yy} - e_{xx})\,\omega \tag{5.8}$$

Let us focus our attention on the stress field (5.1). When the plane continuum is deformed according to (5.3) and point the P_0 displaces to the point P of coordinates ξ, η, the stress at P acquires new values defined by the components

$$\bar{\sigma}_{xx} = S_{xx} + \bar{s}_{xx} \quad \bar{\sigma}_{yy} = S_{yy} + \bar{s}_{yy} \quad \bar{\sigma}_{xy} = S_{xy} + \bar{s}_{xy} \tag{5.9}$$

These components are referred to the fixed directions x, y. The components $\bar{s}_{xx}$, $\bar{s}_{yy}$, $\bar{s}_{xy}$ represent the increment of the stress at the displaced point P of coordinates ξ, η after the additional deformation. They are due not only to the occurrence of strains but also to the rotation ω of the initial stress field when moving from P_0 to P.

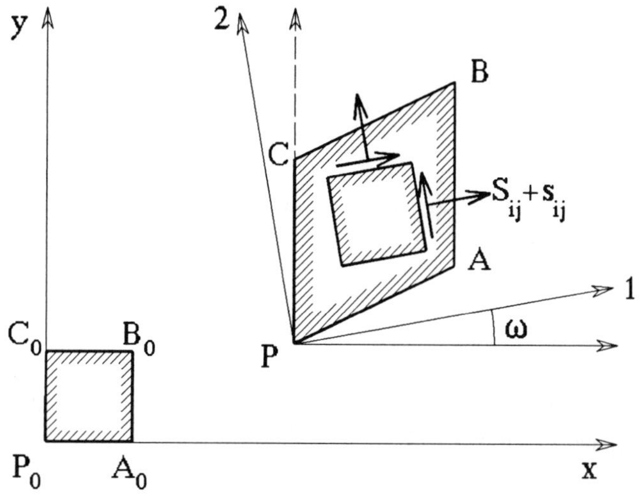

FIGURE 5.1

It is usual at this point, according to Biot,[1-3] to separate geometry from the physics in expressing incremental stress components. This can be done if, rather than referring the stress components to the initial axes x, y we refer them to the new rotated directions 1, 2. The stress components referred to these rotated axes are (Fig. 5.1)

$$\sigma_{xx} = S_{xx} + s_{xx} \quad \sigma_{yy} = S_{yy} + s_{yy} \quad \sigma_{xy} = S_{xy} + s_{xy} \tag{5.10}$$

where $s_{xx,}$ s_{xy}, and s_{yy} are stresses at P referred to the rotated axes.

The material of which the structural continuum is composed, steel for instance, is very rigid because of the corresponding high values of its elastic moduli. In this case, without relevant error, we can evaluate the incremental stresses $s_{xx,}$ s_{xy}, and s_{yy} on the areas of the rotated but not deformed element. Because of the high stiffness of the structural material, changes of the areas due to the deformation of the medium produce, in fact, only negligible changes on the incremental stress components.[9] The incremental stresses $s_{xx,}$ s_{xy}, and s_{yy}, by means the elasticity equations, are thus functions only of the pure strain components.

Let us assume now that the structural element T is further deformed and undergoes an additional virtual deformation moving from configuration C to C′. The virtual displacement is defined by the virtual displacement components

$$\delta u(x,y),\ \delta v(x,y)$$

The volume element undergoes a virtual additional deformation and will take the new virtual deformed configuration P′. The virtual internal work depends only on the virtual increment of the pure deformation. Taking into account the assumed high stiffness of the structural material, this work can be evaluated on the reference element dx dy in C_0. Thus we have

$$\delta L = [(S_{xx} + s_{xx})\,\delta\varepsilon_{xx} + (S_{yy} + s_{yy})\,\delta\varepsilon_{yy} + 2(S_{xy} + s_{xy})\,\delta\varepsilon_{xy}]dxdy \tag{5.11}$$

where $\delta\varepsilon_{xx}$, $\delta\varepsilon_{yy}$, and $\delta\varepsilon_{xy}$ represent the virtual variations of the pure strain components defined by (5.6).

We can immediately extend the previous relations to the tridimensional case. In place of (5.11), we have

$$\delta L = (S_{ij} + s_{ij})\delta\varepsilon_{ij}\ dxdydz \tag{5.12}$$

where, with

$$e_{ij} = \frac{1}{2}\left(u_{i,j} + u_{j,i}\right) \qquad \omega_{ij} = \frac{1}{2}\left(u_{i,j} - u_{j,i}\right) \tag{5.13}$$

$$\varepsilon_{ij}^{(2)} = \frac{1}{2}\left(e_{i\mu}\,\omega_{\mu j} + e_{j\mu}\,\omega_{\mu i}\right) + \frac{1}{2}\,\omega_{i\mu}\,\omega_{\mu j} \tag{5.14}$$

we have

$$\varepsilon_{ij} = e_{ij} + \varepsilon_{ij}^{(2)} \tag{5.15}$$

The components ω_i of the local rigid rotation vector ω are connected to the rotation tensor of components ω_{ij} and given by (5.13) by the following relation

$$\omega_i = -\frac{1}{2}\,e_{ijk}\,\omega_{jk} \tag{5.13′}$$

where e_{ijk} is the alternating tensor.

5.3 ADDITIONAL INTERNAL WORK AND THE NEUTRAL EQUILIBRIUM EQUATION

Let $\delta\Phi^*$ be the internal virtual work for unit volume of T at C_0. Thus

$$\delta\Phi^* = (S_{ij} + s_{ij})\delta\varepsilon_{ij} \tag{5.16}$$

Evaluating the virtual work function up to the second-order terms in the pure strain expansions gives

$$\delta\Phi^* = \delta L_1^* + \delta L_2^* + \delta\Phi \tag{5.17}$$

with

$$\delta L_1^* = S_{ij}\,\delta e_{ij} \quad \delta L_2^* = S_{ij}\,\delta\varepsilon_{ij}^{(2)},\ \delta\Phi = s_{ij}\,\delta e_{ij} \tag{5.18}$$

where the terms $s_{ij}\,\delta\varepsilon_{ij}^{(2)}$ have been neglected because of higher order in the small additional displacement u_i.

Integration of these virtual work functions over the entire region Ω yields

$$\delta\mathcal{V}^* = \int_{\Omega} \delta\Phi^*\, d\Omega \tag{5.19}$$

$$\delta\mathcal{L}_1^* = \int_{\Omega} S_{ij}\delta e_{ij}\, d\Omega;\ \delta\mathcal{L}_2^* = \int_{\Omega} S_{ij}\delta\varepsilon_{ij}^{(2)}\, d\Omega;\ \delta\mathcal{W}_2 = \int_{\Omega} s_{ij}\delta e_{ij}\, d\Omega \tag{5.20}$$

Now we formulate the equilibrium condition of T under the action of the loads F and of the additional loads ΔF that displace T from C_0 to C. Equating the internal virtual work to the virtual work of the external forces yields the variational equilibrium condition of T at C. The virtual work of the external forces can also be written as

$$\delta\mathcal{L}_E = -\,\delta\mathcal{U} + \int_{\partial\Omega} \Delta\mathbf{F}\cdot\delta\mathbf{u}\, d\Omega \tag{5.21}$$

where $\delta\mathcal{U}$ is the virtual variation of the potential energy of the loads **F**. This variation can be expressed as

$$\delta\mathcal{U} = \delta\mathcal{U}_1 + \delta\mathcal{U}_2 + \ldots \tag{5.22}$$

where $\delta\mathcal{U}_1$ and $\delta\mathcal{U}_2$ represent the first- and second-order differentials of the potential energy functional $\mathcal{U}$. The equilibrium condition of T at C, considered in the infinitesimal neighborhood of the principal configuration C_0, yields

$$\delta\mathcal{V}^* = \delta\mathcal{L}_E \tag{5.23}$$

or, more explicitly, according to (5.18), (5.19), (5.20), and (5.22),

$$\delta\mathcal{V}_2 + \delta\mathcal{L}_1^* + \delta\mathcal{L}_2^* + \delta\mathcal{U}_1 + \delta\mathcal{U}_2 = \int_{\partial\Omega} \Delta\mathbf{F}\cdot\delta\mathbf{u}\,d\Omega \tag{5.24}$$

Equilibrium of T at C_0 on the other hand requires

$$\delta\mathcal{L}_1^* + \delta\mathcal{U}_1 = 0 \tag{5.25}$$

and condition (5.24) becomes

$$\delta\mathcal{V}_2 + \delta\mathcal{L}_2^* + \delta\mathcal{U}_2 = \int_{\partial\Omega} \Delta\mathbf{F}\cdot\delta\mathbf{u}\,d\Omega \tag{5.26}$$

The neutral state at C_0 is attained when the equilibrium of the system at C is maintained for $\Delta\mathbf{F} \to \mathbf{0}$. Thus we finally get from (5.26)[10,14]

$$\delta\mathcal{V}_2 + \delta\mathcal{L}_2^* + \delta\mathcal{U}_2 = 0 \tag{5.27}$$

Thus, comparing (1.17) or (3.89) with (5.27) yields

$$\delta\mathcal{E}_2 = \mathcal{E}_{11}(\mathbf{u};\,\delta\mathbf{u}) = \delta\mathcal{V}_2 + \delta\mathcal{L}_2^* = \delta\mathcal{U}_2 = 0 \tag{5.28}$$

The variational condition (5.28) and the expression (5.20) of the internal work show that it is possible to obtain a general and direct expression of the second differential $\mathcal{E}_2(u)$ of the additional potential energy. In fact we recall that the material is assumed elastic and the internal work is path independent. Hence, considering a closed transformation that takes the system T from C_0 to C and then from C to C_0, the total expended internal work is zero. With the assumption of structural material, the internal work increment for unit volume of T at C_0, according to (5.16), is given by

$$d\Phi^* = (S_{ij} + s_{ij})\,d\varepsilon_{ij} \tag{5.29}$$

where $S_{ij} + s_{ij}$ are the stress components, inclusive of the initial stresses S_{ij} and of the incremental stresses s_{ij}, and $d\varepsilon_{ij}$ are the increments of the pure strain components. Thus, according to the assumed elastic constitutive equation, $d\Phi^*$ is an exact differential. If ε_{ij} represent the final values of the additional deformation that takes T from C_0 to C, the internal work function $\Phi^* = \Phi^*(\varepsilon_{ij})$ exists and is given by

$$\Phi^* = \int_0^{\varepsilon_{ij}} (S_{ij} + s_{ij})\,d\varepsilon_{ij} \tag{5.30}$$

Thus

$$S_{ij} + s_{ij} = \frac{\partial \Phi^*}{\partial \varepsilon_{ij}} \tag{5.31}$$

Only the additional stress components depend on the strains. Hence, for the differential to be exact, it is necessary and sufficient that

$$\frac{\partial s_{ij}}{\partial \varepsilon_{pq}} = \frac{\partial s_{pq}}{\partial \varepsilon_{ij}} \tag{5.32}$$

With the assumption of linear elastic material, conditions (5.32) are satisfied if the usual classical elasticity stress-strain law holds. To evaluate the incremental elastic response of the material from C_0 to C, let us assume the linear elastic equation

$$s_{ij} = C_{ijpq}\, \varepsilon_{pq} \tag{5.33}$$

which is the same as was assumed in the passage from the unstressed state C_i to C. Thus, we obtain the additional strain energy for unit volume at C_0, up to the quadratic terms in the displacements,

$$\Phi^* = \Phi + L_1^* = L_2^* \tag{5.34}$$

where

$$\Phi = \frac{1}{2} C_{ijpq} e_{ij} e_{pq} \tag{5.35}$$

or, with reference to an isotropic elastic material,

$$\Phi = \mu\left(e_{xx}^2 + e_{yy}^2 + e_{zz}^2\right) + \frac{\lambda}{2}(e_{xx} + e_{yy} + e_{zz})^2 + 2\mu\left(e_{xy}^2 + e_{xz}^2 + e_{yz}^2\right) \tag{5.35'}$$

$$\lambda L_1^* = \lambda \overline{S}_{ij} e_{ij} \qquad \lambda L_2^* = \lambda \overline{S}_{ij} \varepsilon_{ij}^{(2)} \tag{5.36}$$

where the dependence of initial stresses on the load multiplier λ has been pointed out. According to position (3.64) this dependence has been assumed linear. Integration over the whole volume Ω gives

$$\mathcal{V}^* = \mathcal{V}_2 + \gamma \mathcal{L}_1^* + \gamma \mathcal{L}_2^* \tag{5.34'}$$

with

$$\mathcal{V}_2 = \int_\Omega \Phi \, d\Omega \qquad \mathcal{L}_1^* = \int_\Omega L_1^* \, d\Omega \qquad \mathcal{L}_2^* = \int_\Omega L_2^* \, d\Omega \tag{5.36'}$$

The additional strain energy stored in the system in the passage of T from C_0 to C is composed of three terms. The first term $\mathcal{V}_2$, the double of the energy norm, is the strain energy, not dependent on the initial stresses: it is the strain energy stored in the system, as the primary configuration C_0 is the natural unstressed configuration of T. The second and third terms are, on the contrary, the work of the initial stresses S_{ij} on the pure strain components, respectively of the first and of the second order, in the additional displacement derivatives. Thus

$$\mathcal{E}_2(\mathbf{u}) = \mathcal{V}_2(\mathbf{u}) + \lambda \mathcal{L}_2^*(\mathbf{u}) + \lambda \mathcal{U}_2(\mathbf{u}) \tag{5.28'}$$

Comparing (5.28′) with (3.64) yields

$$\mathcal{V}_2(\mathbf{u}) = {}^1/_2\, |||\mathbf{u}|||^2 \qquad \mathcal{L}_2(\mathbf{u}) = -\,[\mathcal{L}_2^*(\mathbf{u}) + \mathcal{U}_2(\mathbf{u})] \tag{5.37}$$

The destabilizing effect of the initial stresses and of the external loads, globally represented by the functional $\mathcal{L}_2(\mathbf{u})$, thus splits into two terms, the one, $\mathcal{L}_2^*(\mathbf{u})$, representing the effect of the initial stresses, the other, $\mathcal{U}_2(\mathbf{u})$, the effect of the external loads.

In conclusion, equilibrium is neutral if the variational condition (5.27) is satisfied for any admissible virtual variations δu, δv, and δw of the displacement components u, v, and w. In (5.27) $\delta\mathcal{V}_2$, $\delta\mathcal{L}_2^*$, and $\delta\mathcal{U}_2$ are the variations i.e., the first weak differentials of the previously defined functionals $\mathcal{V}_2$, $\mathcal{L}_2^*$, and $\mathcal{U}_2$. When the additional displacements u, v, and w are infinitesimal, all of the same order of magnitude, the functional $\mathcal{U}_2$ is not zero only if the loads **F** acting on the system are not zero and are dependent on u, v and ,w. So, for instance, if the forces acting on the system are dead loads, and consequently don't depend on u, v, and w, and all these last components have the same order of smallness, $\mathcal{U}_2 = \delta\mathcal{U}_2 = 0$.[10,11,14]

5.4 COMPARISONS WITH OTHER FORMULATIONS

Another formulation involving the Green deformation tensor, in place of the pure strains, is frequently met in the literature. In place of (5.29) this other formulation gives

$$d\Phi^* = \left(S_{ij} + s_{ij}\right) d\gamma_{ij} \tag{5.38}$$

or

$$d\Phi^* = \sigma_{ij} d\gamma_{ij} \tag{5.38'}$$

where $d\gamma_{ij}$ are increments of the Green deformation tensor

$$\gamma_{ij} = \frac{1}{2}\,(u_{i,j} + u_{j,i} + u_{k,i}\, u_{k,j}) \tag{5.39}$$

and σ_{ij} are the stress components directed along the tangents to the deformed fibers, initially parallel to the coordinate axes. In spite of its popularity, expression (5.38) is not free from inadequacies, in that it is inconsistent with regard to small terms; some are retained and others that are of the same order of magnitude are rejected. In brief, (5.38) is an imperfect representation of the internal work for structural materials. It comes out from the rigorous formulation of the internal work increment involving the symmetric Piola-Kirchoff stress tensor and the Green deformation tensor[5,6]

$$d\Phi^* = \sigma_{ij}^*\, d\gamma_{ij} \tag{5.40}$$

where

$$\sigma_{ij}^* = \frac{A_i^*}{A_i}\,\frac{\sigma_{ij}}{1 + E_j} \tag{5.41}$$

if A_i and A_i^* are the areas of the faces of the volume element before and after the deformation and E_j are the relative extensions of fibers initially parallel to the coordinate axes. Under small extensions and shears, differences between the dimensions of the element before and after deformation are neglected and the stresses are simply related to the initial areas dimensions of the element. In this context, to obtain (5.38) from (5.40), it is assumed that

$$\frac{A_i^*}{A_i}\frac{1}{1+E_j} \cong 1 \tag{5.42}$$

This approximation in some cases is inconsistent. Let us consider, for example,[8] the virtual work of the component σ_{xx} acting on the faces of an element when deformation is defined by the displacements

$$u = u(x)\ ;\ v = w = 0 \tag{5.43}$$

If the variation $\delta u(x)$ of $u(x)$ occurs, the virtual work of the stress σ_{xx} is

$$\sigma_{xx}\frac{\partial \delta u}{\partial x} \tag{5.44}$$

With the exact expression of the virtual work for finite deformations, taking into account that

$$\frac{A_x^*}{A_x} = 1\ \frac{1}{1+E_{xx}} = \frac{1}{1+\dfrac{\partial u}{\partial x}} \qquad \delta\gamma_{xx} = \frac{\partial \delta u}{\partial x} + \frac{\partial u}{\partial x}\frac{\partial \delta u}{\partial x} \tag{5.45}$$

we get in fact

$$\sigma_{xx}^*\ d\gamma_{xx} = \frac{\sigma_{xx}\ \delta\gamma_{xx}}{1+\dfrac{\partial u}{\partial x}} = \sigma_{xx}\frac{\partial \delta u}{\partial x} \tag{5.46}$$

By using (5.42), on the contrary, we get an inadequate expression of the internal virtual work. In fact we have

$$\sigma_{xx}\ d\gamma_{xx} = \sigma_{xx}\frac{\partial \delta u}{\partial x} + \sigma_{xx}\frac{\partial u}{\partial x}\frac{\partial \delta u}{\partial x} \tag{5.47}$$

On the contrary, by using expression (5.29), involving pure strain components, we obtain

$$\sigma_{xx}\ \delta\varepsilon_{xx} = \sigma_{xx}\frac{\partial \delta u}{\partial x} \tag{5.48}$$

because $\varepsilon_{xx}^{(2)} = 0$.

5.5 PURE STRAINS IN CURVILINEAR COORDINATES

Let us introduce the three parameters α_1, α_2, α_3 related to the rectangular coordinates x, y, and z by the three equations (Fig.5.2)

$$x = f_1(\alpha_1, \alpha_2, \alpha_3)\ y = f_2(\alpha_1, \alpha_2, \alpha_3)\ z = f_3(\alpha_1, \alpha_2, \alpha_3) \tag{5.49}$$

or by the single vectorial equation

$$r = r(\alpha_1, \alpha_2, \alpha_3) \tag{5.50}$$

In these formulas, putting α_2 = constant, α_3 = constant, and considering the parameter α_1 as variable, we obtain a family of curves given in parametric form. We call this family the coordinate lines α_1. Similarly we have the coordinate lines α_2 and α_3. If the correspondence expressed by (5.49) or (5.50) is one-to-one, through every point in the space, or in a particular region of the space, from each of the three families there will only pass one line. If we make

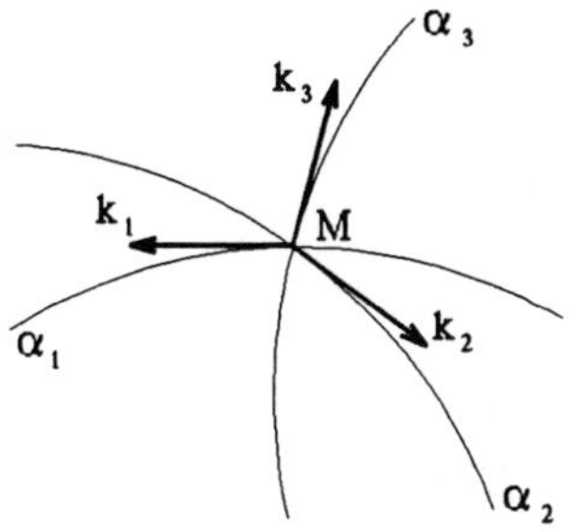

FIGURE 5.2

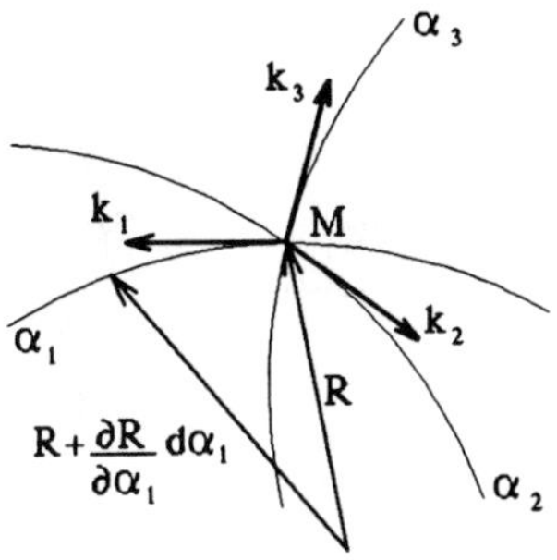

FIGURE 5.3

reference to (5.50) and we put α_2 = constant, α_3 = constant and vary α_1, the tip of the vector $\mathbf{r}$ will be displaced along the coordinate line α_1. Let $\mathbf{k}_1$, $\mathbf{k}_2$, and $\mathbf{k}_3$ be the unit vectors tangent in M to the coordinates lines. Thus (Fig. 5.3)

$$\frac{\partial \mathbf{r}}{\partial \alpha_1} d\alpha_1 = \mathbf{k}_1 \, ds_1 \tag{5.51}$$

i.e., the vector $\frac{\partial \mathbf{r}}{\partial \alpha_1} d\alpha_1$ will be equal in magnitude to the increment of the arc of the line α_1 for a change in the curvilinear coordinate of $d\alpha_1$, and will coincide in direction with $\mathbf{k}_1$. The projections of $\frac{\partial \mathbf{r}}{\partial \alpha_1}$, because of (5.49), will be equal to $\frac{\partial x}{\partial \alpha_1}, \frac{\partial y}{\partial \alpha_1}, \frac{\partial z}{\partial \alpha_1}$; consequently, the length of this vector, called H_1, is

$$H_1 = \frac{ds_1}{d\alpha_1} = \sqrt{\left(\frac{\partial x}{\partial \alpha_1}\right)^2 + \left(\frac{\partial y}{\partial \alpha_1}\right)^2 + \left(\frac{\partial z}{\partial \alpha_1}\right)^2} \tag{5.52}$$

Similarly

$$\mathbf{k}_2 = \frac{1}{H_2}\frac{\partial \mathbf{r}}{\partial \alpha_2} \qquad \mathbf{k}_3 = \frac{1}{H_3}\frac{\partial \mathbf{r}}{\partial \alpha_3} \tag{5.53}$$

where

$$H_2 = \frac{ds_2}{d\alpha_2} = \sqrt{\left(\frac{\partial x}{\partial \alpha_2}\right)^2 + \left(\frac{\partial y}{\partial \alpha_2}\right)^2 + \left(\frac{\partial z}{\partial \alpha_2}\right)^2} \quad H_3 = \frac{ds_3}{d\alpha_3} = \sqrt{\left(\frac{\partial x}{\partial \alpha_3}\right)^2 + \left(\frac{\partial y}{\partial \alpha_3}\right)^2 + \left(\frac{\partial z}{\partial \alpha_3}\right)^2} \tag{5.54}$$

The quantities H_1, H_2, and H_3 are called Lamé parameters. We will limit ourselves to consideration of orthogonal curvilinear coordinates, i.e., those for which the coordinate lines intersect at right angles. The conditions for this are the vector equations

$$\mathbf{k}_1 \cdot \mathbf{k}_2 = \mathbf{k}_2 \cdot \mathbf{k}_3 = \mathbf{k}_3 \cdot \mathbf{k}_1 = 0 \tag{5.55}$$

We recall the rules of differentiation of these vectors

$$\frac{\partial \mathbf{k}_1}{\partial \alpha_1} = -\frac{1}{H_2}\frac{\partial H_1}{\partial \alpha_2}\mathbf{k}_2 - \frac{1}{H_3}\frac{\partial H_1}{\partial \alpha_3}\mathbf{k}_3 ; \quad \frac{\partial \mathbf{k}_1}{\partial \alpha_2} = \frac{1}{H_1}\frac{\partial H_2}{\partial \alpha_1}\mathbf{k}_2 ; \quad \frac{\partial \mathbf{k}_1}{\partial \alpha_3} = \frac{1}{H_1}\frac{\partial H_3}{\partial \alpha_1}\mathbf{k}_3$$

$$\frac{\partial \mathbf{k}_2}{\partial \alpha_1} = \frac{1}{H_2}\frac{\partial H_1}{\partial \alpha_2}\mathbf{k}_1 ; \quad \frac{\partial \mathbf{k}_2}{\partial \alpha_2} = -\frac{1}{H_1}\frac{\partial H_2}{\partial \alpha_1}\mathbf{k}_1 - \frac{1}{H_3}\frac{\partial H_2}{\partial \alpha_3}\mathbf{k}_3 ; \quad \frac{\partial \mathbf{k}_2}{\partial \alpha_3} = \frac{1}{H_2}\frac{\partial H_3}{\partial \alpha_2}\mathbf{k}_3$$

$$\frac{\partial \mathbf{k}_3}{\partial \alpha_1} = \frac{1}{H_3}\frac{\partial H_1}{\partial \alpha_3}\mathbf{k}_1 ; \quad \frac{\partial \mathbf{k}_3}{\partial \alpha_2} = -\frac{1}{H_3}\frac{\partial H_2}{\partial \alpha_3}\mathbf{k}_2; \quad \frac{\partial \mathbf{k}_3}{\partial \alpha_3} = -\frac{1}{H_2}\frac{\partial H_3}{\partial \alpha_2}\mathbf{k}_2 - \frac{1}{H_1}\frac{\partial H_3}{\partial \alpha_1}\mathbf{k}_1 \tag{5.56}$$

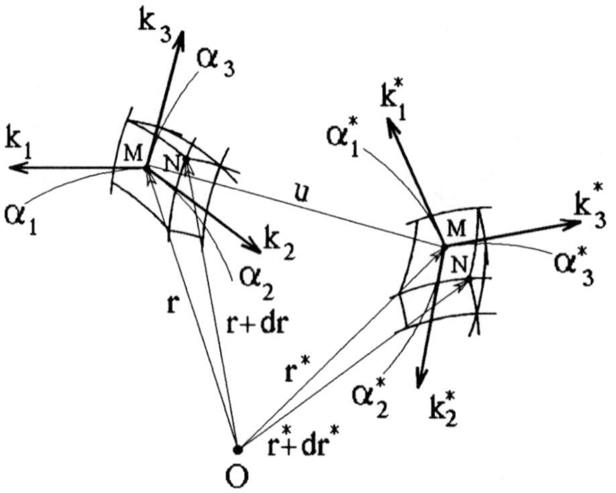

FIGURE 5.4

which enable us to differentiate vectorial functions with respect to the curvilinear coordinates.

To obtain the pure strain components, in orthogonal curvilinear coordinates, let us also recall the essential formulas of the continuum deformation in orthogonal curvilinear coordinates. Hence, suppose that the position of an arbitrary point M is given before deformation by the radius vector $\mathbf{r} = \mathbf{r}(\alpha_1,\alpha_2,\alpha_3)$. Then the position of the same point after deformation will be given by the vector

$$\mathbf{r}^* = \mathbf{r} + \mathbf{u} = \mathbf{r} + u_1\mathbf{k}_1 + u_2\mathbf{k}_2 + u_3\mathbf{k}_3 \tag{5.57}$$

where u_j are the projections of the displacement vector $\mathbf{u}$ on the axes of the local trihedral constructed at the point M. Let us consider also, together with the point M, the close point $N(\alpha_1 + d\alpha_1,\alpha_2 + d\alpha_2,\alpha_3 + d\alpha_3)$. Its position before deformation is given by the vector (Fig. 5.4)

$$\mathbf{r} + d\mathbf{r} = \mathbf{r} + \frac{\partial \mathbf{r}}{\partial \alpha_1}d\alpha_1 + \frac{\partial \mathbf{r}}{\partial \alpha_2}d\alpha_2 + \frac{\partial \mathbf{r}}{\partial \alpha_3}d\alpha_3 = \mathbf{r} + \mathbf{k}_1H_1d\alpha_1 + \mathbf{k}_2H_2d\alpha_2 + \mathbf{k}_3H_3d\alpha_3 \tag{5.58}$$

and after deformation by the vector

$$\mathbf{r}^* + d\mathbf{r}^* = \mathbf{r} + \frac{\partial \mathbf{r}}{\partial \alpha_1}d\alpha_1 + \frac{\partial \mathbf{r}}{\partial \alpha_2}d\alpha_2 + \frac{\partial \mathbf{r}}{\partial \alpha_3}d\alpha_3 + \mathbf{u} + \frac{\partial \mathbf{u}}{\partial \alpha_1}d\alpha_1 + \frac{\partial \mathbf{u}}{\partial \alpha_2}d\alpha_2 + \frac{\partial \mathbf{u}}{\partial \alpha_3}d\alpha_3 \tag{5.59}$$

Taking into account the rule of differentiation of unit vectors $\mathbf{k}_j$ given by (5.56), thus we get

$$d\mathbf{r}^* = [(1 + e_{11})H_1d\alpha_1 + (1/2\ e_{12} - \omega_3)H_2d\alpha_2 + (1/2\ e_{13} + \omega_2)H_3d\alpha_3]\ \mathbf{k}_1$$

$$+ [(1/2\ e_{12} + \omega_3)H_1d\alpha_1 + (1 + e_{22})H_2d\alpha_2 + (1/2\ e_{23} - \omega_1)H_3d\alpha_3]\ \mathbf{k}_2$$

$$+ [(1/2\ e_{13} - \omega_2)H_1d\alpha_1 + (1/2\ e_{23} + \omega_1)H_2d\alpha_2 + (1 + e_{33})H_3d\alpha_3]\ \mathbf{k}_3$$

where

$$e_{11} = \frac{1}{H_1}\frac{\partial u_1}{\partial \alpha_1} + \frac{1}{H_1H_2}\frac{\partial H_1}{\partial \alpha_2}u_2 + \frac{1}{H_1H_3}\frac{\partial H_1}{\partial \alpha_3}u_3$$

$$e_{22} = \frac{1}{H_2}\frac{\partial u_2}{\partial \alpha_2} + \frac{1}{H_2H_3}\frac{\partial H_2}{\partial \alpha_3}u_3 + \frac{1}{H_2H_1}\frac{\partial H_2}{\partial \alpha_1}u_1$$

$$e_{33} = \frac{1}{H_3}\frac{\partial u_3}{\partial \alpha_3} + \frac{1}{H_3H_1}\frac{\partial H_3}{\partial \alpha_1}u_1 + \frac{1}{H_3H_2}\frac{\partial H_3}{\partial \alpha_2}u_2$$

$$e_{12} = e_{21} = \frac{H_2}{H_1}\frac{\partial}{\partial \alpha_1}\left(\frac{u_2}{H_2}\right) + \frac{H_1}{H_2}\frac{\partial}{\partial \alpha_2}\left(\frac{u_1}{H_1}\right)$$

$$e_{13} = e_{31} = \frac{H_3}{H_1}\frac{\partial}{\partial \alpha_1}\left(\frac{u_3}{H_3}\right) + \frac{H_1}{H_3}\frac{\partial}{\partial \alpha_3}\left(\frac{u_1}{H_1}\right)$$

$$e_{23} = e_{32} = \frac{H_2}{H_3}\frac{\partial}{\partial \alpha_3}\left(\frac{u_2}{H_2}\right) + \frac{H_3}{H_2}\frac{\partial}{\partial \alpha_2}\left(\frac{u_3}{H_3}\right) \tag{5.60}$$

$$2\omega_1 = \frac{1}{H_2H_3}\left[\frac{\partial}{\partial \alpha_2}(H_3u_3) - \frac{\partial}{\partial \alpha_3}(H_2u_2)\right]$$

$$2\omega_2 = \frac{1}{H_1H_3}\left[\frac{\partial}{\partial \alpha_3}(H_1u_1) - \frac{\partial}{\partial \alpha_1}(H_3u_3)\right]$$

$$2\omega_3 = \frac{1}{H_2H_1}\left[\frac{\partial}{\partial \alpha_1}(H_2u_2) - \frac{\partial}{\partial \alpha_2}(H_1u_1)\right] \tag{5.61}$$

The pure strain components, with terms up to the second order in the displacements $u(u_1)$, $v(u_2)$, and $w(u_3)$ are expressed as function of the first-order strains and rotations according to (5.14). Thus we have, for instance,

$$\varepsilon_{11} = e_{11} + e_{12}\omega_3 - e_{31}\omega_2 + \frac{1}{2}(\omega_3^2 + \omega_2^2)$$

and

$$\varepsilon_{12} = e_{12} + \frac{1}{2}\omega_3(e_{22} - e_{11}) + \frac{1}{2}\omega_1 e_{31} - \frac{1}{2}\omega_2 e_{23} - \frac{1}{2}\omega_1\omega_2 \tag{5.62}$$

5.6 BUCKLING OF BARS

The principal configuration C_0 is the natural rectilinear configuration of the bar. In the chosen coordinate system the direction z corresponds to the beam axis and x, y the principal directions issued from the cross section center. The compression stresses before buckling are represented by

$$S_{zz} = -\frac{N}{A} \tag{5.63}$$

if N is the applied load and A is the area of the beam cross section. The buckling displacements of the beam are

$$u(x,y,z) = 0 \; ; \; v(x,y,z) = v(z) \; ; \; w(x,y,z) = -yv', \; (\;)' = \frac{d(\;)}{dz} \tag{5.64}$$

corresponding to a flexure in the plane yz.

The second-order pure strain component $\varepsilon_{zz}^{(2)}$, according to (5.14), is given by

$$\varepsilon_{zz}^{(2)} = e_{zz}\omega_y - e_{yz}\omega_x + \frac{1}{2}\left(\omega_x^2 + \omega_y^2\right) \tag{5.65}$$

On the other hand, with (5.64) and taking into account (5.13), we have

$$e_{zx} = 0;\ e_{yz} = 0;\ \omega_y = 0\ ;\ \omega_x = -\ v' \tag{5.66}$$

and

$$\varepsilon^{(2)}_{zz} = \frac{1}{2}\, v'^2 \tag{5.67}$$

The second-order internal work $\mathcal{L}_2^*$ is[10,11]

$$\mathcal{L}_2^* = \int_\Omega S_{zz}\, \varepsilon^{(2)}_{zz}\, d\Omega = -\frac{N}{2}\int_0^L v'^2\, dz \tag{5.68}$$

The strain energy $\mathcal{V}_2$ (5.37) with

$$e_{zz} = -\ yv'' \quad e_{xx} = e_{yy} = \nu yv'' \quad e_{xy} = e_{xz} = e_{yz} = 0 \tag{5.69}$$

thus becomes

$$\mathcal{V}_2 = \frac{1}{2} EI_x \int_0^L v''^2\, dz \tag{5.70}$$

The force N is a dead load, and it keeps its direction when the rod bends. The buckling components (5.64) are all of the first order in the displacement function v. Thus, the second order work of the load N is zero and $\mathcal{U}_2 = 0$, $\delta\mathcal{U}_2 = 0$. The variational equation of the buckling of the rod is

$$\delta\left(\frac{1}{2} EI_x \int_0^L v''^2\, dz - \frac{N}{2}\int_0^L v'^2\, dz\right) = 0 \tag{5.71}$$

and we find again (1.19). Components (5.64), all infinitesimal of the same order, involve higher-order strains along the beam axis. Thus $\mathcal{U}_2 = 0$ while $\mathcal{L}_2^*$ is given by (5.68).

In place of (5.64) we could assume, on the contrary, the following displacement components

$$u(x,y,z) = 0 \quad v(x,y,z) = v(z)$$

$$w(x,y,z) = -\ yv' + \int_0^z \left[\sqrt{1 - v'^2} - 1\right] d\zeta = -\ yv' + \int_0^z \left[-\frac{v'^2}{2} - \frac{v'^4}{8} - \ldots\right] d\zeta$$

These displacements, containing higher-order terms, correspond to bending without axial extension of the rod. We have

$$\frac{\partial w}{\partial z} = -\ yv'' - \frac{v'^2}{2} - \frac{v'^4}{8} - \ldots$$

and the second-order pure strain component $\varepsilon^{(2)}_{zz}$ vanishes. In fact

$$\varepsilon_{zz} = \frac{\partial w}{\partial z} + \frac{1}{2}\omega_x^2 + \ldots = -yv'' - \frac{v'^2}{2} + \frac{v'^2}{2} + \ldots$$

from which we get that the second-order pure strain along the beam axis is zero. Hence, now $\mathcal{L}_2^* = 0$. On the contrary, in this case, as from (4.28), we have

$$\mathcal{U}_2 = -\frac{N}{2}\int_0^L v'^2\,dz$$

5.7 BUCKLING BY TORSION AND FLEXURE OF THIN-WALLED OPEN CROSS SECTIONS

In the general case of a column of thin-walled open cross section, buckling occurs by a combination of torsion and bending. Let us consider the unsymmetrical cross section shown in Fig. 5.5. In this figure the buckling displacement of the cross section is sketched in the plane Cxy.

The beam, inextensible, is rectilinear in the principal configuration C_0, and the initial stress state S_{ij} is represented by an uniform compression, given by (5.63). The buckling displacement field can be represented by the well known formulas[12]

$$u(x,y,z) = u(z) + (y_0 - y)\varphi(z)$$

$$v(x,y,z) = v(z) - (x_0 - x)\varphi(z)$$

$$w(x,y,z) = -xu'(z) - yv'(z) - \omega(x,y)\varphi'(z) \tag{5.72}$$

where u(z) and v(z) define the flexure of the shear center's axis along directions x and y and $\varphi(z)$ is the torsional rotation of the sections. The coordinate $\omega(x,y)$, the warping function, represents the doubled sectorial area corresponding to the curvilinear abscissas of the middle line of the cross section.

Functions (5.72) give only a simplified description of the double bending and torsion of the bar. The strain energy corresponding to the classical torsional rigidity C cannot be taken into account from (5.72). Thus, to evaluate the strain energy $\mathcal{V}$, we have to simply add to the bending and warping strain energies, depending on (5.72), the torsional strain energy, depending on the torsional stiffness C.

Evaluating the components of the infinitesimal deformation according to (5.13) and (5.72), we have

$$e_{xx} = \frac{\partial u}{\partial x} = 0 \quad e_{yy} = \frac{\partial v}{\partial y} = 0 \quad e_{xy} = \frac{1}{2}\left(\frac{\partial u}{\partial y} + \frac{\partial v}{\partial x}\right) = 0 \quad e_{zz} = \frac{\partial w}{\partial z} = -xu'' - yv'' - \omega\phi'' \tag{5.73}$$

Approximately, we also write

$$e_{xx} = e_{yy} = -\nu e_{zz} \tag{5.73'}$$

where ν is the Poisson ratio and

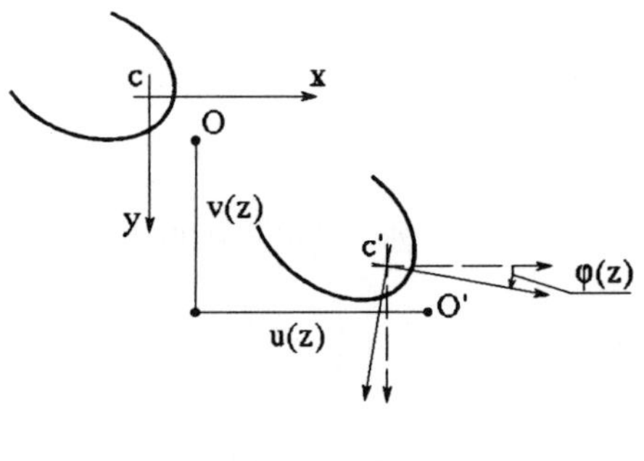

FIGURE 5.5

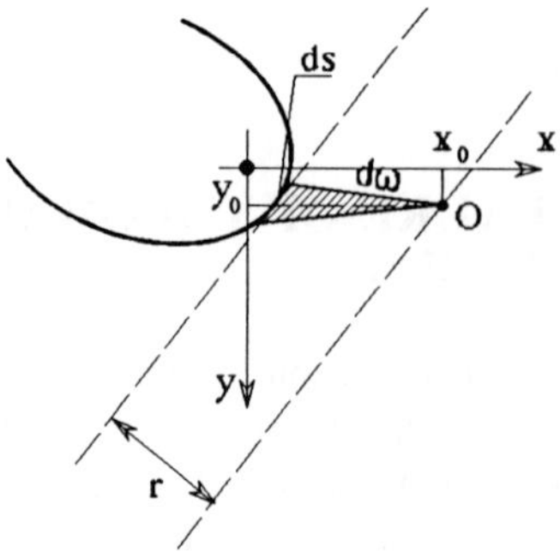

FIGURE 5.6

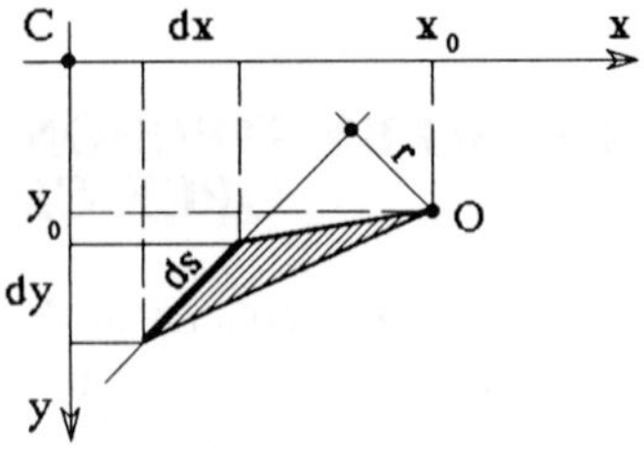

FIGURE.5.7

$$e_{xz} = \frac{1}{2}\left(\frac{\partial u}{\partial z} + \frac{\partial w}{\partial x}\right) = \frac{1}{2}\left[(y_0 - y)\phi' - \frac{\partial \omega}{\partial x}\phi'\right] \quad e_{yz} = \frac{1}{2}\left(\frac{\partial v}{\partial z} + \frac{\partial w}{\partial y}\right) = \frac{1}{2}\left[-(x_0 - x)\phi' - \frac{\partial \omega}{\partial y}\phi'\right] \quad (5.73'')$$

To obtain a more explicit form of (5.73) let us evaluate the derivatives

$$\frac{\partial \omega}{\partial x}, \ \frac{\partial \omega}{\partial y} \tag{5.74}$$

From the theory of thin-walled sections and with reference to Figs. 5.6 and 5.7 we have

$$d\omega = r\, ds = (y_0 - y)dx - (x_0 - x)dy \tag{5.75}$$

and we obtain

$$\frac{\partial \omega}{\partial x} = r\frac{\partial s}{\partial x} = (y_0 - y)\,; \quad \frac{\partial \omega}{\partial y} = r\frac{\partial s}{\partial y} = -(x_0 - x) \tag{5.76}$$

Thus

$$e_{xz} = e_{yz} = 0 \tag{5.77}$$

Let us now evaluate the local rotation components. From (5.13) and (5.72) then we have

$$\omega_x = \frac{1}{2}\left(\frac{\partial w}{\partial y} - \frac{\partial v}{\partial z}\right) = -v' + (x_0 - x)\phi'$$

$$\omega_y = \frac{1}{2}\left(\frac{\partial u}{\partial z} - \frac{\partial w}{\partial x}\right) = u' + (y_0 - y)\phi'$$

$$\omega_z = \frac{1}{2}\left(\frac{\partial v}{\partial x} - \frac{\partial u}{\partial y}\right) = \phi \tag{5.78}$$

Thus we can evaluate the second-order pure strain component $\varepsilon_{zz}^{(2)}$. We have

$$\varepsilon_{zz}^{(2)} = e_{zx}\omega_y + e_{yz}\omega_x + \frac{1}{2}\left(\omega_y^2 + \omega_x^2\right) = \frac{1}{2}\left[u' + (y_o - y)\phi\right]^2 + \frac{1}{2}\left[-v' + (x_o - x)\phi\right]^2 \tag{5.79}$$

and we immediately obtain the second order internal works $\mathcal{L}_2^*$ and $\mathcal{V}$. In fact[14]

$$\overset{*}{\mathcal{L}}_2 = -\frac{N}{2}\int_0^L \left(u'^2 + v'^2\right) dz + \frac{I_o}{A}\int_0^L \phi'^2\, dz + 2y_o\int_0^L u'\phi'\, dz - 2x_o\int_0^L v'\phi'\, dz \qquad (5.80)$$

$$\mathcal{V}_2 = \frac{EI_y}{2}\int_0^L u''^2\, dz + \frac{EI_x}{2}\int_0^L v''^2\, dz + \frac{C}{2}\int_0^L \phi'^2\, dz + \frac{C_1}{2}\int_0^L \phi''^2\, dz \qquad (5.81)$$

In expression (5.81)

$$C = GJ \qquad (5.82)$$

is the torsional rigidity of the bar, where G is the shearing modulus of elasticity and J the torsion constant,

$$C_1 = EC_\omega \qquad (5.83)$$

is the warping rigidity of the section. In (5.83) E is the Young modulus and

$$C_\omega = \int_0^m \left(\overline{\omega}_s - \omega_s\right)^2 t\, ds \qquad (5.84)$$

the warping constant, if m is the length of the middle line of the cross section and $\overline{\omega}_s$ the average value of ω_s. By means the variational condition (5.27) we immediately obtain the buckling differential equations[12,13]

$$EI_y u^{IV} + Nu'' + Ny_o\phi'' = 0$$

$$EI_x v^{IV} + Nv'' - Nx_o\phi'' = 0$$

$$C_1\phi^{IV} - \left(C - \frac{I_o}{A}\right)N\phi'' - Nx_o v'' + Ny_o u'' = 0 \qquad (5.85)$$

and the corresponding boundary conditions of thin-walled beams. If the bar has simple supports so that the ends of the bar are free to warp and to rotate about the x and the y axes, we obtain the following cubic equation for calculating the critical loads

$$\frac{I_o}{A}\left(N - N_{E_y}\right)\left(N - N_{E_x}\right)\left(N - N_\phi\right) - N^2y_o^2\left(N - N_{E_x}\right) - N^2x_o^2\left(N - N_{E_y}\right) = 0 \qquad (5.86)$$

where

$$N_{E_x} = \pi^2\frac{EI_x}{L^2} \qquad N_{E_y} = \pi^2\frac{EI_y}{L^2} \qquad N_\Phi = \frac{A}{I_o}\left(C + C_1\frac{\pi^2}{L^2}\right) \qquad (5.86')$$

5.8 LATERAL BUCKLING OF DEEP BEAMS

Let us consider the deep beam of Fig. 5.8 loaded by transversal forces acting along the center axis. The cross section of the beam has two planes of symmetry and the beam is assumed to be subjected to arbitrary transversal loads p(z) acting in the yz plane, which is the plane of maximum rigidity. The loads are distributed along the centroid axis.

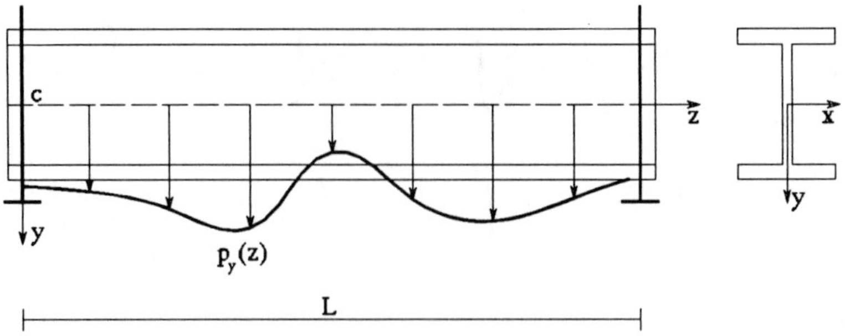

FIGURE 5.8

The beam can buckle with a sudden lateral deflection involving combined flexure and torsion. The lateral buckling of the beam can thus be defined by the following displacement field[12]

$$u(x,y,z) = u(z) - y\phi(z) \; ; \; v(x,y,z) = x\phi(z) \; ; \; w(x,y,z) = - xu'(z) - \omega(x,y)\phi'(z) \qquad (5.87)$$

where, with reference to Fig. 5.9, u(z) is the lateral deflection of the beam axis and $\phi(z)$ the rotation angle of the cross sections.

The stress state at the fundamental equilibrium state is represented by the following components

$$S_{zz}\, S_{zx}\, S_{zy} \qquad (5.88)$$

where

$$S_{zz} = \frac{M_x(z)}{I_x}\, y(s) \qquad (5.89)$$

if y(s) is the distance of the generic point P on the cross section from the x axis. The shear stresses are then given by (Fig. 5.10)

$$S_{zx} = \tau(s)\cos\psi \qquad S_{zy} = \tau(s)\sin\psi \qquad (5.90)$$

where $\tau(s)$ is the shear stress uniformly distributed over the thickness t of the wall of the beam and parallel to the tangent to the middle line of the cross section, and ψ is the angle between the tangent to s and the x axis.

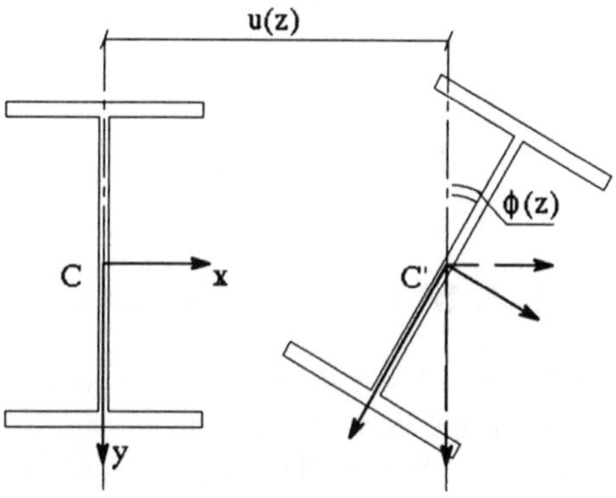

FIGURE 5.9

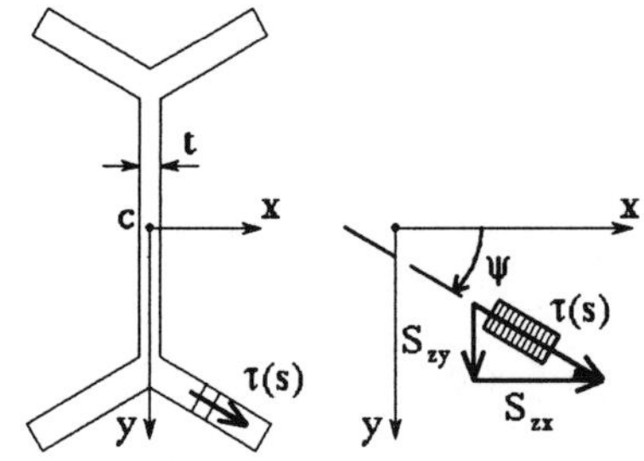

FIGURE 5.10

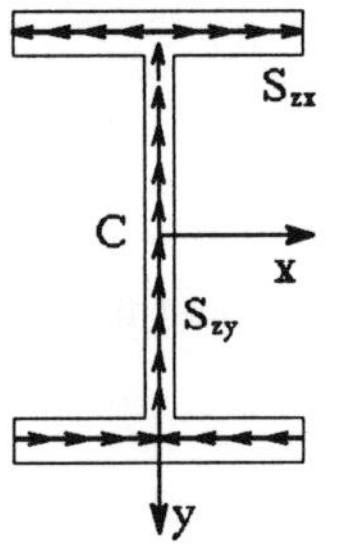

FIGURE 5.11

Along the middle line of the cross section the stresses $\tau(s)$ vary with the distance s from the edge of the section according to

$$\tau(s) = -\frac{M'_x(z)}{t\, I_x} S(s) \tag{5.91}$$

where S(s) is the static moment about neutral axis of cross-sectional area between the free edge and the chord normal to the line s and

$$M'(z) = T_y \tag{5.92}$$

the shear force. Let us evaluate the second order work $\mathcal{L}_2^*$ of primary stresses (5.88) for the second-order buckling pure strains. With the position

$$\mathcal{L}_2^* = \mathcal{L}_{2\sigma}^* + \mathcal{L}_{2\tau}^* \tag{5.93}$$

we distinguish the work of the normal stresses S_{zz} from the work of the shear stresses S_{zx} and S_{zx}. We have (Fig. 5.11)

$$\mathcal{L}_{2\tau}^* = 2\int_0^L \int_A \left(S_{zx}\, \varepsilon_{zx}^{(2)} + S_{zy}\, \varepsilon_{zy}^{(2)}\right) dA\, dz \tag{5.94}$$

where, according to (5.14), we can assume

$$\varepsilon_{xz}^{(2)} \approx -\frac{1}{2}\omega_z \omega_x \qquad \varepsilon_{yz}^{(2)} \approx -\frac{1}{2}\omega_z \omega_y \tag{5.95}$$

and, consequently, with (5.87)

$$\varepsilon_{xz}^{(2)} = \frac{1}{2} x\phi\phi' \qquad \varepsilon_{yz}^{(2)} = -\frac{1}{2}\phi(u' - y\phi') \tag{5.96}$$

Thus we get

$$\mathcal{L}_{2\tau}^* = 2\int_0^L \int_A S_{zx} \frac{1}{2} x\phi\phi'\, dA\, dz - 2\int_0^L \int_A S_{zy} \frac{1}{2}\phi(u' - y\phi')\, dA\, dz \tag{5.97}$$

On the other hand, because of double symmetry of the cross sections of the beam we have

$$\int_A S_{zx}\, x\, dA = 0 \qquad \int_A S_{zy}\, y\, dA = 0 \tag{5.98}$$

and

$$\mathcal{L}_{2\tau}^{*} = -\int_{0}^{L} M_x^{'} \, \phi \, u' \, dz \tag{5.99}$$

Taking into account that

$$\varepsilon_{zz}^{(2)} = \frac{1}{2}(u' - y\phi')^2 + \frac{1}{2} x^2 \phi'^2 \tag{5.100}$$

the second order work $\mathcal{L}_{2\sigma}^{*}$ is given by[10,14]

$$\mathcal{L}_{2\sigma}^{*} = -\int_{0}^{L} M_x \, u' \, \phi' \, dz \tag{5.101}$$

Summing (5.99) with (5.101), we obtain

$$\mathcal{L}_{2\sigma}^{*} = -\int_{0}^{L} M_x \, u' \, \phi' \, dz - \int_{0}^{L} M_x^{'} \, \phi \, u' \, dz \tag{5.102}$$

On the other hand, the strain energy $\mathcal{V}_2$ is

$$\mathcal{V}_2 = \frac{EI_y}{2}\int_{0}^{L} u''^2 \, dz + \frac{C}{2}\int_{0}^{L} \phi'^2 \, dz + \frac{C_1}{2}\int_{0}^{L} \phi''^2 \, dz \tag{5.103}$$

Thus, from the variational condition (5.27) we obtain the following differential equations

$$EI_y u^{IV} + (M_x\phi)'' = 0 \qquad C_1\phi^{IV} - C\phi'' + M_x u'' = 0 \tag{5.104}$$

and the boundary conditions

$$\left|EI_y u'' \delta u'\right|_0^L = 0 \qquad \left|\left[EI_y u''' + (M_x\phi)'\right]\delta u\right|_0^L = 0$$

$$\left|C_1\phi'' \delta\phi'\right|_0^L = 0 \qquad \left|(C_1\phi''' - C\phi' + M_x u')\delta\phi\right|_0^L = 0 \tag{5.105}$$

that control the lateral buckling of deep beams. If the loads are applied on the top flange of the beam, the primary normal stresses S_{yy} will act along the web of the beam (Fig. 5.12). These stresses can be evaluated as

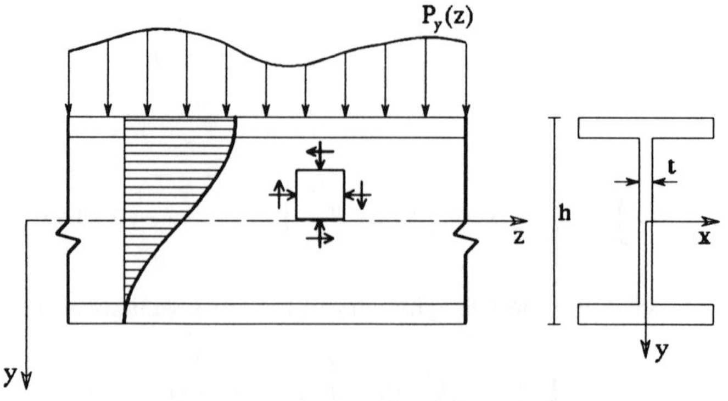

FIGURE 5.12

$$S_{yy} = -\frac{p_y}{2t} + \frac{p_y}{t}\left[\frac{3}{2}\frac{y}{h} - 2\frac{y^3}{h^3}\right] \tag{5.106}$$

The corresponding second-order pure strain component $\varepsilon_{yy}^{(2)}$ is given by

$$\varepsilon_{yy}^{(2)} \approx \frac{1}{2}\left(x^2\phi'^2 + \phi^2\right) \tag{5.107}$$

and, consequently,

$$\mathcal{L}^*_{2S_{yy}} = \int_V S_{yy}\,\varepsilon_{yy}^{(2)}\,dV = -\frac{h}{4}\int_0^L p_y\phi^2\,dz \tag{5.108}$$

Taking this new term into account, the buckling differential equation of the beam becomes, in place of (5.104),

$$EI_y u^{IV} + (M_x\phi)'' = 0 \qquad C_1\phi^{IV} - C\phi'' + M_x u'' - \frac{h}{2}p_y\phi = 0 \tag{5.109}$$

while the boundary conditions are still given by (5.105). If the transversal load p_y is applied at the lower flange of the beam, the second equation of (5.109) changes because, in place of $-\frac{h}{2}p_y\phi$, we have $+\frac{h}{2}p_y\phi$.

If, in particular, the cross section of the beam is rectangular and the external loads are only represented by two bending couples M at the end sections, we have

$$M_x = \text{constant} \tag{5.110}$$

and the differential equations (5.104) become

$$EI_y u^{IV} + M_x\phi'' = 0 \qquad -C\phi'' + M_x u'' = 0 \tag{5.111}$$

Fig. 5.13 gives the critical values of various loads distribution. In this figure the critical couples are indicated with M_c, the critical forces by F_c, and the resultant of the critical load distributions by $(qL)_c$.

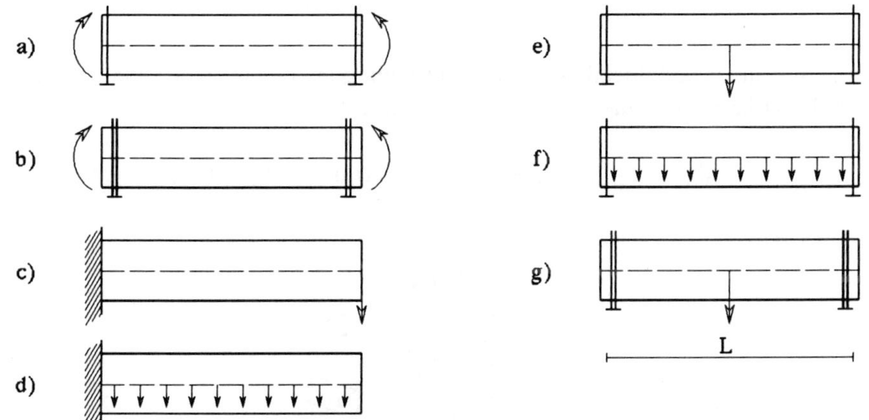

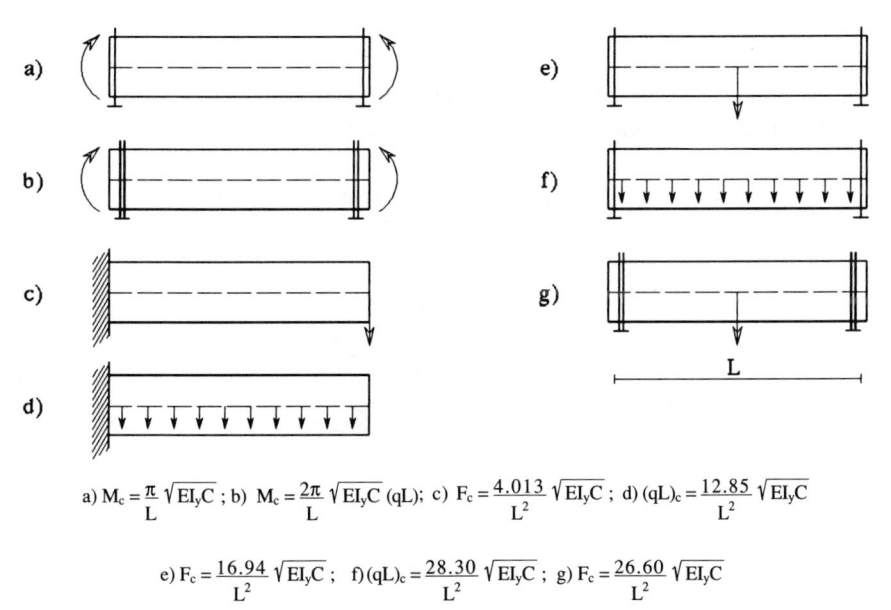

a) $M_c = \frac{\pi}{L}\sqrt{EI_yC}$; b) $M_c = \frac{2\pi}{L}\sqrt{EI_yC}$ (qL); c) $F_c = \frac{4.013}{L^2}\sqrt{EI_yC}$; d) $(qL)_c = \frac{12.85}{L^2}\sqrt{EI_yC}$

e) $F_c = \frac{16.94}{L^2}\sqrt{EI_yC}$; f) $(qL)_c = \frac{28.30}{L^2}\sqrt{EI_yC}$; g) $F_c = \frac{26.60}{L^2}\sqrt{EI_yC}$

FIGURE 5.13

5.9 BUCKLING OF SHELLS

5.9.1 SOME RECALLS FROM THE THEORY OF SURFACES

Let us recall some basic principles from the theory of surfaces and from the theory of deformation of elastic shells. We will make reference to the simple Novozhilov formulation.[15]

A smooth surface can be represented in cartesian coordinates by three equations

$$x = f_1(\alpha_1, \alpha_2) \qquad y = f_2(\alpha_1, \alpha_2) \qquad z = f_3(\alpha_1, \alpha_2) \tag{5.112}$$

where f_1, f_2, and f_3 are single-valued continuous functions of the two variable parameters α_1, and α_2, defined in a region F, called curvilinear coordinates of the surface.[16] Letting the parameter α_1 take successively different constant values and varying α_2, one obtains on the surface a family of curves, the coordinate lines α_1. Likewise, giving α_1 different constant values and varying α_2, a second family of curves, the coordinate lines α_2, is obtained. Only one curve of each of these two families of curves will pass through each point of the surface. Every point on the surface is the intersection of two coordinate lines. The simplest description of a surface is obtained when the two families of curves are simultaneously lines of principal curvature of the surface. Since the directions of principal curvature are mutually perpendicular, this system of coordinates is orthogonal. The three scalar equations (5.112) correspond to a single vectorial equation (Fig. 5.14)

$$\mathbf{r} = \mathbf{r}\,(\alpha_1,\alpha_2) \tag{5.113}$$

whose projections on the fixed axes x, y, and z are given by (5.112). The derivatives

$$\mathbf{r}_{\alpha_1} = \frac{\partial \mathbf{r}}{\partial \alpha_1} \quad ; \quad \mathbf{r}_{\alpha_2} = \frac{\partial \mathbf{r}}{\partial \alpha_2} \tag{5.114}$$

will be tangential to the α_1 and α_2 lines, respectively, at each point of the surface. Thus, any increase of one of the curvilinear coordinates corresponds to a shift of the tip of the vector on the surface in the direction corresponding to this coordinate line. The vector

$$ds = \frac{\partial \mathbf{r}}{\partial \alpha_1}\, d\alpha_1 + \frac{\partial \mathbf{r}}{\partial \alpha_2}\, d\alpha_2 \tag{5.115}$$

equals in magnitude and direction the segment joining the points (α_1,α_2) and $(\alpha_1 + d\alpha_1,\alpha_2 + d\alpha_2)$ on the surface. The square length of this segment is

$$|ds|^2 = |\mathbf{r}_{\alpha_1}|^2 (d\alpha_1)^2 + 2(\mathbf{r}_{\alpha_1} \cdot \mathbf{r}_{\alpha_2}) d\alpha_1 d\alpha_2 + |\mathbf{r}_{\alpha_2}|^2 (d\alpha_2)^2 \tag{5.116}$$

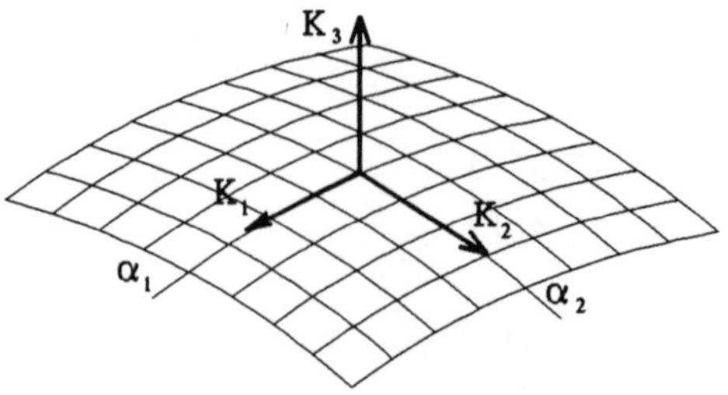

FIGURE 5.14

where the symbol $|\mathbf{ds}|^2$ indicates the product scalar of the vector **ds** with itself. Similarly, the symbol $(\mathbf{r}_{\alpha_1} \cdot \mathbf{r}_{\alpha_2})$ represents the scalar product of the two vectors $\mathbf{r}_{\alpha_1}$ and $\mathbf{r}_{\alpha_2}$. The projections of $\mathbf{r}_{\alpha_1}$ and $\mathbf{r}_{\alpha_2}$ on the coordinate axes are

$$\frac{\partial x}{\partial \alpha_1} \; ; \; \frac{\partial y}{\partial \alpha_1} \; ; \; \frac{\partial z}{\partial \alpha_1} \qquad \frac{\partial x}{\partial \alpha_2} \; ; \; \frac{\partial y}{\partial \alpha_2} \; ; \; \frac{\partial z}{\partial \alpha_2} \tag{5.117}$$

and we obtain the lengths of the vectors $\mathbf{r}_{\alpha_1}$ and $\mathbf{r}_{\alpha_2}$, i.e., the Lamé parameters H_1 and H_2

$$H_1 = \left|\mathbf{r}_{\alpha_1}\right| = \sqrt{\left(\frac{\partial x}{\partial \alpha_1}\right)^2 + \left(\frac{\partial y}{\partial \alpha_1}\right)^2 + \left(\frac{\partial z}{\partial \alpha_1}\right)^2} \quad H_2 = \left|\mathbf{r}_{\alpha_1}\right| = \sqrt{\left(\frac{\partial x}{\partial \alpha_2}\right)^2 + \left(\frac{\partial y}{\partial \alpha_2}\right)^2 + \left(\frac{\partial z}{\partial \alpha_2}\right)^2} \tag{5.118}$$

On the basis of the assumed orthogonality of the curvilinear coordinates we have

$$\left(\mathbf{r}_{\alpha_1} \cdot \mathbf{r}_{\alpha_2}\right) = 0 \tag{5.119}$$

and (5.116) becomes

$$|ds|^2 = H_1^2 (d\alpha_1)^2 + H_2^2 (d\alpha_2)^2 \tag{5.120}$$

In the particular case when only a curvilinear coordinate is varied, we get

$$ds_1 = H_1 \, d\alpha_1 \qquad ds_2 = H_2 \, d\alpha_2 \tag{5.121}$$

All vectors that are functions of the points on the surface can be given in terms of their projections on the directions of the tangent to the coordinate lines α_1 and α_2 and on the normal to the surface at the point considered. For example, for the vector **F** we have

$$\mathbf{F} = F_{\alpha_1} \mathbf{k}_1 + F_{\alpha_2} \mathbf{k}_2 + F_n \mathbf{k}_n \tag{5.122}$$

where F_{α_1}, F_{α_2}, and F_n are scalars and represent the projections of the vector **F** on the local reference system $\mathbf{k}_1$, $\mathbf{k}_2$, $\mathbf{k}_n$ given by

$$\mathbf{k}_1 = \frac{1}{H_1} \frac{\partial \mathbf{r}}{\partial \alpha_1} \qquad \mathbf{k}_2 = \frac{1}{H_2} \frac{\partial \mathbf{r}}{\partial \alpha_2} \qquad \mathbf{k}_n = \mathbf{k}_1 \wedge \mathbf{k}_2 \tag{5.123}$$

The vector $\mathbf{k}_n$ is directed to the convex side of the surface or to the sides of the centers of negative curvature if the signs of the principal curvatures are not equal. Fundamental to the study of the surfaces are the derivatives of the unit vectors $\mathbf{k}_1$, $\mathbf{k}_2$, $\mathbf{k}_n$ with respect to the coordinate variables α_1,α_2. We have, if R_1 and R_2 denote the principal radii of curvature,

$$\begin{aligned}
\frac{\partial \mathbf{k}_1}{\partial \alpha_1} &= -\frac{1}{H_2} \frac{\partial H_1}{\partial \alpha_2} \mathbf{k}_2 - \frac{H_1}{R_1} \mathbf{k}_n & \frac{\partial \mathbf{k}_1}{\partial \alpha_2} &= \frac{1}{H_1} \frac{\partial H_2}{\partial \alpha_1} \mathbf{k}_2 \\
\frac{\partial \mathbf{k}_2}{\partial \alpha_1} &= -\frac{1}{H_2} \frac{\partial H_1}{\partial \alpha_2} \mathbf{k}_1 & \frac{\partial \mathbf{k}_2}{\partial \alpha_2} &= -\frac{1}{H_1} \frac{\partial H_2}{\partial \alpha_1} \mathbf{k}_1 - \frac{H_2}{R_2} \mathbf{k}_n \\
\frac{\partial \mathbf{k}_n}{\partial \alpha_1} &= -\frac{H_1}{R_1} \mathbf{k}_1 & \frac{\partial \mathbf{k}_n}{\partial \alpha_2} &= -\frac{H_2}{R_2} \mathbf{k}_2
\end{aligned} \tag{5.124}$$

These formulas permit differentiation of vectors with given projections on to the local axes of coordinates $\mathbf{k}_1$, $\mathbf{k}_2$, $\mathbf{k}_n$. These equations lead immediately to important relations between the

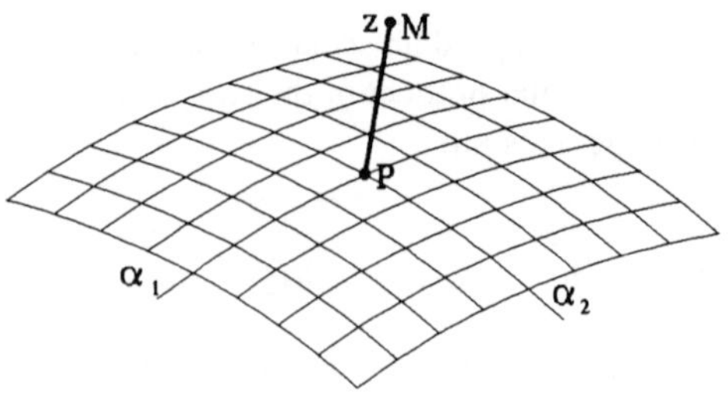

FIGURE 5.15

parameters H_1, and H_2 and the principal radii of curvature R_1, and R_2, i.e., the well-known Codazzi-Gauss equations[5,6]

$$\frac{\partial}{\partial \alpha_1}\left(\frac{H_2}{R_2}\right) = \frac{1}{R_1}\frac{\partial H_2}{\partial \alpha_1} \qquad \frac{\partial}{\partial \alpha_2}\left(\frac{H_1}{R_1}\right) = \frac{1}{R_2}\frac{\partial H_1}{\partial \alpha_2}$$

$$\frac{\partial}{\partial \alpha_1}\left(\frac{1}{H_1}\frac{\partial H_2}{\partial \alpha_1}\right) + \frac{\partial}{\partial \alpha_2}\left(\frac{1}{H_2}\frac{\partial H_1}{\partial \alpha_2}\right) = -\frac{H_1 H_2}{R_1 R_2} \tag{5.125}$$

It follows, as is well known, that the four functions H_1, H_2 and R_1, R_2 of the two parameters α_1, α_2, if chosen arbitrarily, will not, as a rule, define a surface. They may be considered as Lamé parameters and principal radii of curvature of the surface only if they satisfy the above Codazzi-Gauss conditions.

5.9.2 THE LAMÉ PARAMETERS OF A THIN SHELL

Let the position of a point P on the middle surface of a shell be defined by the two curvilinear coordinates α_1, and α_2, which will be assumed to be the principal coordinates, so that the coordinates of the lines are also lines of principal curvature of the middle surface. At point P erect the normal to the middle surface and consider a second point M in the interior of the shell and located on this normal at a distance z from the middle surface; z will be considered positive if the segment PM projects on the positive direction of k_n (Fig. 5.15). We can consider a tridimensional system of orthogonal coordinates (Fig. 5.16)

$$\alpha_1^{(z)}, \quad \alpha_2^{(z)}, \quad \alpha_3 = z \tag{5.126}$$

running inside the shell and of immediate geometrical significance. So, with reference to Fig. 5.17 we get

$$\frac{ds_1^{(z)}}{ds_1} = \frac{R_1 + z}{R_1} \tag{5.127}$$

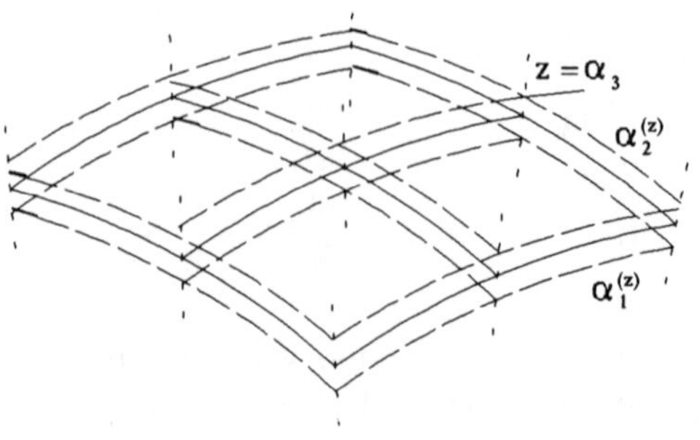

FIGURE 5.16

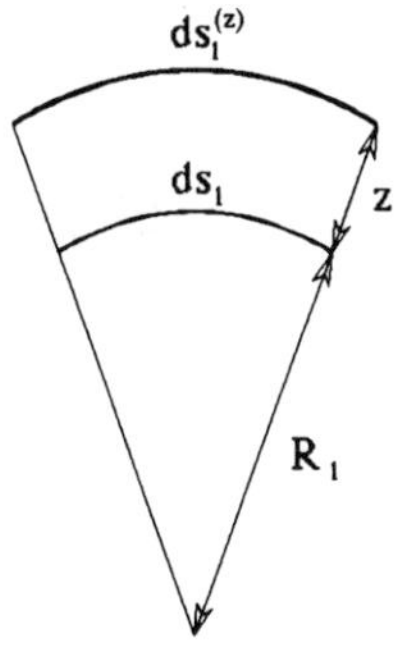

FIGURE 5.17

Taking into account that $ds_1 = H_1\, d\alpha_1$ we have

$$ds_1^{(z)} = H_1\left(1 + \frac{z}{R_1}\right)d\alpha_1 \tag{5.128}$$

On the other hand,

$$ds_1^{(z)} = H_1^{(z)} d\alpha_1 \tag{5.129}$$

and

$$H_1^{(z)} = H_1\left(1 + \frac{z}{R_1}\right) \tag{5.130}$$

Likewise we have

$$H_2^{(z)} = H_2\left(1 + \frac{z}{R_2}\right) \qquad H_3 = 1 \tag{5.131}$$

5.9.3 DISPLACEMENTS ACROSS THE THICKNESS OF THE THIN SHELL

When a thin shell is deformed, a generic point P on its middle surface will displace of **u**. This vector will have components

$$u(\alpha_1,\alpha_2) \qquad v(\alpha_1,\alpha_2) \qquad w(\alpha_1,\alpha_2) \tag{5.132}$$

along the directions of the unit reference vectors $\mathbf{k}_1$, $\mathbf{k}_2$, $\mathbf{k}_n$ passing through the point P. The displacement of the point M, located on the normal to the surface and having distance z from it, will be indicated by $\mathbf{u}^{(z)}$ and will have components

$$u^{(z)}(\alpha_1,\alpha_2) \qquad v^{(z)}(\alpha_1,\alpha_2) \qquad w^{(z)}(\alpha_1,\alpha_2) \tag{5.133}$$

The deformation of a thin shell is based on the following assumptions, proposed by Kirchhoff for thin plates and extended by Love[17] to shells:

1. The straight fibers of a shell that are perpendicular to the middle surface before deformation remain so after deformation and do not change their length.
2. The normal stress acting on planes parallel to the middle surface may be neglected in comparison with the other stresses.

According to these assumptions, we can immediately formulate the law of variation of displacements across the thickness of the shell.[15] With reference to Fig. 5.18, because of the shell deformation, the point P displaces in P′ and M in M′. The segment P′M′ will be

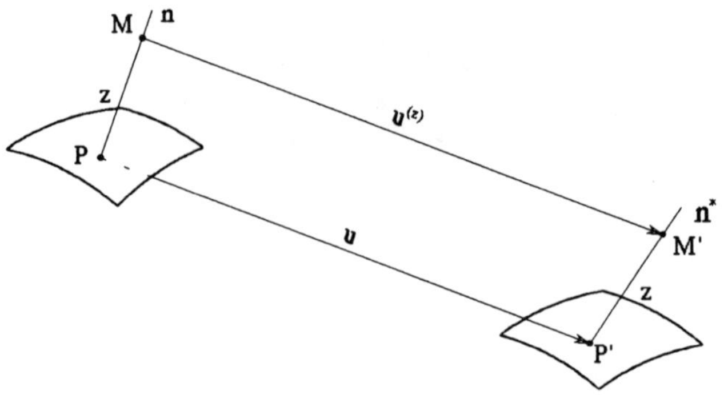

FIGURE 5.18

perpendicular to the middle surface of the deformed shell and the length of P′M′ will still be equal to z. Thus we have the vectorial equation

$$z\mathbf{k}_n + \mathbf{u}^{(z)} = z\mathbf{k}_n^* + \mathbf{u} \tag{5.134}$$

where $\mathbf{k}_n^*$ is the unit vector normal at P′ to the deformed middle surface of the shell. This vector can be easily obtained as

$$\mathbf{k}_n^* = \mathbf{k}_1^* \wedge \mathbf{k}_2^* \tag{5.135}$$

if $\mathbf{k}_1^*$ and $\mathbf{k}_2^*$ are the unit vectors tangential at P′ to the lines α_1, α_2 on the deformed middle surface. The equation of the deformed middle surface, on the other hand, is

$$\mathbf{r}^* = \mathbf{r} + \mathbf{u} = \mathbf{r} + u\mathbf{k}_1 + v\mathbf{k}_2 + w\mathbf{k}_n \tag{5.136}$$

The unit vector $\mathbf{k}_1^*$ can be obtained as

$$\mathbf{k}_1^* = \frac{1}{H_1^*}\frac{\partial \mathbf{r}^*}{\partial \alpha_1} = \frac{1}{H_1^*}\left[\frac{\partial \mathbf{r}}{\partial \alpha_1} + \frac{\partial(u\mathbf{k}_1)}{\partial \alpha_1} + \frac{\partial(v\mathbf{k}_2)}{\partial \alpha_1} + \frac{\partial(w\mathbf{k}_n)}{\partial \alpha_1}\right] \tag{5.137}$$

where

$$H_1^* = \sqrt{\frac{\partial \mathbf{r}^*}{\partial \alpha_1}\cdot\frac{\partial \mathbf{r}^*}{\partial \alpha_1}} \tag{5.138}$$

On the other hand, we have

$$\frac{\partial \mathbf{r}^*}{\partial \alpha_1} = H_1\mathbf{k}_1 + u\frac{\partial \mathbf{k}_1}{\partial \alpha_1} + v\frac{\partial \mathbf{k}_2}{\partial \alpha_1} + w\frac{\partial \mathbf{k}_n}{\partial \alpha_1} + \frac{\partial u}{\partial \alpha_1}\mathbf{k}_1 + \frac{\partial v}{\partial \alpha_1}\mathbf{k}_2 + \frac{\partial w}{\partial \alpha_1}\mathbf{k}_n \tag{5.139}$$

Thus, taking into account (5.124), we have

$$\frac{\partial \mathbf{r}^*}{\partial \alpha_1} = H_1\left[1 + \frac{1}{H_1}\frac{\partial u}{\partial \alpha_1} + \frac{1}{H_1H_2}\frac{\partial H_1}{\partial \alpha_2}v + \frac{w}{R_1}\right]\mathbf{k}_1$$

$$+\left[\frac{\partial v}{\partial \alpha_1} - \frac{1}{H_2}\frac{\partial H_1}{\partial \alpha_2}u\right]\mathbf{k}_2 + \left[\frac{\partial w}{\partial \alpha_1} - \frac{H_1}{R_1}u\right]\mathbf{k}_n \tag{5.140}$$

Evaluating the Lamé parameter H_1^* by means (5.138), taking into account that

$$\mathbf{k}_1 \cdot \mathbf{k}_2 = 0 \qquad \mathbf{k}_1 \cdot \mathbf{k}_n = 0 \qquad \mathbf{k}_2 \cdot \mathbf{k}_n = 0 \tag{5.141}$$

we get, keeping only terms of the first order in u, v, and w,

$$H_1^{*2} = \frac{\partial \mathbf{r}^*}{\partial \alpha_1} \cdot \frac{\partial \mathbf{r}^*}{\partial \alpha_1} = H_1^2 \left[1 + \frac{1}{H_1} \frac{\partial u}{\partial \alpha_1} + \frac{1}{H_1 H_2} \frac{\partial H_1}{\partial \alpha_2} v + \frac{w}{R_1} \right]^2 \tag{5.142}$$

Thus we have

$$H_1^* = H_1(1 + e_1) \tag{5.143}$$

where

$$e_1 = \frac{1}{H_1} \frac{\partial u}{\partial \alpha_1} + \frac{1}{H_1 H_2} \frac{\partial H_1}{\partial \alpha_2} v + \frac{w}{R_1} \tag{5.144}$$

is the first-order elongation along line α_1. From (5.137) we get

$$\mathbf{k}_1^* = \frac{1}{H_1^*} \frac{\partial \mathbf{r}^*}{\partial \alpha_1} \approx \mathbf{k}_1 + \left[\frac{1}{H_1} \frac{\partial v}{\partial \alpha_1} - \frac{1}{H_1 H_2} \frac{\partial H_1}{\partial \alpha_2} u \right] \mathbf{k}_2 + \left[\frac{1}{H_1} \frac{\partial w}{\partial \alpha_1} - \frac{u}{R_1} \right] \mathbf{k}_n \tag{5.145}$$

Likewise

$$\mathbf{k}_2^* = \frac{1}{H_2^*} \frac{\partial \mathbf{r}^*}{\partial \alpha_2} \approx \mathbf{k}_2 + \left[\frac{1}{H_2} \frac{\partial u}{\partial \alpha_2} - \frac{1}{H_1 H_2} \frac{\partial H_2}{\partial \alpha_1} u \right] \mathbf{k}_1 + \left[\frac{1}{H_2} \frac{\partial w}{\partial \alpha_2} - \frac{v}{R_2} \right] \mathbf{k}_n \tag{5.146}$$

We can obtain now the unit vector $\mathbf{k}_n^*$, normal to the middle deformed surface. We have

$$\mathbf{k}_n^* = \mathbf{k}_n + A\mathbf{k}_1 + B\mathbf{k}_2 \tag{5.147}$$

where

$$A = -\frac{1}{H_1} \frac{\partial w}{\partial \alpha_1} + \frac{u}{R_1} \qquad B = -\frac{1}{H_2} \frac{\partial w}{\partial \alpha_2} + \frac{v}{R_2} \tag{5.148}$$

As can be seen by inspection of expression (5.147), A and B are the projections of the vector $\mathbf{k}_n^*$ on the direction $\mathbf{k}_1$ and $\mathbf{k}_2$. Since $\mathbf{k}_n^*$ is of unit length and small displacements are assumed, these projections are equal to the angles by which during deformation the normal to the middle surface rotates about the axes $\mathbf{k}_2$ and $\mathbf{k}_1$, respectively. Likewise, we can say that A and B are the angles by which during deformation the tangents to the lines α_1, and α_2 of the middle surface rotate about the axes $\mathbf{k}_2$ and $\mathbf{k}_1$. These angles will be equal, in fact, to the angles of rotation of the normal. Returning to (5.134) we can write

$$\mathbf{u}^{(z)} = \mathbf{u} + z(A\mathbf{k}_1 + B\mathbf{k}_2) \tag{5.149}$$

or, projecting on the axes of the local reference system, we have the equations

$$u^{(z)} = u + zA \quad v^{(z)} = v + zB \quad w^{(z)} = w \tag{5.150}$$

which give the law of variation of the displacements across the thickness of the shell.

5.9.4 PURE DEFORMATIONS AND ROTATIONS IN THIN SHELLS

Equations (5.60) and (5.61) give the expressions of the infinitesimal strain and rotation components in a system of orthogonal curvilinear coordinates with Lamé parameters H_1, H_2, and H_3. We can immediately apply these formulas to thin shells considering the system of curvilinear coordinates $\alpha_1^{(z)}$, $\alpha_2^{(z)}$, $\alpha_3 = z$. In fact, taking into account the expressions of the Lamé parameters for thin shells and of the Codazzi-Gauss relations we have

$$e_{11}^{(z)} = \frac{1}{1+\frac{z}{R_1}}\left[\frac{1}{H_1}\frac{\partial u^{(z)}}{\partial \alpha_1} + \frac{1}{H_1 H_2}\frac{\partial H_1}{\partial \alpha_2} v^{(z)} + \frac{w^{(z)}}{R_1}\right]$$

$$e_{22}^{(z)} = \frac{1}{1+\frac{z}{R_2}}\left[\frac{1}{H_2}\frac{\partial v^{(z)}}{\partial \alpha_2} + \frac{1}{H_1 H_2}\frac{\partial H_2}{\partial \alpha_1} u^{(z)} + \frac{w^{(z)}}{R_2}\right]$$

$$e_{zz}^{(z)} = \frac{\partial w}{\partial z}$$

$$2e_{12}^{(z)} = \frac{1}{1+\frac{z}{R_1}}\left[\frac{1}{H_1}\frac{\partial v^{(z)}}{\partial \alpha_1} - \frac{1}{H_1 H_2}\frac{\partial H_1}{\partial \alpha_2} u^{(z)}\right] + \frac{1}{1+\frac{z}{R_2}}\left[\frac{1}{H_2}\frac{\partial u^{(z)}}{\partial \alpha_2} - \frac{1}{H_1 H_2}\frac{\partial H_2}{\partial \alpha_1} v^{(z)}\right]$$

$$2e_{1z}^{(z)} = \frac{\partial u^{(z)}}{\partial z} + \frac{1}{1+\frac{z}{R_1}}\left[\frac{1}{H_1}\frac{\partial w^{(z)}}{\partial \alpha_1} - \frac{u^{(z)}}{R_1}\right]$$

$$2e_{2z}^{(z)} = \frac{\partial v^{(z)}}{\partial z} + \frac{1}{1+\frac{z}{R_2}}\left[\frac{1}{H_2}\frac{\partial w^{(z)}}{\partial \alpha_2} - \frac{v^{(z)}}{R_2}\right] \tag{5.151}$$

At the same time the local rotation components are

$$2\omega_1^{(z)} = -\frac{\partial v^{(z)}}{\partial z} + \frac{1}{1+\frac{z}{R_2}}\left[\frac{1}{H_2}\frac{\partial w^{(z)}}{\partial \alpha_2} - \frac{v^{(z)}}{R_2}\right]$$

$$2\omega_2^{(z)} = \frac{\partial u^{(z)}}{\partial z} - \frac{1}{1+\frac{z}{R_1}}\left[\frac{1}{H_1}\frac{\partial w^{(z)}}{\partial \alpha_1} - \frac{u^{(z)}}{R_1}\right]$$

$$2\omega_z^{(z)} = \frac{1}{1+\frac{z}{R_1}}\left[\frac{1}{H_1}\frac{\partial v^{(z)}}{\partial \alpha_1} - \frac{1}{H_1 H_2}\frac{\partial H_1}{\partial \alpha_2} u^{(z)}\right] - \frac{1}{1+\frac{z}{R_2}}\left[\frac{1}{H_2}\frac{\partial u^{(z)}}{\partial \alpha_2} - \frac{1}{H_1 H_2}\frac{\partial H_2}{\partial \alpha_1} v^{(z)}\right] \tag{5.152}$$

where H_1, H_2, and H_3 are the Lamé parameters of the middle surface and $u^{(z)}$, $v^{(z)}$, and $w^{(z)}$ are the displacements of a point in the thickness of the shell having distance z from the middle surface. Relations (5.151) and (5.152), on the other hand, are not the final expressions of the small strains and small rotations because displacements $u^{(z)}$, $v^{(z)}$, and $w^{(z)}$ have to be expressed by means of (5.150). Substitution of (5.150) into (5.151) thus gives first of all

$$e_{zz}^{(z)} = e_{1z}^{(z)} = e_{2z}^{(z)} = 0 \tag{5.153}$$

The other components of deformation for a thin shell

$$e_{11}^{(z)} \quad e_{22}^{(z)} \quad e_{12}^{(z)}$$

can be simply formulated introducing the following quantities, representing the relative changes in length and in shear of the middle surface

$$e_1 = \frac{1}{H_1}\frac{\partial u}{\partial \alpha_1} + \frac{1}{H_1 H_2}\frac{\partial H_1}{\partial \alpha_2} v + \frac{w}{R_1} \qquad e_2 = \frac{1}{H_2}\frac{\partial v}{\partial \alpha_2} + \frac{1}{H_1 H_2}\frac{\partial H_2}{\partial \alpha_1} u + \frac{w}{R_2}$$

$$\gamma = \frac{H_2}{H_1}\frac{\partial}{\partial \alpha_1}\left(\frac{v}{H_2}\right) + \frac{H_1}{H_2}\frac{\partial}{\partial \alpha_2}\left(\frac{u}{H_1}\right) \tag{5.154}$$

and the middle surface flexural and torsional changes of curvatures

$$\chi_1 = -\frac{1}{H_1}\frac{\partial}{\partial \alpha_1}\left(\frac{1}{H_1}\frac{\partial w}{\partial \alpha_1} - \frac{u}{R_1}\right) - \frac{1}{H_1 H_2}\frac{\partial H_1}{\partial \alpha_2}\left(\frac{1}{H_2}\frac{\partial w}{\partial \alpha_2} - \frac{v}{R_2}\right)$$

$$\chi_2 = -\frac{1}{H_2}\frac{\partial}{\partial \alpha_2}\left(\frac{1}{H_2}\frac{\partial w}{\partial \alpha_2} - \frac{v}{R_2}\right) - \frac{1}{H_1 H_2}\frac{\partial H_2}{\partial \alpha_1}\left(\frac{1}{H_1}\frac{\partial w}{\partial \alpha_1} - \frac{u}{R_1}\right)$$

$$\tau = -\frac{1}{H_1 H_2}\left(\frac{\partial^2 w}{\partial \alpha_1 \partial \alpha_2} - \frac{1}{H_1}\frac{\partial H_1}{\partial \alpha_2}\frac{\partial w}{\partial \alpha_1} - \frac{1}{H_2}\frac{\partial H_2}{\partial \alpha_1}\frac{\partial w}{\partial \alpha_2}\right)$$

$$+ \frac{1}{R_1}\left(\frac{1}{H_2}\frac{\partial u}{\partial \alpha_2} - \frac{1}{H_1 H_2}\frac{\partial H_1}{\partial \alpha_2} u\right) + \frac{1}{R_2}\left(\frac{1}{H_1}\frac{\partial v}{\partial \alpha_1} - \frac{1}{H_1 H_2}\frac{\partial H_2}{\partial \alpha_1} v\right) \tag{5.155}$$

Thus we have, across the thickness of the shell and retaining only linear terms in z

$$e_{11}^{(z)} \approx e_1 + z\left(\chi_1 - \frac{e_1}{R_1}\right);\quad e_{22}^{(z)} \approx e_2 + z\left(\chi_2 - \frac{e_2}{R_2}\right);\quad 2e_{12}^{(z)} \approx \gamma + 2z\left[\tau - \left(\frac{1}{R_1} + \frac{1}{R_2}\right)\frac{\gamma}{2}\right] \tag{5.156}$$

The changes of curvature of the middle surface in the direction α_1, α_2, and the change of twist are therefore

$$\chi_1' = \frac{1}{R_1^*} - \frac{1}{R_1} = \chi_1 - \frac{e_1}{R_1};\quad \chi_2' = \frac{1}{R_2^*} - \frac{1}{R_2} = \chi_2 - \frac{e_2}{R_2};\quad \tau' = \tau - \left(\frac{1}{R_1} + \frac{1}{R_2}\right)\frac{\gamma}{2} \tag{5.157}$$

On the other hand, for simplicity and without relevant error, we can assume, according to many authors,

$$\chi_1' \approx \chi_1 \;;\qquad \chi_2' \approx \chi_2 \;;\qquad \tau' \approx \tau \tag{5.158}$$

and across the thickness of the shell we have

$$e_{11}^{(z)} \approx e_1 + z\chi_1 \;;\quad e_{22}^{(z)} \approx e_2 + z\chi_2 \;;\quad 2e_{12}^{(z)} \approx \gamma + 2z\tau \tag{5.159}$$

Simple expressions can also be given for the components of the local rotation. First of all we have, according to the definition of the above quantities A and B given by (5.148),

$$\omega_1^{(z)} = -B \qquad \omega_2^{(z)} = A \tag{5.148$'$}$$

For the third component $\omega_z^{(z)}$, introducing the quantities

$$\gamma_1 = \frac{1}{H_1}\frac{\partial v}{\partial \alpha_1} - \frac{1}{H_1 H_2}\frac{\partial H_1}{\partial \alpha_2} u\;;\quad \gamma_2 = \frac{1}{H_2}\frac{\partial u}{\partial \alpha_2} - \frac{1}{H_1 H_2}\frac{\partial H_2}{\partial \alpha_1} v \tag{5.160}$$

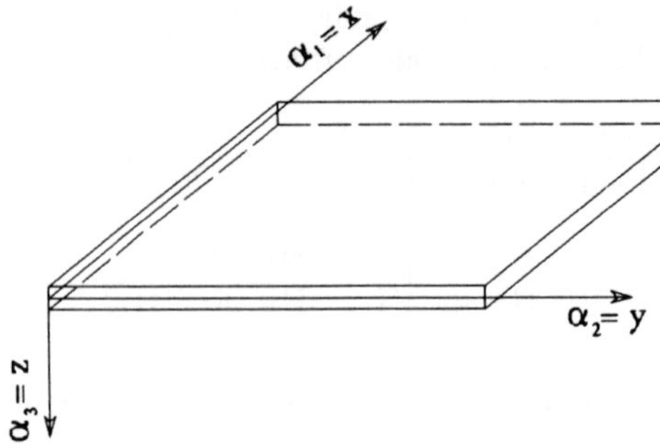

FIGURE 5.19

$$\tau_1 = \frac{1}{H_1}\frac{\partial B}{\partial \alpha_1} - \frac{1}{H_1 H_2}\frac{\partial H_1}{\partial \alpha_2} A\,; \qquad \tau_2 = \frac{1}{H_2}\frac{\partial A}{\partial \alpha_2} - \frac{1}{H_1 H_2}\frac{\partial H_2}{\partial \alpha_1} B \tag{5.161}$$

we have

$$2\omega_z^{(z)} = \gamma_1 - \gamma_2 + z(\tau_1 - \tau_2) + z\left(\frac{\gamma_2}{R_2} - \frac{\gamma_1}{R_1}\right) \tag{5.162}$$

The pure strain components, up to the second order in the displacement functions u, v, and w, can be immediately obtained, according to (5.62). We have in fact, with (5.153)

$$\varepsilon_{11}^{(2)} = e_{12}^{(z)}\omega_z^{(z)} + \frac{1}{2}\left(\omega_z^{(z)2} + \omega_2^{(z)2}\right);\ \varepsilon_{22}^{(2)} = -\, e_{12}^{(z)}\omega_z^{(z)} + \frac{1}{2}\left(\omega_1^{(z)2} + \omega_z^{(z)2}\right);$$

$$\varepsilon_{12}^{(2)} = \frac{1}{2}\,\omega_z^{(z)}\left(e_{22}^{(z)} - e_{11}^{(z)}\right) - \frac{1}{2}\,\omega_1^{(z)}\omega_2^{(z)} \tag{5.163}$$

In many cases we can retain only the most relevant terms in (5.163). Thus we have

$$\varepsilon_{11}^{(2)} = \frac{1}{2}\left(\omega_z^{(z)^2} + \omega_2^{(z)^2}\right); \qquad \varepsilon_{22}^{(2)} = \frac{1}{2}\left(\omega_1^{(z)^2} + \omega_z^{(z)^2}\right); \qquad \varepsilon_{12}^{(2)} = -\frac{1}{2}\omega_1^{(z)}\,\omega_2^{(z)} \tag{5.163'}$$

We have now all the expression that will be used in the analysis of buckling of shells. From the previous results we can obtain also a description of the deformation of plates. In fact, in the particular case when (Fig. 5.19)

$$R_1 \to \infty \qquad R_2 \to \infty \tag{5.164}$$

$$\alpha_1 = x \qquad \alpha_2 = y \qquad H_2 = 1 \qquad H_1 = 1 \tag{5.165}$$

the shell becomes a plate. From (5.154) and (5.163′) we thus have

$$e_{xx}^{(z)} = \frac{\partial u}{\partial x} - z\frac{\partial^2 w}{\partial x^2}; \qquad e_{yy}^{(z)} = \frac{\partial v}{\partial y} - z\frac{\partial^2 w}{\partial y^2}; \qquad 2e_{xy}^{(z)} = \frac{\partial u}{\partial y} + \frac{\partial v}{\partial x} - 2z\frac{\partial^2 w}{\partial x \partial y}$$

$$\varepsilon_{xx}^{(2)} = \frac{1}{2}\left(\frac{\partial w}{\partial x}\right)^2; \qquad \varepsilon_{yy}^{(2)} = \frac{1}{2}\left(\frac{\partial w}{\partial y}\right)^2; \qquad \varepsilon_{xy}^{(2)} = \frac{1}{2}\frac{\partial w}{\partial x}\frac{\partial w}{\partial y} \tag{5.166}$$

The case of shallow shells is not too different from that of the plates. In this case the more relevant displacement component is the component w, orthogonal to the middle surface of the shell. Thus, in defining the changes of curvature of the middle surface we can retain only terms depending on w and, in the framework of Donnell theory,[21-23] in place of (5.155), we have

$$\chi_1 = -\frac{1}{H_1}\frac{\partial}{\partial\alpha_1}\left(\frac{1}{H_1}\frac{\partial w}{\partial\alpha_1}\right) - \frac{1}{H_1H_2^2}\frac{\partial H_1}{\partial\alpha_2}\frac{\partial w}{\partial\alpha_2}$$

$$\chi_2 = -\frac{1}{H_2}\frac{\partial}{\partial\alpha_2}\left(\frac{1}{H_2}\frac{\partial w}{\partial\alpha_2}\right) - \frac{1}{H_1^2H_2}\frac{\partial H_2}{\partial\alpha_1}\frac{\partial w}{\partial\alpha_1}$$

$$\tau = -\frac{1}{H_1H_2}\left(\frac{\partial^2 w}{\partial\alpha_1\partial\alpha_1} - \frac{1}{H_1}\frac{\partial H_1}{\partial\alpha_2}\frac{\partial w}{\partial\alpha_1} - \frac{1}{H_2}\frac{\partial H_2}{\partial\alpha_1}\frac{\partial w}{\partial\alpha_2}\right) \tag{5.167}$$

Also, the expressions of the second-order pure strains can be conveniently simplified. To evaluate the component $\omega_z^{(z)}$, fundamental in the expressions of the second-order pure strains, we can write

$$A \approx -\frac{1}{H_1}\frac{\partial w}{\partial\alpha_1} \qquad B \approx -\frac{1}{H_2}\frac{\partial w}{\partial\alpha_2} \tag{5.168}$$

and consequently

$$\omega_1^{(z)} \approx -\frac{1}{H_2}\frac{\partial w}{\partial\alpha_2} \qquad \omega_2^{(z)} \approx -\frac{1}{H_1}\frac{\partial w}{\partial\alpha_1} \tag{5.169}$$

On the other hand, consistently with the assumption of shallow shell, we have also

$$2\omega_z^{(z)} \approx z\,(\tau_1 - \tau_2) \tag{5.170}$$

In the same context, thus we have

$$\tau_1 \approx -\frac{1}{H_1}\frac{\partial}{\partial\alpha_1}\left(\frac{1}{H_2}\frac{\partial w}{\partial\alpha_2}\right) + \frac{1}{H_1^2H_2}\frac{\partial H_1}{\partial\alpha_2}\frac{\partial w}{\partial\alpha_1}$$

$$\tau_2 \approx -\frac{1}{H_2}\frac{\partial}{\partial\alpha_2}\left(\frac{1}{H_1}\frac{\partial w}{\partial\alpha_1}\right) + \frac{1}{H_1H_2^2}\frac{\partial H_2}{\partial\alpha_1}\frac{\partial w}{\partial\alpha_2} \tag{5.171}$$

and

$$\tau_1 - \tau_2 \approx 0 \tag{5.172}$$

Consequently

$$\omega_z^{(z)} \approx 0 \tag{5.173}$$

Thus, the destabilizing effect of the local rotation component $\omega_z^{(z)}$ is negligible with respect to the effect due to $\omega_1^{(z)}$ and $\omega_2^{(z)}$. The rotation $\omega_z^{(z)}$ substantially comes out, in fact, from the stretching of the middle surface of the shell. We now analyze some particular cases.

5.9.5 SHALLOW CYLINDRICAL SHELLS

With reference to Fig. 5.20 we have

$$\alpha_1 = x \qquad \alpha_2 = s \tag{5.174}$$

Consequently, taking into account that dx = dx and ds = ds we get

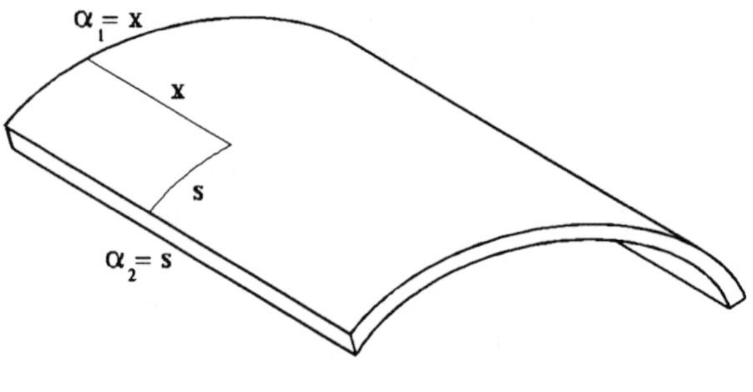

FIGURE 5.20

$$H_1 = H_x = 1 \quad H_2 = H_s = 1 \quad H_x^{(z)} = 1 \quad H_s^{(z)} = 1 + \frac{z}{R_2} \tag{5.175}$$

If the directrix of the shell is an arc of circle we have also R_2 = constant and we get

$$e_x = \frac{\partial u}{\partial x}; \quad e_s = \frac{\partial v}{\partial s} + \frac{w}{R}; \quad \gamma = \frac{\partial v}{\partial x} + \frac{\partial u}{\partial s}; \quad \chi_x = -\frac{\partial^2 w}{\partial x^2}; \quad \chi_s = -\frac{\partial^2 w}{\partial s^2}; \quad \tau = -\frac{\partial^2 w}{\partial w \partial s} \tag{5.176}$$

and

$$e_x^{(z)} = \frac{\partial u}{\partial x} - z\frac{\partial^2 w}{\partial x^2}; \quad e_s^{(z)} = \left(\frac{\partial v}{\partial s} + \frac{w}{R}\right) - z\frac{\partial^2 w}{\partial s^2}; \quad 2e_{xs}^{(z)} = \left(\frac{\partial v}{\partial x} + \frac{\partial u}{\partial s}\right) - 2z\frac{\partial^2 w}{\partial x \partial s} \tag{5.177}$$

At the same time we have

$$\omega_x = \frac{\partial w}{\partial s} - \frac{v}{R}; \quad \omega_s = -\frac{\partial w}{\partial x} \quad \omega_z = \frac{1}{2}\left(\frac{\partial v}{\partial x} - \frac{\partial u}{\partial s}\right) \tag{5.178}$$

The second-order pure strain components thus are[14]

$$\varepsilon_{xx}^{(2)} = \frac{1}{2}\left(\frac{\partial w}{\partial x}\right)^2; \quad \varepsilon_{ss}^{(2)} = \frac{1}{2}\left(\frac{\partial w}{\partial s}\right)^2; \quad \varepsilon_{xs}^{(2)} = \frac{1}{2}\frac{\partial w}{\partial x}\frac{\partial w}{\partial s} \tag{5.179}$$

and turn out to be very similar to the analogous expressions for the thin plates.

5.9.6 THE CASE OF NONSHALLOW CYLINDRICAL SHELLS

It is useful to evaluate the expressions of curvatures and of second-order pure strain components in the case of nonshallow cylindrical shells. According to the general expressions (5.152) and (5.155) we have

$$\omega_x = \frac{\partial w}{\partial s} - \frac{v}{R}; \quad \omega_s = -\frac{\partial w}{\partial x}; \quad \omega_z \approx \frac{1}{2}\left(\frac{\partial v}{\partial x} - \frac{\partial u}{\partial s}\right);$$

$$\chi_x = -\frac{\partial^2 w}{\partial x^2}; \quad \chi_s = -\frac{\partial^2 w}{\partial s^2} + \frac{1}{R}\frac{\partial v}{\partial s}; \quad \tau = -\frac{\partial^2 w}{\partial x \partial s} + \frac{1}{R}\frac{\partial v}{\partial x}; \tag{5.180}$$

and[14]

$$\varepsilon_{xx}^{(2)} = \frac{1}{2}\left(\frac{\partial w}{\partial x}\right)^2 + \frac{1}{8}\left(\frac{\partial v}{\partial x} - \frac{\partial u}{\partial s}\right)^2; \quad \varepsilon_{ss}^{(2)} = \frac{1}{2}\left(\frac{\partial w}{\partial s} - \frac{v}{R}\right)^2 + \frac{1}{8}\left(\frac{\partial v}{\partial x} - \frac{\partial u}{\partial s}\right)^2$$

$$\varepsilon_{xs}^{(2)} = \frac{1}{2}\frac{\partial w}{\partial x}\frac{\partial w}{\partial s} - \frac{1}{2R}v\frac{\partial w}{\partial x} \tag{5.181}$$

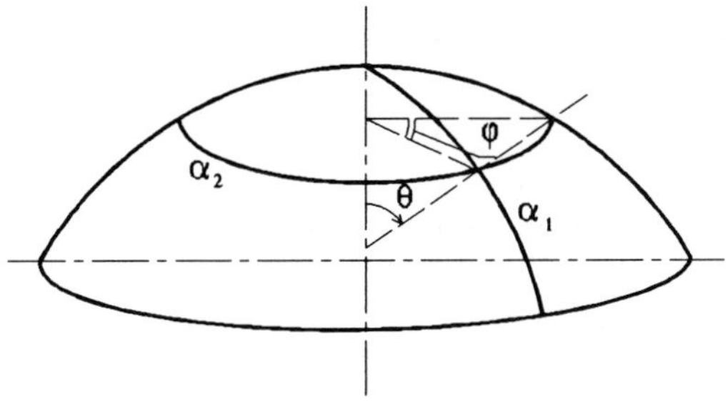

FIGURE 5.21

It is useful to remark the difference between the expression of $\varepsilon_{ss}^{(2)}$ for the shallow and for the nonshallow cylindrical shell. The first term of the second-order circumferential pure strain $\varepsilon_{ss}^{(2)}$ for the non-shallow cylindrical shell is

$$\frac{1}{2}\left(\frac{\partial w}{\partial s} - \frac{v}{R}\right)^2 \tag{5.182}$$

while the corresponding term for the shallow shell has the more simplified expression

$$\frac{1}{2}\left(\frac{\partial w}{\partial s}\right)^2 \tag{5.183}$$

The term v/R contained in (5.182) has relevant influence in the case of buckling of long cylinders under external pressure. In this case, in fact, buckling occurs only with two circumferential waves and it is not possible to use the shallow shell approximation. This argument will be examined later.

5.9.7 THE SHALLOW SPERICAL CUP

In the case of shells of revolution the principal curvature lines are their meridians and parallel circles. Accordingly, one may take as the principal curvilinear coordinates of the middle surface the angle θ, between the normal to the middle surface and the axis of the shell, and the angle φ, determining the position of a point on the corresponding parallel circle (Fig. 5.21).

Let R_1 be the radius of curvature of the meridian. The second radius of curvature will be the length of the intercept of the normal to the middle surface between this surface and the axis of the shell. In fact, considering two adjacent points on the same parallel circle, the normals from these points intersect on the axis of the shell. Thus we have

$$\alpha_1 = \theta \qquad \alpha_2 = \varphi \tag{5.184}$$

The elements of arc of meridian and of parallel circle are respectively

$$ds_1 = R_1\, d\theta \quad ds_2 = R_2 \sin\theta\, d\varphi \tag{5.185}$$

Correspondingly, the Lamé parameters of meridians and parallel circles are

$$H_1 = R_1 \quad H_2 = R_2 \sin\theta \tag{5.186}$$

The first two Codazzi-Gauss relations (5.125) are satisfied identically, while the third gives

$$\frac{\partial(R_2 \sin\theta)}{\partial\theta} = R_1 \cos\theta \tag{5.187}$$

Thus the Lamé parameters of the shells of revolution are

$$H_\theta^{(z)} = R_1\left(1+\frac{z}{R_1}\right); \quad H_\varphi^{(z)} = R_2 \sin\theta\left(1+\frac{z}{R_2}\right) \tag{5.188}$$

Assuming shallowness of the spherical shell, we can thus write

$$H_1 = R; \quad H_2 = R\sin\theta \approx R\,\theta \tag{5.189}$$

if R is the radius of the spherical cup. The first-order strain components are therefore

$$e_\theta = \frac{1}{R}\frac{\partial u}{\partial\theta}+\frac{w}{R}; \quad e_\varphi = \frac{1}{R\theta}\frac{\partial v}{\partial\varphi}+\frac{u}{R\theta}+\frac{w}{R}; \quad \gamma = \frac{1}{R}\frac{\partial v}{\partial\theta}-\frac{v}{R\theta}+\frac{1}{R\theta}\frac{\partial u}{\partial\varphi} \tag{5.190}$$

and the curvature changes are

$$\chi_\varphi = -\frac{1}{R^2\theta^2}\frac{\partial^2 w}{\partial\varphi^2}-\frac{1}{R^2\theta}\frac{\partial w}{\partial\theta}; \quad \chi_\theta = -\frac{1}{R^2}\frac{\partial^2 w}{\partial\theta^2}; \quad \tau = -\frac{1}{R^2\theta}\frac{\partial^2 w}{\partial\theta\partial\varphi}+\frac{1}{R^2\theta^2}\frac{\partial w}{\partial\varphi} \tag{5.191}$$

The second-order pure strain components then are

$$\varepsilon_{\theta\theta}^{(2)} = \frac{1}{2R^2}\left(\frac{\partial w}{\partial\theta}\right)^2; \quad \varepsilon_{\varphi\varphi}^{(2)} = \frac{1}{2R^2\theta^2}\left(\frac{\partial w}{\partial\varphi}\right)^2; \quad \varepsilon_{\theta\varphi}^{(2)} = \frac{1}{2R^2\theta}\frac{\partial w}{\partial\varphi}\frac{\partial w}{\partial\theta} \tag{5.192}$$

On the other hand, taking into account the shallowness of the cup, we can write

$$R\theta = s \qquad R\,d\theta = ds \tag{5.193}$$

$$e_\theta = \frac{\partial u}{\partial s}+\frac{w}{R}; \quad e_\varphi = \frac{1}{s}\frac{\partial v}{\partial\varphi}+\frac{u}{s}+\frac{w}{R}; \quad \gamma = \frac{\partial v}{\partial s}-\frac{v}{s}+\frac{1}{s}\frac{\partial u}{\partial\varphi} \tag{5.194}$$

$$\chi_\theta = -\frac{\partial^2 w}{\partial s^2}; \quad \chi_\varphi = -\frac{1}{s^2}\frac{\partial^2 w}{\partial\varphi^2}-\frac{1}{s}\frac{\partial w}{\partial s}; \quad \tau = -\frac{1}{s}\frac{\partial^2 w}{\partial s\partial\varphi}+\frac{1}{s^2}\frac{\partial w}{\partial\varphi} \tag{5.195}$$

$$\varepsilon_{\theta\theta}^{(2)} = \frac{1}{2}\left(\frac{\partial w}{\partial s}\right)^2; \quad \varepsilon_{\varphi\varphi}^{(2)} = \frac{1}{2s^2}\left(\frac{\partial w}{\partial\varphi}\right)^2; \quad \varepsilon_{\theta\varphi}^{(2)} = \frac{1}{2s}\frac{\partial w}{\partial\varphi}\frac{\partial w}{\partial s} \tag{5.196}$$

For the axialsymmetric case, i.e., when (Fig. 5.22)

$$u = u(s) \qquad w = w(s) \qquad v = 0 \tag{5.197}$$

in particular we have

$$e_\theta = \frac{\partial u}{\partial s}+\frac{w}{R}; \quad e_\varphi = \frac{u}{s}+\frac{w}{R}; \quad \gamma = 0 \qquad \chi_\theta = -\frac{\partial^2 w}{\partial s^2}; \quad \chi_\varphi = -\frac{1}{s}\frac{\partial w}{\partial s}; \quad \tau = 0$$

$$\varepsilon_{\theta\theta}^{(2)} = \frac{1}{2}\left(\frac{\partial w}{\partial s}\right)^2; \quad \varepsilon_{\varphi\varphi}^{(2)} = \varepsilon_{\theta\varphi}^{(2)} = 0 \tag{5.198}$$

5.9.8 THE VARIATIONAL EQUATION OF BUCKLING OF THIN SHELLS

According to (5.27) the variational condition of the neutral equilibrium is

$$\delta(\mathcal{V}_2 + \mathcal{L}_2^* + \mathcal{U}_2) = 0 \tag{5.199}$$

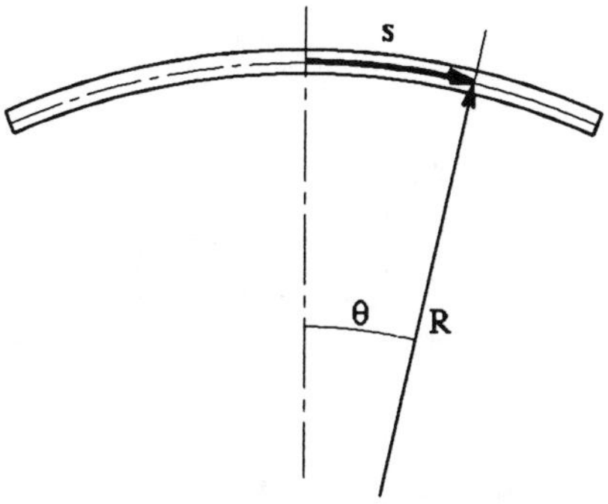

FIGURE 5.22

We analyze the various terms of this condition. First of all the strain energy of the shell is

$$\mathcal{V}_2 = \int_V \left[G\left(e_{11}^{(z)2} + e_{22}^{(z)2} + e_{33}^{(z)2}\right) + \frac{\lambda}{2}\left(e_{11}^{(z)} + e_{22}^{(z)} + e_{33}^{(z)}\right)^2 + 2G\, e_{12}^{(z)2} \right] dV \qquad (5.200)$$

where G is the modulus of tangential rigidity and λ the first Lamé modulus. In this expression the component $e_{33}^{(z)}$ takes origin from the Poisson effect. By assuming $\sigma_{33} = 0$ we get

$$e_{33}^{(z)} = -\frac{\nu}{1-\nu}\left(e_{11}^{(z)} + e_{22}^{(z)}\right) \qquad (5.201)$$

where ν is the Poisson coefficient. The second-order internal work of the membrane stresses S_{11}, S_{22}, and S_{12} acting at the equilibrium configuration C_0 of the shell (Fig. 5.23) is

$$\overset{*}{\mathcal{L}}_2 = \int_V (S_{11}\, \varepsilon_{11}^{(2)} + S_{22}\, \varepsilon_{22}^{(2)} + 2S_{12}\, \varepsilon_{12}^{(2)}) dV \qquad (5.202)$$

where $\varepsilon_{11}^{(2)}$, $\varepsilon_{22}^{(2)}$, $\varepsilon_{12}^{(2)}$ are the second-order pure strain components of the buckling deformation, whose expressions have been previously obtained. Finally

$$\mathcal{U}_2 \qquad (5.203)$$

is the second-order increment of potential energy of the external loads. If these loads are not dependent on the deformation of the shell and the buckling displacement components are all of the same order of magnitude, we will have $\mathcal{U}_2 = 0$. Very important, on the contrary, is the

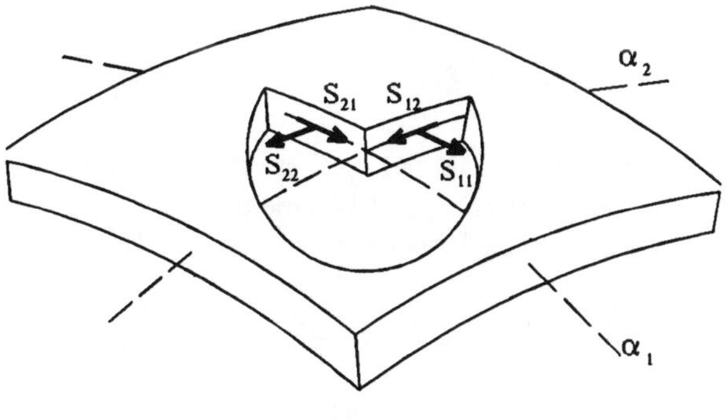

FIGURE 5.23

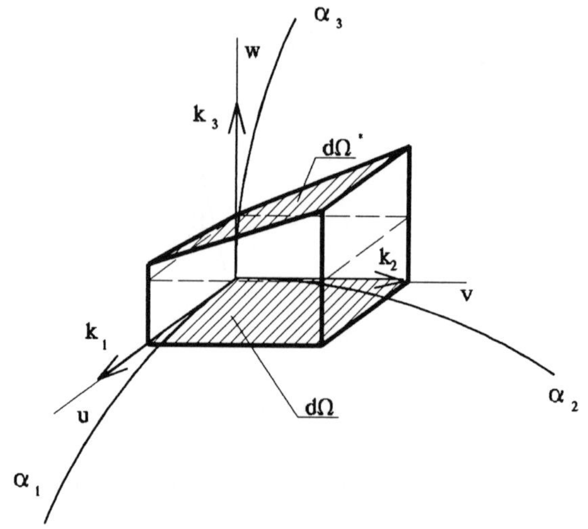

FIGURE 5.24

case in which the shell is dipped in a fluid and the forces acting on the shell are due to an hydrostatic pressure p. In this case the second-order term $\mathcal{U}_2$ is very significant and is given by[25]

$$\mathcal{U}_2 = -\int_{\Omega_0} p\, d_2 V \tag{5.204}$$

where d_2V is the second-order volume change of the closed dipped shell in the passage from a boundary Ω_0 to a boundary Ω. In expression (5.204) the pressure p is taken positive if acting along the outward normal and d_2V is positive if it represents an expansion of the internal volume of the shell. Let us consider now a surface element $d\Omega = ds_1\, ds_2$ where ds_1 and ds_2 are taken along the principal coordinates lines α_1 and α_2. Thus d_2V represents the second order variation of the elementary volume traced by the surface element when moving from the initial position $d\Omega$ at C_0 to the buckled configuration $d\Omega^*$. Before the buckling, the element of surface $d\Omega$ is given by

$$d\Omega = H_1 H_2 d\alpha_1\, d\alpha_2 \tag{5.205}$$

and after the buckling deformation is

$$d\Omega^* = H_1^* H_2^* \sin\left(\mathbf{k}_1^*, \mathbf{k}_2^*\right) d\alpha_1 d\alpha_2 \tag{5.206}$$

where E_{11}, and E_{22} are the strain components along the lines α_1 and α_2. The volume spanned by $d\Omega$ during its motion to the buckled configuration $d\Omega^*$, up to the second order terms is given by

$$dV = d\Omega\, w + \frac{d\Omega^* - d\Omega}{2} w + \frac{d_v\Omega^* v}{2} + \frac{d_u\Omega^* u}{2} \tag{5.207}$$

where with reference to Fig. 5.24, $d_v\Omega^*$ and $d_u\Omega^*$ are the projections of $d\Omega^*$ on the planes normal to v and u. To evaluate the change of volume dV, up to second-order terms in u, v, and w, it suffices to consider only the first-order terms of $d\Omega^*$. Thus we get, by using (5.154)

$$d\Omega^* = H_1 H_2 d\alpha_1 d\alpha_2 \left[1 + \frac{1}{H_1}\frac{\partial u}{\partial \alpha_1} + \frac{1}{H_1 H_2}\frac{\partial H_1}{\partial \alpha_2} v + \frac{w}{R_1}\right.$$

$$\left. + \frac{1}{H_2}\frac{\partial v}{\partial \alpha_2} + \frac{1}{H_1 H_2}\frac{\partial H_2}{\partial \alpha_1} u + \frac{w}{R_2}\right] \tag{5.208}$$

To evaluate the projections of dΩ on the normal planes, i.e., on the planes $\mathbf{k}_2$, $\mathbf{k}_3$ and $\mathbf{k}_1$, $\mathbf{k}_3$ we have

$$d_u\Omega^* = AH_1H_2d\alpha_1d\alpha_2 = \left(-\frac{1}{H_1}\frac{\partial w}{\partial \alpha_1} + \frac{u}{R_1}\right)H_1H_2d\alpha_1d\alpha_2;$$

$$d_v\Omega^* = BH_1H_2d\alpha_1d\alpha_2 = \left(-\frac{1}{H_2}\frac{\partial w}{\partial \alpha_2} + \frac{v}{R_2}\right)H_1H_2d\alpha_1d\alpha_2; \qquad (5.209)$$

taking into account the significance of functions A and B. Hence we have[19]

$$d_1V = H_1H_2d\alpha_1d\alpha_2\, w \qquad (5.210)$$

$$d_2V = H_1H_2\frac{d\alpha_1d\alpha_2}{2}\left[w^2\left(\frac{1}{R_1} + \frac{1}{R_2}\right) + \frac{u^2}{R_1} + \frac{v^2}{R_2} + \frac{w}{H_1}\frac{\partial u}{\partial \alpha_1}\right.$$

$$\left. + \frac{w}{H_2}\frac{\partial v}{\partial \alpha_2} + \frac{vw}{H_1H_2}\frac{\partial H_1}{\partial \alpha_2} - \frac{u}{H_1}\frac{\partial w}{\partial \alpha_1} - \frac{v}{H_2}\frac{\partial w}{\partial \alpha_2}\right] \qquad (5.211)$$

In case of cylindrical coordinates we thus have

$$d_2V = \frac{1}{2}\left[\frac{w^2}{R} + \frac{v^2}{R} + w\frac{\partial u}{\partial x} + w\frac{\partial v}{\partial s} - u\frac{\partial w}{\partial x} - v\frac{\partial w}{\partial s}\right]dxds \qquad (5.212)$$

and in the case of spherical coordinates

$$d_2V = \frac{1}{2}\left[2w^2 + u^2 + v^2 + w\frac{\partial u}{\partial \theta} + \frac{w}{\sin\theta}\frac{\partial v}{\partial \varphi}\right.$$

$$\left. + \frac{uw}{\mathrm{tg}\theta} - u\frac{\partial w}{\partial \theta} - \frac{v}{\sin\theta}\frac{\partial w}{\partial \varphi}\right]R\sin\theta\, d\theta d\varphi \qquad (5.213)$$

A complete analysis of the volume changes of a deforming surface has been given by Ferrarese.[19]

5.9.9 BUCKLING OF CYLINDRICAL SHELLS UNDER EXTERNAL PRESSURE

5.9.9.1 Cylinders of Indefinite Length

Only a uniform circumferential stress S_{ss} is acting at the fundamental state C_0. In this case the buckling mode is not dependent on the abscissa x (Fig. 5.25).

Buckling of very long cylindrical shells is characterized only by circumferential waves. Thus let us evaluate the stretching and the curvature changes of the middle surface of the shell from (5.180), relative to the non-shallow shell equations. We have

$$e_x = 0; \quad e_{ss} = \frac{dv}{ds} + \frac{w}{R}; \quad \gamma = 0 \qquad \chi_x = 0; \quad \chi_s = -\frac{d^2w}{ds^2} + \frac{1}{R}\frac{dv}{ds}; \quad \tau = 0 \qquad (5.214)$$

The second-order pure strain component is

$$\varepsilon_{ss}^{(2)} = \frac{1}{2}\left(\frac{dw}{ds} - \frac{v}{R}\right)^2 \qquad (5.215)$$

According to (5.200) we obtain the following expression of the strain energy of the shell

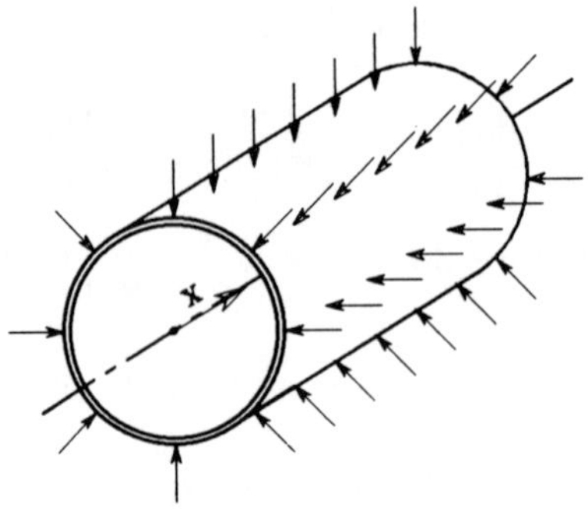

FIGURE 5.25

$$\mathcal{V}_2 = \frac{D_e}{2}\int_0^{2\pi R}\left(\frac{dv}{ds} - \frac{w}{R}\right)^2 ds + \frac{D_f}{2}\int_0^{2\pi R}\left(-\frac{d^2w}{ds^2} + \frac{1}{R}\frac{dv}{ds}\right)^2 ds \tag{5.216}$$

where D_e and D_f are the extensional and flexural stiffnesses of the cylindrical wall

$$D_e = \frac{Et}{(1-\nu^2)} \qquad D_f = \frac{Et^3}{12(1-\nu^2)} \tag{5.217}$$

When buckling occurs, the principal circumferential stresses S_{ss} do the following work on the second-order pure strain component (5.215)

$$\mathcal{L}_2^* = \int_V S_{ss}\, \varepsilon_{ss}^{(2)}\, dV \tag{5.218}$$

Then we have

$$\mathcal{L}_2^* = -\frac{N_{ss}}{2}\int_0^{2\pi R}\left(\frac{dw}{ds} - \frac{v}{R}\right)^2 ds \tag{5.219}$$

In order to evaluate the second-order variation of the potential energy of the external pressure, from (5.212) we get

$$d_2V = \frac{1}{2}\left[\frac{w^2}{R} + \frac{v^2}{R} + w\frac{dv}{ds} - v\frac{dw}{ds}\right]ds \tag{5.220}$$

With the inextensional buckling mode (Fig. 5.26)

$$v(s) = \frac{C}{2}\cos\frac{2ns}{R} \qquad w(s) = C\sin\frac{2ns}{R} \qquad (n = 1, 2, \ldots) \tag{5.221}$$

and we have

$$\mathcal{V}_2 = C^2\frac{\pi}{R^3}D_f\, 2n^2\left(2n - \frac{1}{2}\right)^2 \quad \mathcal{L}_2^* = -C^2\frac{\pi}{2R}N_{ss}\left(2n - \frac{1}{2}\right)^2 \quad \mathcal{U}_2 = -p\, C^2\frac{\pi}{4}\left(2n - \frac{1}{2}\right) \tag{5.222}$$

The smallest value of the critical pressure is obtained taking n = 1. Correspondingly we obtain

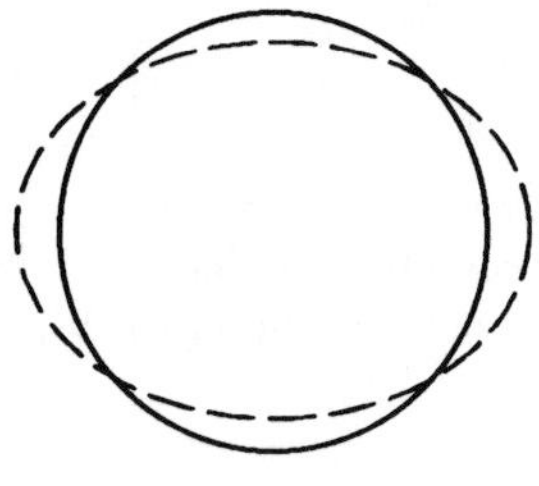

FIGURE 5.26

$$\mathcal{V}_2 = \frac{9}{2} C^2 \frac{\pi}{R^3} D_f \qquad \mathrm{L}\mathcal{L}_2^* = -\frac{9}{8} C^2 \pi p \qquad \iota\, \mathcal{U}_2 = -\frac{3}{8} C^2 \pi p \tag{5.223}$$

and by using the variational condition (5.199) we get the value of the critical pressure

$$p_{cr} = \frac{3D_f}{R^3} \tag{5.224}$$

first obtained by Boussinesq.[20] We remark that, without considering the destabilizing effect of the work of the external pressure on the second-order volume change of the ovalizing cylinder, in place of (5.224) we have

$$p_{cr} = \frac{4D_f}{R^3} \tag{5.225}$$

with an error of 33%. When the number n of half-wavelengths increases, the influence of the work d_2U becomes gradually less relevant. In the next section, where we will formulate the Donnell equation, we will comment further on this question.

5.9.9.2 Buckling of Complete Cylinders of Finite Length Under External Pressure and Axial/Torsional Loadings Applied at End Sections; the Donnell Equation

For complete cylinders of finite length (Fig. 5.27) buckling of the cylindrical wall occurs with several circumferential waves. In this case the shallow shell equations are particularly useful. According to (5.200) the strain energy $\mathcal{V}$ is given by

$$\mathcal{V}_2 = \mathcal{V}_{2e} + \mathcal{V}_{2f} \tag{5.226}$$

where

$$\mathcal{V}_{2e} = \frac{D_e}{2} \int_\Omega \left[\left(\frac{\partial u}{\partial x}\right)^2 + \left(\frac{\partial v}{\partial s}\right)^2 + 2\nu \frac{\partial u}{\partial x}\frac{\partial v}{\partial s} + \frac{1-\nu}{2}\left(\frac{\partial u}{\partial s}\right)^2 + \frac{1-\nu}{2}\left(\frac{\partial v}{\partial x}\right)^2 \right.$$

$$\left. + (1-\nu)\frac{\partial u}{\partial s}\frac{\partial v}{\partial x} + \frac{w^2}{R^2} + 2\frac{w}{R}\frac{\partial v}{\partial s} + 2\nu \frac{w}{R}\frac{\partial u}{\partial x} \right] d\Omega \tag{5.227}$$

is the stretching energy. The flexural energy $\mathcal{V}_2$ on the other hand is

$$\mathcal{V}_{2f} = \frac{D_f}{2} \int_\Omega \left\{ \left(\frac{\partial^2 w}{\partial x^2} + \frac{\partial^2 w}{\partial y^2} \right)^2 - 2(1-\nu)\left[\frac{\partial^2 w}{\partial x^2}\frac{\partial^2 w}{\partial y^2} - \left(\frac{\partial^2 w}{\partial x \partial y}\right)^2 \right] \right\} d\Omega \tag{5.228}$$

Unlike the buckling of plates, here we have coupling between the normal displacement w and inplane components u and v. The second-order work of the internal stresses in fact is

$$\mathcal{L}_2^* = \int_V \left[-\frac{N_{ss}}{2}\left(\frac{\partial w}{\partial s}\right)^2 - \frac{N_{xx}}{2}\left(\frac{\partial w}{\partial x}\right)^2 - 2N_{xs}\frac{\partial w}{\partial x}\frac{\partial w}{\partial s}\right] dV \tag{5.229}$$

In this case we can neglect the second-order change of the potential energy of the external pressure. Then, by using the variational condition (5.199), we obtain the following system of partial differential equations

$$-D_e\left[\frac{\partial^2 u}{\partial x^2} + \frac{1-\nu}{2}\frac{\partial^2 u}{\partial s^2} + \frac{1+\nu}{2}\frac{\partial^2 v}{\partial x \partial s} + \frac{\nu}{R}\frac{\partial w}{\partial x}\right] = 0 \tag{5.230}$$

$$-D_e\left[\frac{\partial^2 v}{\partial s^2} + \frac{1-\nu}{2}\frac{\partial^2 v}{\partial x^2} + \frac{1+\nu}{2}\frac{\partial^2 u}{\partial x \partial s} + \frac{1}{R}\frac{\partial w}{\partial s}\right] = 0 \tag{5.231}$$

$$D_f \Delta\Delta w + \frac{D_e}{R}\left(\frac{w}{R} + \frac{\partial v}{\partial s} + \nu\frac{\partial u}{\partial x}\right) + N_{xx}\frac{\partial^2 w}{\partial x^2} + N_{ss}\frac{\partial^2 w}{\partial s^2} + 2N_{xs}\frac{\partial^2 w}{\partial x \partial s} = 0 \tag{5.232}$$

governing the problem of the buckling of cylindrical shells under external pressure. The three equations are coupled because contain derivatives of the three displacements u, v, and w. The corresponding boundary conditions are

$$\left(\frac{\partial u}{\partial x} + \nu\frac{\partial v}{\partial s} + \frac{\nu}{2}\frac{w}{R}\right)\delta u = 0 \qquad \left(\frac{\partial v}{\partial x} + \frac{\partial u}{\partial s}\right)\delta v = 0 \tag{5.233}$$

$$D_f\left(\frac{\partial^2 w}{\partial x^2} + \nu\frac{\partial^2 w}{\partial s^2}\right)\frac{\partial \delta w}{\partial x} = 0 \qquad \left\{D_f\left[\frac{\partial^3 w}{\partial w^3} + (2-\nu)\frac{\partial^3 w}{\partial x \partial s^2}\right] - N_x\frac{\partial w}{\partial x} - N_{xy}\frac{\partial w}{\partial s}\right\}\delta w = 0 \tag{5.233'}$$

The unknown displacement functions, cyclic with the variable s, have to satisfy conditions (5.233) and (5.233′) at the end sections x = 0 and x = L. The last two conditions are those typical of the plates. The first two conditions, on the contrary, typical of the curved shells, require that the stresses N_{xx} and N_{ss} be zero at the end sections of the complete cylinder.

The system of three equations (5.230), (5.231), and (5.232) can be uncoupled and placed in terms of the displacement w only by using a scheme introduced by Donnell.[21] In this form, solutions for various buckling problems can be readily effected. Consider the three operators

$$\text{(a)}\ \frac{\partial^2}{\partial x^2} \qquad \text{(b)}\ \frac{\partial^2}{\partial s^2} \qquad \text{(c)}\ \frac{\partial^2}{\partial x \partial s}$$

Upon operating with (a) on (5.231), (b) on (5.231), and (c) on (5.230), we obtain

$$\frac{\partial^4 u}{\partial x^3 \partial s} = -\frac{2}{1+\nu}\left[\frac{\partial^4 v}{\partial x^2 \partial s^2} + \frac{1-\nu}{2}\frac{\partial^4 v}{\partial x^4} + \frac{1}{R}\frac{\partial^3 w}{\partial x^2 \partial s}\right] \tag{5.234}$$

$$\frac{\partial^4 u}{\partial x \partial s^3} = -\frac{2}{1+\nu}\left[\frac{\partial^4 v}{\partial s^4} + \frac{1-\nu}{2}\frac{\partial^4 v}{\partial s^2 \partial x^2} + \frac{1}{R}\frac{\partial^3 w}{\partial s^3}\right] \tag{5.235}$$

$$\frac{\partial^4 u}{\partial s \partial x^3} + \frac{1-\nu}{2}\frac{\partial^4 u}{\partial x \partial s^3} + \frac{1+\nu}{2}\frac{\partial^4 v}{\partial x^2 \partial s^2} + \frac{\nu}{R}\frac{\partial^3 w}{\partial x^2 \partial s} = 0 \tag{5.236}$$

Substituting (5.234) and (5.235) in (5.236) and multiplying each term by -[(1 + ν)/(1 - ν)] gives

$$- R\,\Delta\Delta v = (2 + \nu)\frac{\partial^3 w}{\partial s \partial x^2} + \frac{\partial^3 w}{\partial s^3} \tag{5.237}$$

We can proceed in the same manner by operating on (5.230) and then using (5.229) to solve for the v derivatives. In this manner we obtain

$$- R\,\Delta\Delta u = \nu\frac{\partial^3 w}{\partial x^3} - \frac{\partial^3 w}{\partial x \partial s^2} \tag{5.238}$$

Turning now to (5.232), we apply the operator ΔΔ with the result

$$D_f\,\Delta\Delta\Delta\Delta w + \Delta\Delta\left(N_{xx}\frac{\partial^2 w}{\partial x^2} + 2N_{xs}\frac{\partial^2 w}{\partial x \partial s} + N_{ss}\frac{\partial^2 w}{\partial s^2}\right) + \frac{D_e}{R}\Delta\Delta\left(\frac{w}{R} + \frac{\partial v}{\partial s} + \nu\frac{\partial u}{\partial x}\right) = 0 \tag{5.239}$$

To eliminate the u derivative in the last term of (5.239) we utilize (5.238) in the form

$$\nu\Delta\Delta\,\frac{\partial u}{\partial x} = \frac{1}{R}\left(-\nu^2\frac{\partial^4 w}{\partial x^4} + \nu\frac{\partial^4 w}{\partial x^2 \partial s^2}\right) \tag{5.240}$$

Similarly, for the v derivative term, we utilize eq.(5.237)

$$\Delta\Delta\frac{\partial v}{\partial s} = \frac{1}{R}\left[- (2 + \nu)\,\frac{\partial^4 w}{\partial s^2 \partial x^2} - \frac{\partial^4 w}{\partial s^4}\right] \tag{5.241}$$

Finally, taking into account that

$$\frac{\Delta\Delta w}{R} = \frac{1}{R}\left[\frac{\partial^4 w}{\partial x^4} + 2\,\frac{\partial^4 w}{\partial s^2 \partial x^2} + \frac{\partial^4 w}{\partial s^4}\right] \tag{5.242}$$

collecting all the coefficients of the w derivatives in (5.240), (5.241), and (5.242), and using (5.239), we arrive at the Donnell equation for buckling of cylindrical shells in terms of the lateral displacement w only

$$D_f\,\Delta\Delta\Delta\Delta w + \frac{Et}{R^2}\frac{\partial^4 w}{\partial x^4} + \Delta\Delta\left(N_{xx}\frac{\partial^2 w}{\partial x^2} + 2N_{xs}\frac{\partial^2 w}{\partial x \partial s} + N_{ss}\frac{\partial^2 w}{\partial s^2}\right) = 0 \tag{5.243}$$

The Donnell equation has been obtained in the "shallow shell" approximation. Among the various simplifications assumed, particularly in the second-order work of the external pressure, certain terms that have relevant influence only in the presence of several buckling waves have been neglected. Thus the Donnell equation remains generally valid only for buckling problems in which there are several buckle wavelengths in the circumferential direction.

We now analyze, in detail, the buckling of the cylinder under external pressure normal to the cylindrical surface (Fig. 5.27). In this case the Donnell equation becomes

$$D_f\,\Delta\Delta\Delta\Delta w + \frac{Et}{R^2}\frac{\partial^4 w}{\partial x^4} + \Delta\Delta\left(N_{ss}\frac{\partial^2 w}{\partial s^2}\right) = 0 \tag{5.244}$$

A nontrivial solution of this equation that satisfies the boundary conditions of simple support at the ends of the cylinder of length L and radius R is

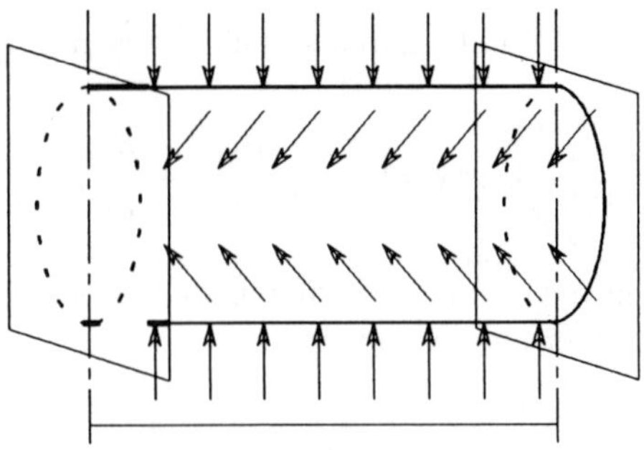

FIGURE 5.27

$$w(x,s) = w_{mn} \sin \frac{m\pi x}{L} \sin \frac{ns}{R} \tag{5.245}$$

In this expression m is the number of half-wavelengths axially, and n is the number of half-wavelengths circumferentially. To (5.245) are associated the functions

$$v(x,s) = V_{mn} \sin \frac{m\pi x}{L} \cos \frac{ns}{R} \qquad u(x,s) = U_{mn} \cos \frac{m\pi x}{L} \sin \frac{ns}{R} \tag{5.246}$$

corresponding to (5.237) and (5.238). Functions (5.245) and (5.246) satisfy at x = 0 and x = L the following boundary conditions

$$w = 0 \qquad \frac{\partial^2 w}{\partial x^2} = 0 \qquad v = 0 \qquad \frac{\partial u}{\partial x} = 0 \tag{5.247}$$

The assumed displacement functions (5.245) and (5.246) satisfy the first of the boundary conditions (5.233)

$$\frac{\partial u}{\partial x} + v\frac{\partial v}{\partial s} + \frac{v}{2}\frac{w}{R} = 0 \tag{5.248}$$

which implies the vanishing of stresses N_{xx} at the end sections of the cylinder. Substituting the appropriate derivatives of (5.245) into (5.244), the following solution is obtained

$$S_{sscr} = \frac{\pi^2 E}{12(1 - v^2)} \left(\frac{t}{L}\right)^2 k_p \tag{5.249}$$

where

$$k_p = \frac{(m^2+\beta^2)^2}{\beta^2} + \frac{12Z^2}{\pi^4\beta^2\left(1 + \frac{\beta^2}{m^2}\right)^2} \tag{5.250}$$

with

$$\beta = \frac{nL}{\pi R} \qquad Z = \frac{L^2}{Rt}\sqrt{1 - v^2} \tag{5.251}$$

It now becomes evident that, in terms of m, a minimum value of k_p is obtained when m has its smallest value, namely m = 1. Thus, for m = 1, (5.250) becomes

$$k_p = \frac{(1+\beta^2)^2}{\beta^2} + \frac{12Z^2}{\pi^4\beta^2(1 + \beta^2)^2} \tag{5.252}$$

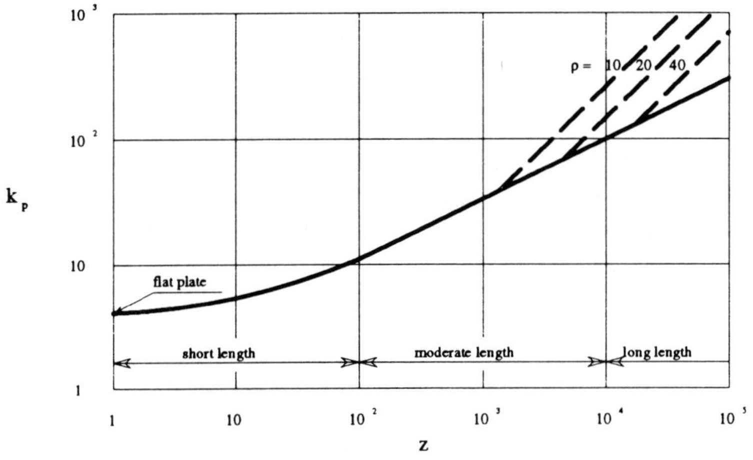

FIGURE 5.28

As consequence, a circular cylinder of any length under external pressure buckles into a single half-wavelength in the axial direction. A more convenient form of (5.249) is the following

$$p_{cr} = \pi^2 k_p \frac{D_f}{L^2 R} \tag{5.253}$$

which is useful for comparisons with (5.224). When $Z \approx 0$, we have

$$k_p = \frac{(1+\beta^2)^2}{\beta^2} \tag{5.254}$$

When this equation is minimized, it is found that $\beta = 1$ and the equation gives $k_p = 4$. Consequently, from (5.251) for $\beta = 1$ we get

$$n = \frac{\pi R}{L} \tag{5.255}$$

Thus, in the case of a very short cylinder under lateral pressure, the behavior of the cylinder approaches that of a very long flat plate in compression that buckles in half-waves, the length of which approaches the width of the plate. The axial length corresponds to the short length of a very long flat plate, and the circumferential direction is that of the long dimension of the flat plate. Minima values of k_p for small values of Z and as a function of Z can be found using (5.252). According to the results obtained by Batdorf,[22] Fig. 5.28 plots the values of k_p versus the values of the parameter Z. As Z increases, the value of k_p rises as well as the value of β. For cylinders of moderate length, i.e., for values of Z beyond 100, the relation between k_p and Z appears linear on the log-log plot, and $\beta^2 >> 1$.Under this condition $1 + \beta^2 \approx \beta^2$ and (5.252) simplifies to

$$k_p = \beta^2 + \frac{12Z^2}{\pi^4 \beta^6} \tag{5.256}$$

Upon minimizing this equation with

$$\frac{\partial k_p}{\partial \beta} = 0 \tag{5.257}$$

we obtain $k_p = 1,038\sqrt{Z}$ and, correspondingly,

$$S_{sscr} = \frac{0.855E}{(1-\nu^2)^{3/4}}\left(\frac{t}{R}\right)^{3/2}\frac{R}{L} \tag{5.258}$$

When the parameter Z further increases, i.e., in the case of long cylinders, the number of circumferential half-waves decreases until it reaches the value n = 2. At this point the circular cross section at the buckling becomes oval, i.e., buckles into an elliptical form. For n = 2

$$\beta = \frac{2L}{\pi R} \tag{5.259}$$

and from (5.252) we obtain

$$k_p = \beta^2 = \frac{4L^2}{\pi^2 R^2} = \frac{4}{\pi^2\sqrt{1-\nu^2}} Z\frac{t}{R} \tag{5.260}$$

Substituting (5.260) in (5.253) gives

$$S_{sscr} = \frac{E}{3(1-\nu^2)}\left(\frac{t}{R}\right)^2 \tag{5.261}$$

We observe that for very long cylinders (5.224) gives without any approximation the exact value of the critical pressure. The corresponding exact circumferential critical stress S_{ss} is

$$S_{sscr} = \frac{E}{4(1-\nu^2)}\left(\frac{t}{R}\right)^2 \tag{5.262}$$

Comparing (5.261) with (5.262), we have to remark again that the Donnell equation is generally valid for buckling problems in which several buckle wavelengths occur in the circumferential direction.

In Section 5.9.10.1 we pointed out that, disregarding (as in the Donnell equation) the destabilizing second-order work of the external pressure, the critical value of the pressure was approximately given by (5.225) in place of the exact value (5.224). To this approximate value of the critical pressure corresponds the critical value (5.262) of the circumferential stress S_{ss}.

5.9.9.3 Buckling of Cylinders Under Axial Compression

In the case of circular cylinders under axial compression (Fig. 5.29) the general Donnell equation reduces to

$$D_f\,\Delta\Delta\Delta\Delta w + \frac{Et}{R^2}\frac{\partial^4 w}{\partial x^4} + \Delta\Delta\left(N_{xx}\frac{\partial^2 w}{\partial x^2}\right) = 0 \tag{5.263}$$

A solution of this equation that satisfies the boundary conditions of simple supports at the ends of the cylinder is

$$w(x,s) = w_{mn}\sin\frac{m\pi x}{L}\sin\frac{ns}{R} \tag{5.264}$$

Upon substituting (5.264) into (5.263) the following nontrivial solution is obtained

$$S_{xxcr} = \frac{\pi^2 E}{12(1-\nu^2)}\left(\frac{t}{L}\right)^2 k_c \tag{5.265}$$

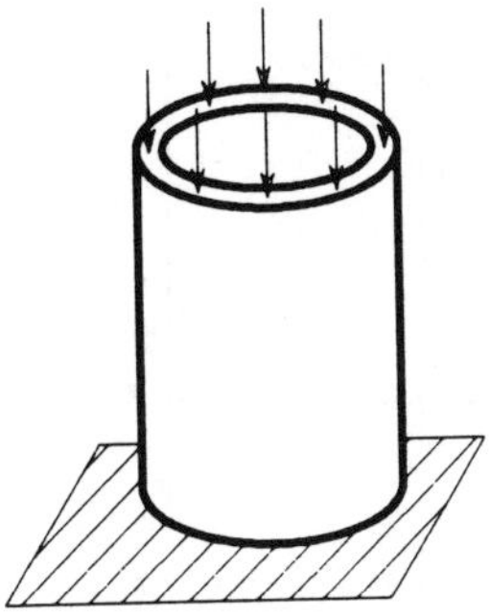

FIGURE 5.29

where

$$k_c = \frac{(m^2+\beta^2)^2}{m^2} + \frac{12Z^2m^2}{\pi^4\left(m^2+\beta^2\right)^2} \tag{5.266}$$

with the notation given by (5.251). The wavelength parameter

$$\frac{(m^2+\beta^2)^2}{m^2} \tag{5.267}$$

appears in both terms of (5.266). By minimizing (5.266) with respect to this parameter, we have

$$\frac{(m^2+\beta^2)^2}{m^2} = \sqrt{\frac{12Z^2}{\pi^4}} \tag{5.268}$$

Substituting into (5.266) we obtain the simple result

$$S_{xxcr} = 0.605E\frac{t}{R} \tag{5.269}$$

We return to (5.268) to establish the range of validity of (5.269). Solving (5.268) for β we obtain

$$\beta = \sqrt{\frac{\left(12Z^2\right)^{1/4}}{\pi}m - m^2} \tag{5.270}$$

Since, as a minimum, m = 1, it is apparent that for real values of β

$$Z \geq \frac{\pi^2}{\sqrt{12}} \approx 2.85 \tag{5.271}$$

Thus the moderate length cylinder solution applies for values of Z greater than 2.85. For short cylinders, i.e., when Z < 2.85, an approximated solution of the problem is obtained by assuming, for the search of the minimum of k_c, the values $\beta = 0$ and m = 1. Substituting these last values into (5.266) we get

$$k_c = 1 + \frac{12Z^2}{\pi^4} \tag{5.272}$$

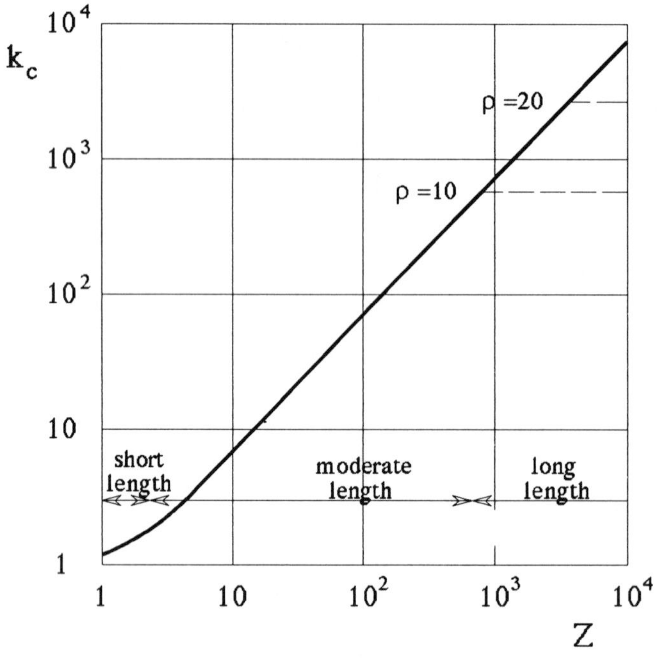

FIGURE 5.30

In this region the short cylinder behaves in the same manner as a wide simply supported column, which buckles into one half-wavelength in the loaded axial direction, because m = 1 and none in the unloaded direction, i.e., with n = 0. The results obtained are plotted in Fig. 5.30. For larger values of Z, long cylinders buckle as the Euler column with no distortion of the circular cross section. On the other hand, for practical values of the ratio R/t, Euler buckling occurs at values of Z beyond those shown in Fig. 5.30.[24]

In the next chapter the postbuckling behavior of cylinders and their imperfection sensitivity under external pressure will be examined.

REFERENCES

1. Biot, M. A., Nonlinear theory of elasticity and the linearized case for a body under initial stress, Phil. Magaz. 27(7), 1939.
2. Biot, M. A., Theory of elasticity with large displacements and rotations, Proc. 5th Intern. Congr. Appl. Mech., John Wiley & Sons, New York, 1939.
3. Biot, M. A., Mechanics of incremental deformations, John Wiley & Sons, New York, 1965.
4. Trefftz, E., Über die ableitung der stabilitäts — Kriterien des elastischen glechgewichtes aus der elastizitättheorie endlicher deformationen, Proc. 3th Int. Congr. Appl. Mech. 3, 1931.
5. Novozhilov, V. V., Foundations of a Nonlinear Theory of Elasticity. Graylock, NY, 1953.
6. Novozhilov, V. V., Theory of Elasticity, Pergamon Press, New York, 1961.
7. Föppl, A., Drang und Zwang, R. Oldembourg, Berlin, 1924.
8. Como, M., Considerazioni critiche sulla formulazione dell'energia di deformazione nella stabilità dell'equilibrio elastico, Costruzioni Metalliche 1, 1966.
9. Grimaldi, A. and Romano, G., On the formulation of the nonlinear analysis of slender elastic structures, Dept. of Struct. University of Calabria, Report no. 6, 1974.
10. Krall, G., Stabilità dell'equilibrio elastico, Annali di Matematica Pura e Applicata, Zanichelli, Bologna, 1939.
11. Galli, A., Scienza delle Costruzioni, Vol. III. Pellerano Del Gaudio, Napoli, 1954.
12. Timoshenko, S. and Gere, J. M., Theory of elastic stability McGraw-Hill, New York, 1961.
13. Vlasov, V. Z., Thin Walled Elastic Beams. Pergamon Press, New York, 1961.
14. Como, M., Teoria della Stabilità dell'equilibrio elastico, Liguori Editore, Napoli, 1967.
15. Novozhilov, V. V., The Theory of Thin Shells. Noordhoff Groningen, The Netherlands, 1959.

16. Willmore, I. J., An Introduction to Differential Geometry, Oxford at Clarendon Press, 1964.
17. Love, A. E. H., A treatise on the mathematical theory of elasticity, 4th Ed., Dover Publ., New York, 1960.
18. Flügge, W., Stresses in shells, Springer-Verlag, Berlin, 1962.
19. Ferrarese, G., Sull'incremento di volume di un corpo per deformazioni finite, Rend. Acc. Naz. le Lincei, Cl. Sci. Fis. Mat. 1, 1965.
20. Boussinesq, J., Resistance d'un anneau à la flexion quand sa surface exterieure supporte une pression normale, C. R, 1883.
21. Donnell, L. H., A new theory for the buckling of thin cylinders under raxial compression and bending, Trans. ASME 56, 1934.
22. Batdorf, S. B., A simplified method of elastic stability analysis for thin cylindrical shells Donnell Equation, N.A.C.A., T.N. 1341, 1947.
23. Koiter, W. T., General Equations of Elastic Stability for Thin Shells. Donnell Anniversary Volume, Houston, 1966.
24. Gerard, G., Introduction to Structural Stability Theory. McGraw-Hill, New York, 1962.
25. Krall, G., Caligo, D., Moltiplicatore critico λ_c per volte autoportante, Rend. Acc. Naz. le Lincei, Cl. Sci. Fis. Mat. 30, 1962.

Chapter 6

BIFURCATION OR SNAPPING AT THE CRITICAL STATE OF CONTINUOUS ELASTIC STRUCTURES

6.1 INTRODUCTION

All the common elastic systems can be divided into two different groups. The first group is composed of general systems with dominant nonlinear behavior from the beginning of loading. A failure state is frequently met at the critical state, reached step-by-step during the loading with a progressive reduction of stiffness. Let us consider, for example, the simple structural scheme of Fig. 6.1.

In this case the potential energy function of all external and internal forces acting on the system is given by

$$\mathcal{E}(\alpha,\lambda) = kL^2\left(\frac{1}{\cos\alpha_o} - \frac{1}{\cos\alpha}\right)^2 - \lambda PL(\tan\alpha_o - \tan\alpha)$$

where k is the axial stiffness of the struts. The fundamental equilibrium branch, for relatively small values of the angles α_o and α, is represented by

$$D\mathcal{E}(\alpha,\lambda;\,\delta\alpha) = \left[kL^2\alpha(\alpha^2 - \alpha_o^2) + \lambda PL(1+\alpha^2)\right]\delta\alpha = 0$$

and the relation between the load λP and the angle α is

$$\lambda P \cong kL\ \alpha(\alpha_o^2 - \alpha^2)$$

A failure is attained along this branch when

$$\frac{d(\lambda P)}{d\alpha} = 0$$

to which correspond a critical value α_c of the angle α and a critical load, respectively given by

$$\alpha_c = \frac{\alpha_o}{\sqrt{3}}, \qquad \lambda_c\ P = kL^2\frac{2}{3\sqrt{3}}\alpha_o^3$$

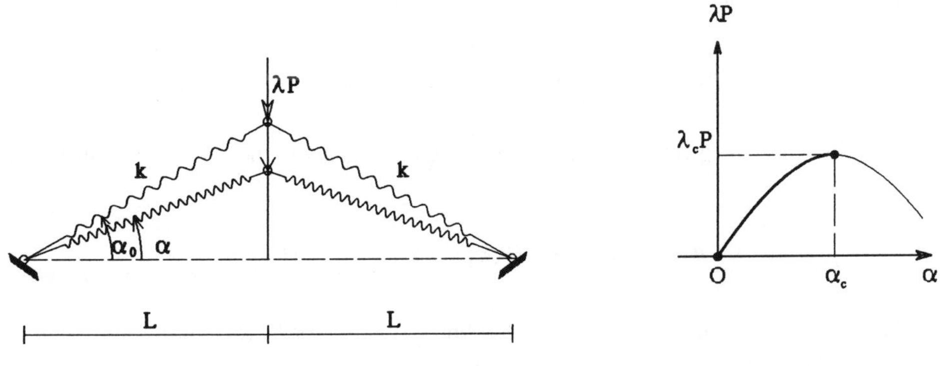

FIGURE 6.1

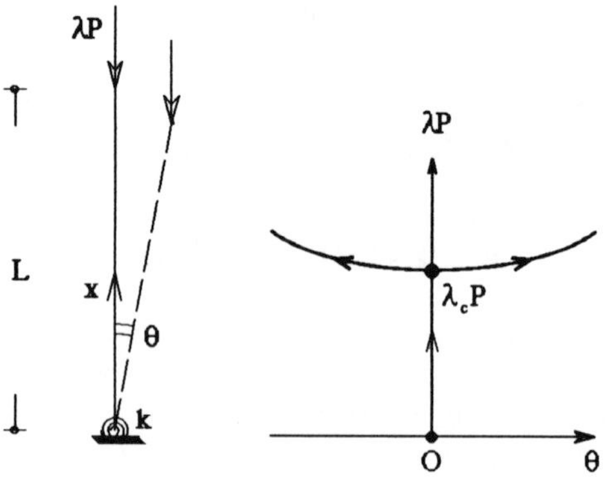

FIGURE 6.2

The behavior of this scheme is strongly nonlinear, and snap-through occurs at the critical state.

The second group of structures is composed of systems that, because of their geometry and load distribution, maintain their stiffness in a first stage of loading and when they reach the critical state a new deformation suddenly occurs. Frequently a linear equation between displacements and the load multiplier λ describes the progress of their first deformations.

Let us also consider in this case some instructive simple examples. A first structural scheme is sketched in Fig. 6.2. The bar, axially deformable but rigid in flexure, is hinged at the base, but it is there constrained by a linear rotational elastic spring of stiffness k. The axial load is λP and the extensional stiffness of the bar is EA.

The initial length of the unstressed bar is L_0 and

$$L = L_o\left(1 - \frac{\lambda P}{EA}\right)$$

is the length of the bar under the axial load λP. Under the action of the gradually increasing axial load λP the bar thus will exhibit an elastic behavior characterized by linearly increasing axial stresses and strains. A principal branch of equilibrium

$$C_\lambda = \{\ \lambda u_0(x),\ \sigma = \lambda\sigma_0,\ \sigma_0 = P/A,\ \lambda \geq 0\}$$

is so defined, if $u_0(x)$ defines the shortening displacement of the bar when $\lambda = 1$.

The lateral rotation θ induces in the bar a state of stress and strain that is completely different from that corresponding to the principal equilibrium branch C_λ. If we displace the bar from C_λ to a a rotated configuration C, the increment of the potential energy is

$$\Delta\mathcal{E} = \mathcal{E}(C) - \mathcal{E}(C_\lambda) = \mathcal{E}(\theta,\lambda) = \frac{1}{2}k\theta^2 - \lambda P L_o(1-\cos\theta)\left(1 - \frac{\lambda P}{EA}\right)$$

If we can disregard the difference between the length of the bar in the stressed and unstressed states, we have also

$$\mathcal{E} = \mathcal{E}(C) - \mathcal{E}(C_\lambda) = \mathcal{E}(\theta,\lambda) = \frac{1}{2}k\theta^2 - \lambda P L_o(1-\cos\theta)$$

and the additional potential energy becomes a linear function of λ. The equilibrium branches are defined by the equation

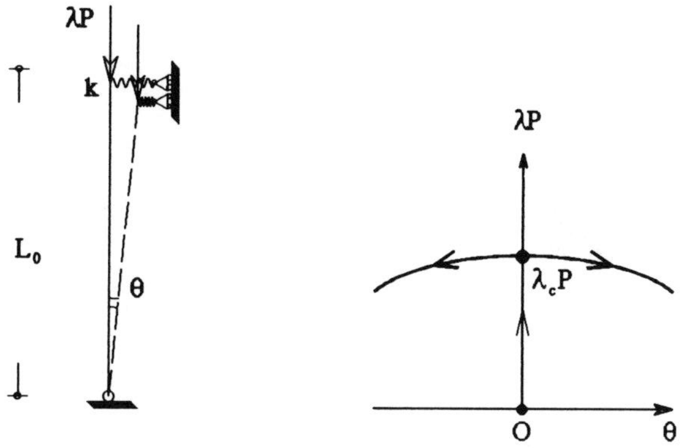

FIGURE 6.3

$$D\mathcal{E}(\theta,\lambda;\,\delta\theta)=\left[k\theta-\lambda PL_o\,\sin\theta\right]\delta\theta=0$$

The principal equilibrium path, corresponding to the bar in the vertical position, is defined by

$$\theta=0\ \ \forall\,\lambda\ge 0$$

Along this branch the differential $D\mathcal{E}(0,\lambda;\,\delta\theta)$ is in fact identically equal to zero, i.e.,

$$D\mathcal{E}(\theta,\lambda;\,\delta\theta)\equiv 0\quad \theta=0\,\forall\,\lambda\ge 0$$

Bifurcation occurs at the critical state. The critical load is

$$\lambda_c P=\frac{k}{L_o}$$

disregarding the small quantity $\frac{\lambda P}{EA}$ with respect to the unity. The second equilibrium path is thus defined by

$$\lambda P=\lambda_c P\frac{\theta}{\sin\theta}$$

and intersects the first branch $\theta=0\ \forall\,\lambda\ge 0$ at $\lambda P=\lambda_c P$. In Fig. 6.2 are sketched the two equilibrium branches of the system.

For the case represented in Fig. 6.3, disregarding the small difference between the lengths of the bar in the stressed and unstressed states, the potential function is linear in λ and is given by

$$\mathcal{E}(\theta,\lambda)=\frac{1}{2}kL_o^2\,\sin^2\theta-\lambda PL_o(1-\cos\theta)$$

The equilibrium paths are described by the implicit equation

$$D\mathcal{E}(\theta,\lambda;\,\delta\theta)=\left(\frac{1}{2}kL_o^2\,\sin 2\theta-\lambda PL_o\sin\theta\right)\delta\theta=0$$

The fundamental branch C_λ is still given by $\theta=0\ \ \forall\,\lambda\ge 0$, while the second, defined by

$$\lambda P=\frac{1}{2}\frac{kL_o\,\sin 2\theta}{\sin\theta}$$

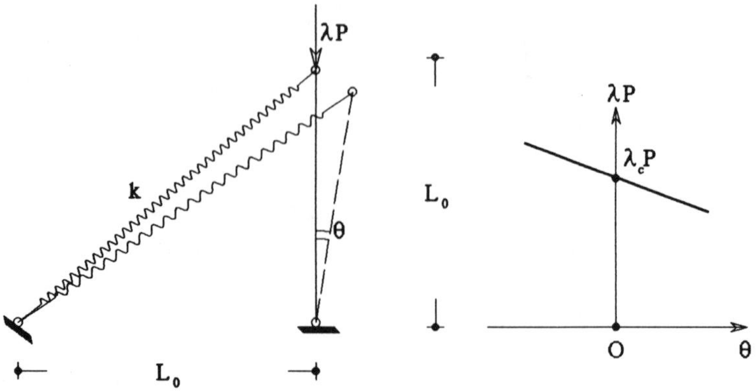

FIGURE 6.4

intersects the first at the critical load

$$\lambda_c P = kL_0$$

In this case the critical state, still symmetric, is unstable. The other scheme (Fig. 6.4) describes, on the contrary, a nonsymmetric bifurcation. In this case the potential energy is

$$\mathcal{E}(\theta,\lambda) = kL_o^2\left[2\left(1-\sqrt{1+\sin\theta}\right)+\sin\theta\right] - \lambda PL_o(1-\cos\theta)$$

and the equilibrium branches are obtained by solving the equation

$$D\mathcal{E}(\theta,\lambda;\,\delta\theta) = \left\{-kL_o^2\left[(1+\sin\theta)^{-1/2}\cos\theta\right] + kL_o^2\cos\theta - \lambda PL_o\sin\theta\right\}\delta\theta = 0$$

The fundamental equilibrium branch is still represented by $\theta = 0 \quad \forall\ \lambda \geq 0$, while the second branch

$$\lambda P(\theta) = \frac{kL_o}{\tan\theta}\left[1 - \frac{1}{\sqrt{1+\sin\theta}}\right]$$

intersects the first at $\lambda_c P = kL_o/2$. A nonsymmetric bifurcation occurs and the critical state is unstable. Structural systems that belong to this last group possess a fundamental branch C_λ that bifurcates when the critical state is attained. As a rule, along this principal path C_λ the system exhibits linear behavior. This type of structural system has been called by Koiter "linear" or "perfect".

In the framework of the general analysis previously developed, in the next sections we will study these systems whose behavior is similar to that of the last three schemes now examined.

In mathematics branching and snapping are problems proper to the framework of perturbation methods applied to the solution of nonlinear differential equations and, in particular, to the context of bifurcation theory.[1-6] Bifurcation analysis was first developed by Poincaré,[5] who first studied the stability exchanges occurring at the critical state of finite degrees of freedom systems. In structural engineering a fundamental work was the Koiter thesis, where the postbuckling behavior of perfect and imperfect structures was first defined and studied in detail.[8,9] In this field, Thompson and others[10-14] made further contributions to the snapping and bifurcation analysis of discrete elastic systems. A complementary contribution to the understanding of evolutionary instabilities was the Catastrophe Theory of René Thom[15] that, developed independently in a more abstract context, is remarkably similar to the above quoted formulations of the nonlinear stability. Main developments to the postbuckling theory of structures, particularly in shell analysis, are due to Koiter,[16] and Budiansky and Hutchinson.[17,18,20] A wide bibliography in postbuckling theory was given by Koiter and Hutchinson[19] and in books by Thompson, Hunt, Huseyn, and others.[21-26]

The aim of this chapter is to provide a unified analysis of the equilibrium states that occur in the neighborhood of the critical state of elastic continuous structural systems in the framework of the results achieved in the previous part of the book. Stability analysis, performed in the "energy" Hilbert space, and the assumption of a smooth potential energy functional are the necessary tools. Some of the results given in this chapter replicate, in a more general mathematical context, the above recalled Koiter achievements, others extend to continuous elastic systems results obtained by Thompson for elastic systems with finite degrees of freedom.

The equilibrium evolution under varying loads of a general structural elastic system is analyzed in the first part of the chapter. As a rule, snapping occurs at the critical state. At the critical state a general elastic system, because of nonlinear geometrical effects, in fact, becomes incapable of sustaining any further small increase of load and fails. The study of the strong correlation existing between stability and uniqueness and between stability exchanges and bifurcation, is the main subject of the first part of the chapter.

The second part is dedicated to the study of the "linear" or "perfect" elastic structures, represented by those elastic systems that exhibit a linear behavior during loading up to the critical state. For these systems branching occurs at the critical state where the system "buckles", seeking for new equilibrium states. The peculiar aspect of the critical state of these perfect systems is, in fact, the occurrence of buckling with stable or unstable branching. The general analysis of the equilibrium paths of nonlinear elastic systems, performed in the first part of the chapter, allows us to obtain a synthetic description of the behavior of these "linear" or "perfect" elastic structures in the neighborhood of the critical state.

Perfect, or "linear", systems represent, on the other hand, only ideal models of the real structures. According to Koiter, the behavior of a real structural system can be described by means of perturbation tecniques, adding small "imperfections" to the perfect system. The study of the behavior of the "quasi-linear" or "quasi-perfect" elastic structures ends the first part of the chapter. Some examples are then developed and the equilibrium bifurcations of elastic rods, deep beams, and frames are investigated. The analysis of the imperfection sensitivity of thin cylinders under external pressure ends the chapter.

6.2 EQUILIBRIUM PATHS OF GENERAL NONLINEAR ELASTIC SYSTEMS

6.2.1 DEFINITIONS

Consider an elastic continuous system T and let H_A be the corresponding energy space of its configurations. Let us analyze the behavior of T under the action of the loads $\lambda(\alpha)\mathbf{q}$, where $\lambda(\alpha)$ is the load multiplier and α a variable defining the progress of the loading. Under the loads $\lambda_0\mathbf{q}$ the system T is in equilibrium at the principal configuration C_0 defined by the displacement field $\mathbf{u}_0$ and evaluated with respect to the reference unstressed configuration C_i. We intend to determine the equilibrium configurations of T under the loads $\lambda\mathbf{q}$, with λ close to λ_0. We first give the following definitions:

1. *Path or branch of equilibrium states*: This represents the set of the equilibrium configurations of T defined by the continuous functions in the energy space H_A (Fig. 6.5)

$$\mathbf{u} = \mathbf{u}(\alpha) \; \lambda = \lambda(\alpha) \; \forall \; \alpha \in [\alpha_1, \alpha_2] \tag{6.1}$$

   Equations (6.1) — as a rule not invertible — specify the path of the equilibrium states crossed by T under the loads $\lambda(\alpha)\mathbf{q}$ for α varying in the interval $[\alpha_1, \alpha_2]$. All the

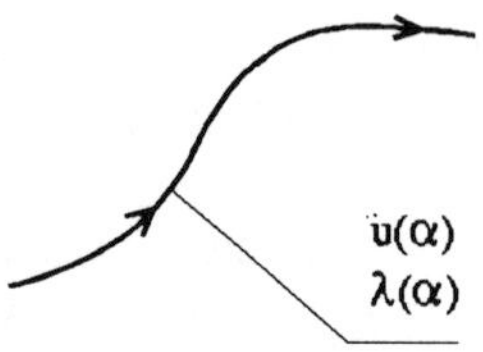

FIGURE 6.5

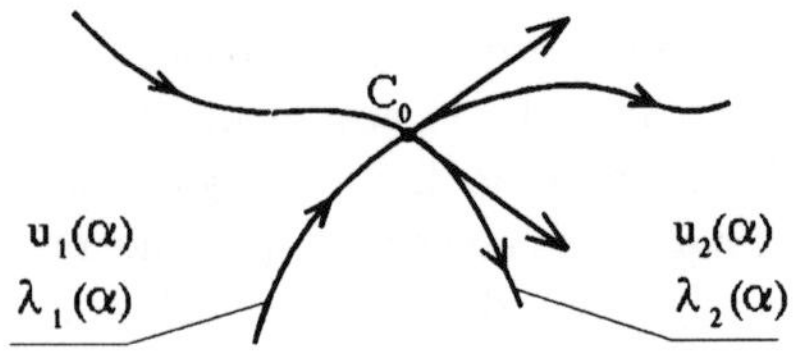

FIGURE 6.6

smoothness properties for the abstract function $\mathbf{u} = \mathbf{u}(\alpha)$, such as continuity, differentiability, etc., must be intended in the sense of the metric of the space H_A.

2. *Branching point:* The equilibrium configuration C_0, defined by $\alpha = \alpha_0$ on the path (6.1), is a branching point (Fig. 6.6) if at C_0 two or more branches intersect each other. At a branching point the unit tangent vectors to the various branches will therefore be distinct from each other. Thus at a bifurcation point two different branches, respectively defined by the equations

$$\mathbf{u}_1 = \mathbf{u}_1(\alpha)\ \lambda_1 = \lambda_1(\alpha) \qquad \mathbf{u}_2 = \mathbf{u}_2(\alpha) \qquad \lambda_2 = \lambda_2(\alpha) \qquad \forall\ \alpha \in [\alpha_1, \alpha_2] \tag{6.2}$$

will satisfy the following conditions

$$\mathbf{u}_1(\alpha_0) = \mathbf{u}_2(\alpha_0),\ \lambda_1(\alpha_0) = \lambda_2(\alpha_0),\ \frac{\dot{\mathbf{u}}_1(\alpha_0)}{|||\dot{\mathbf{u}}_1(\alpha_0)|||} \neq \frac{\dot{\mathbf{u}}_1(\alpha_0)}{|||\dot{\mathbf{u}}_1(\alpha_0)|||} \tag{6.3}$$

and with $\dot{\lambda}_1(\alpha_0) \neq \dot{\lambda}_2(\alpha_0)$ and at least one of the two derivatives $\dot{\lambda}_1$, $\dot{\lambda}_2$ different from zero.

3. *Snapping point:* The configuration of equilibrium C_0 on the path (6.1) is a snapping point if at C_0 (Fig. 6.7)

$$\dot{\mathbf{u}}(\alpha_0) \neq 0 \qquad \dot{\lambda}(\alpha_0) = 0 \tag{6.4}$$

At a snapping point the system T exhibits deformation increments without any increment of load. If the load that can be sustained reaches at the snapping its maximum value, the snapping load corresponds to the *failure load* of the elastic system.

6.2.2 THE IMPLICIT FUNCTION PROBLEM OF EQUILIBRIUM BRANCHES

The potential energy functional $\mathcal{E}(\mathbf{u})$ of all the external and internal forces acting on T is defined in a suitable set of H_A for every configuration $\mathbf{u} = \mathbf{u}(\alpha)$. We will assume $\mathcal{E}(\mathbf{u})$ to be n times Fréchet differentiable at $\mathbf{u} = \mathbf{u}(\alpha)$ in H_A. The coordinate points of the path (6.1) will satisfy the equilibrium equations, which, according to (3.8), are

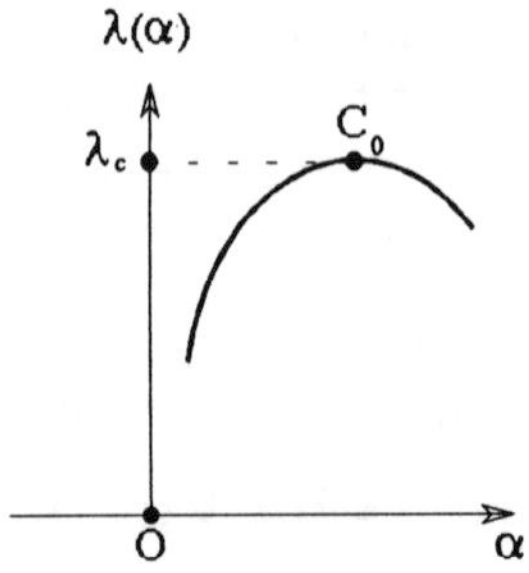

FIGURE 6.7

$$\mathcal{E}_1[\mathbf{u}(\alpha),\lambda(\alpha);\ \delta\mathbf{u}] = 0 \qquad \forall\,\delta\mathbf{u} \in H_A,\ \forall\,\alpha \in [\alpha_1,\ \alpha_2] \tag{6.5}$$

where $\mathcal{E}_1[\mathbf{u}(\alpha),\lambda(\alpha);\ \delta\mathbf{u}]$ is the Fréchet differential of $\mathcal{E}(\mathbf{u})$ at $[\mathbf{u}(\alpha),\lambda(\alpha)]$ and along the direction $\delta\mathbf{u}$. With a different notation we can also write

$$D\mathcal{E}[\mathbf{u}(\alpha),\lambda(\alpha);\ \delta\mathbf{u}\ ] = 0 \qquad \forall\,\delta\mathbf{u} \in H_A\ \forall\,\alpha \in [\alpha_1,\ \alpha_2] \tag{6.5$'$}$$

Assuming for simplicity that $\alpha_0 = 0$, we have

$$\mathbf{u}(0) = \mathbf{u}_0 \qquad \lambda(0) = \lambda_0 \tag{6.6}$$

and

$$D\mathcal{E}[\mathbf{u}_0,\ \lambda_0;\ \delta\mathbf{u}] = 0 \qquad \forall\,\delta\mathbf{u} \in H_A \tag{6.5$''$}$$

On the contrary, there is the problem of the *existence* of equilibrium branches passing through $C_0\,(\mathbf{u}_0,\lambda_0)$, i.e., the problem of the existence of solutions in a neighborhood of (C_0,λ_0) of the implicit problem (6.5). The assumed Fréchet differentiability properties of $\mathcal{E}(\mathbf{u},\lambda)$ at (C_0,λ_0) permit us to prove the local existence of smooth functions (6.1), which branch off the trivial solution (6.6) of (6.5′).[1-4]

The mutual relation existing for a general elastic system between "infinitesimal" stability and the existence and uniqueness of the equilibrium branch (6.1), will be analyzed in Section 6.2.3. Only at the critical state is more than one equilibrium branch possible. Therefore, taking for granted the existence of these equilibrium states, we are going to characterize and analyze them in detail. We will admit that functions (6.1), at least at C_0, have continuous derivatives with respect to α up to the required order n. Thus the equilibrium branch passing through (C_0,λ_0), i.e., the solution of the implicit equation (6.5), is obtained by means of the Taylor theorem

$$\mathbf{u}(\alpha) - \mathbf{u}(0) = \alpha\dot{\mathbf{u}}(0) + \frac{\alpha^2}{2}\ddot{\mathbf{u}}(0) + \ldots + \frac{\alpha^n}{n!}\mathbf{u}^{(n)}(0) + o(\alpha^n) \tag{6.7}$$

$$\lambda(\alpha) = \lambda(0) + \alpha\dot{\lambda}(0) + \frac{\alpha^2}{2}\ddot{\lambda}(0) + \ldots + \frac{\alpha^n}{n!}\lambda^n(0) + o(\alpha^n) \tag{6.8}$$

where

$$(\dot{\ }) = \frac{d(\ )}{d\alpha} \qquad \lim_{\alpha\to 0}\frac{o(\alpha^n)}{\alpha^n} = 0 \tag{6.9}$$

Expressions (6.7) and (6.8) can also be written as

$$\mathbf{u}(\alpha) - \mathbf{u}(0) = \alpha\mathbf{u}_1 + \alpha^2\mathbf{u}_2 + \ldots + \alpha^n\mathbf{u}_n + o(\alpha^n) \tag{6.7$'$}$$

$$\lambda(\alpha) = \lambda(0) + \alpha\lambda_1 + \alpha^2\lambda_2 + \ldots + \alpha^n\lambda_n + o(\alpha^n) \tag{6.8$'$}$$

which are frequently used in postbuckling analysis of structures.

Functions $\dot{\mathbf{u}}(0), \ddot{\mathbf{u}}(0), \dddot{\mathbf{u}}(0), \ldots \mathbf{u}^{(n)}(0), \ldots$ will be generated as solutions of a sequence of variational equations, as a rule corresponding to seeking minima of suitable quadratic functionals. The equilibrium path $[\mathbf{u}(\alpha),\lambda(\alpha)]$ is thus expressed in the neighborhood of the state (C_0,λ_0) by means of the Taylor polynomials (6.7) and (6.8) where the coefficient derivatives $\dot{\mathbf{u}}(0), \ddot{\mathbf{u}}(0), \ldots; \dot{\lambda}(0), \ddot{\lambda}(0), \ldots$ are unknowns. All the functions $\dot{\mathbf{u}}(\alpha), \ddot{\mathbf{u}}(\alpha), \ldots; \dot{\lambda}(\alpha), \ddot{\lambda}(\alpha), \ldots$, will thus be evaluated by solving a sequence of equations obtained equating to zero all the subsequent derivatives with respect to α of the first differential $D\mathcal{E}[\mathbf{u}(\alpha),\lambda(\alpha);\,\delta\mathbf{u}]$ of the potential energy $\mathcal{E}[\mathbf{u}(\alpha),\lambda(\alpha)]$. Along this path, in fact, $D\mathcal{E}[\mathbf{u}(\alpha),\lambda(\alpha);\,\delta\mathbf{u}]$ is identically equal to zero. Hence we have the subsequent equations

$$\frac{d}{d\alpha}D\mathcal{E}\left[\mathbf{u}(\alpha),\,\lambda(\alpha);\delta\mathbf{u}\right]=0,\qquad \frac{d^2}{d\alpha^2}D\mathcal{E}\left[\mathbf{u}(\alpha),\,\lambda(\alpha);\delta\mathbf{u}\right]=0,\ldots \tag{6.10}$$

expressing continuity of the equilibrium along the path. Evaluation of the first derivative of the differential $D\mathcal{E}[\mathbf{u}(\alpha),\lambda(\alpha);\,\delta\mathbf{u}]$ gives

$$\frac{d}{d\alpha}D\mathcal{E}\left[\mathbf{u}(\alpha),\,\lambda(\alpha);\delta\mathbf{u}\right]=\dot\lambda(\alpha)D\mathcal{E}'\left[\mathbf{u}(\alpha),\lambda(\alpha);\,\delta\mathbf{u}\right]+D^2\mathcal{E}\left[\mathbf{u}(\alpha),\,\lambda(\alpha);\,\dot{\mathbf{u}}(\alpha),\delta\mathbf{u}\right] \tag{6.11}$$

where

$$\frac{\partial(\;)}{\partial\lambda}=(\;)' \tag{6.12}$$

The first equation of group (6.10), more expressively, can be written in the following form

$$D^2\mathcal{E}\left[\mathbf{u}(\alpha),\lambda(\alpha);\,\dot{\mathbf{u}}(\alpha),\delta\mathbf{u}\right]+\dot\lambda(\alpha)D\mathcal{E}'\left[\mathbf{u}(\alpha),\lambda(\alpha);\delta\mathbf{u}\right]=0,\quad \forall\;\delta\mathbf{u}\in H_A,\;\forall\;\alpha\in\left[\alpha_1,\alpha_2\right] \tag{6.13}$$

Likewise we have, with the same procedure

$$\begin{aligned}&D^3\mathcal{E}\left[\mathbf{u}(\alpha),\lambda(\alpha);\dot{\mathbf{u}}^2(\alpha),\delta\mathbf{u}\right]+D^2\mathcal{E}\left[\mathbf{u}(\alpha),\lambda(\alpha);\ddot{\mathbf{u}}(\alpha),\delta\mathbf{u}\right]+2\dot\lambda(\alpha)D^2\mathcal{E}'\left[\mathbf{u}(\alpha),\lambda(\alpha);\dot{\mathbf{u}}(\alpha),\delta\mathbf{u}\right]\\&+\ddot\lambda(\alpha)D\mathcal{E}'\left[\mathbf{u}(\alpha),\lambda(\alpha);\delta\mathbf{u}\right]+\dot\lambda^2(\alpha)D\mathcal{E}''\left[\mathbf{u}(\alpha),\lambda(\alpha);\delta\mathbf{u}\right]=0\end{aligned} \tag{6.14}$$

which will be satisfied $\forall\,\delta\mathbf{u}\in H_A$, $\forall\;\alpha\in[\alpha_1,\alpha_2]$. Differentiate again identity (6.13). We get, $\forall\;\alpha\in[\alpha_1,\alpha_2]$,

$$\begin{aligned}&3D^3\mathcal{E}\left[\mathbf{u}(\alpha),\lambda(\alpha);\dot{\mathbf{u}}(\alpha),\ddot{\mathbf{u}}(\alpha),\delta\mathbf{u}\right]+D^4\mathcal{E}\left[\mathbf{u}(\alpha),\lambda(\alpha);\dot{\mathbf{u}}^3(\alpha),\delta\mathbf{u}\right]\\&+3\dot\lambda(\alpha)D^3\mathcal{E}'\left[\mathbf{u}(\alpha),\lambda(\alpha);\dot{\mathbf{u}}^2(\alpha),\delta\mathbf{u}\right]+D^2\mathcal{E}\left[\mathbf{u}(\alpha),\lambda(\alpha);\dddot{\mathbf{u}}(\alpha),\delta\mathbf{u}\right]\\&+3\dot\lambda(\alpha)D^2\mathcal{E}'\left[\mathbf{u}(\alpha),\lambda(\alpha);\ddot{\mathbf{u}}(\alpha),\delta\mathbf{u}\right]+3\ddot\lambda(\alpha)D^2\mathcal{E}'\left[\mathbf{u}(\alpha),\lambda(\alpha);\dot{\mathbf{u}}(\alpha),\delta\mathbf{u}\right]\\&+3\dot\lambda^2(\alpha)D^2\mathcal{E}''\left[\mathbf{u}(\alpha),\lambda(\alpha);\dot{\mathbf{u}}(\alpha),\delta\mathbf{u}\right]+\dddot\lambda(\alpha)D\mathcal{E}'\left[\mathbf{u}(\alpha),\lambda(\alpha);\delta\mathbf{u}\right]\\&+3\dot\lambda(\alpha)\ddot\lambda(\alpha)D\mathcal{E}''\left[\mathbf{u}(\alpha),\lambda(\alpha);\delta\mathbf{u}\right]+\dot\lambda^3D\mathcal{E}'''\left[\mathbf{u}(\alpha),\lambda(\alpha);\delta\mathbf{u}\right]=0\end{aligned} \tag{6.15}$$

Equations (6.13), (6.14), and (6.15), and so on, yield the incremental solutions of the nonlinear elastic problems. Particularly, the coefficient derivatives $\dot{\mathbf{u}}(\alpha)$, $\ddot{\mathbf{u}}(\alpha)$, …; $\dot\lambda(\alpha)$, $\ddot\lambda(\alpha)$, … , at the fundamental state $\alpha = 0$ describe the incremental solutions (6.7) and (6.8) of the implicit problem (6.5′).

6.2.3 UNIQUENESS AND STABILITY

We can prove[27] the following Theorem 6.1: The configuration C_0 on the path (6.1) cannot be a branching or a snapping point if the equilibrium at C_0 is stable in the sense that

$$D^2\mathcal{E}\left[\mathbf{u}(0),\lambda(0);\mathbf{u}^2\right]\geq\omega|||\mathbf{u}|||^2,\;\;\omega>0,\;\forall\;\mathbf{u}\in H_A \tag{6.16}$$

Equation (6.13) at $\alpha = 0$ in fact becomes

$$D^2\mathcal{E}[\mathbf{u}(0),\lambda(0);\dot{\mathbf{u}}(0),\delta\mathbf{u}] \geq \dot{\lambda}(0)D\mathcal{E}'[\mathbf{u}(0),\lambda(0);\delta\mathbf{u}] = 0 \tag{6.17}$$

and is solved by

$$\dot{\mathbf{u}}(0) = \dot{\lambda}(0)\mathbf{u}_1 \tag{6.18}$$

where $\mathbf{u}_1$ is the unique solution of the equation

$$D^2\mathcal{E}[\mathbf{u}(0),\lambda(0);\mathbf{u}_1,\delta\mathbf{u}] + D\mathcal{E}'[\mathbf{u}(0),\lambda(0);\delta\mathbf{u}] = 0 \quad \forall\ \delta\mathbf{u} \in H_A \tag{6.17$'$}$$

Equation (6.17′), coupled to condition (6.16), is the variational equation corresponding to the special problem of finding the minimum of the definite positive quadratic functional

$$\mathbf{Q}_2(\mathbf{u}) = \frac{1}{2}D^2\mathcal{E}[\mathbf{u}(0),\lambda(0);\mathbf{u},\mathbf{u}] + D\mathcal{E}'[\mathbf{u}(0),\lambda(0);\mathbf{u}] \tag{6.19}$$

which we discussed in Section 2.7.1.2.The unit tangent vector to the equilibrium path (6.1) is then unique at $\mathbf{u}(0)$. Branching thus cannot occur at C_0. On the other hand C_0 cannot be a snapping point because, on account of (6.17), $\dot{\lambda}(0) = 0$, yields $D^2\mathcal{E}[\mathbf{u}(0),\lambda(0);\dot{\mathbf{u}}(0),\delta\mathbf{u}] = 0$, $\forall\delta\mathbf{u} \in H_A$, contradicting assumption (6.16) unless $\dot{\mathbf{u}}(0) = 0$. In conclusion, assumption (6.16) implies that to every increment $\dot{\lambda}$ of the load multiplier corresponds an unique increment $\dot{\mathbf{u}}$ of the deformation of the system.

Theorem 6.1 comes from the theorem of implicit functions which, with assumption (6.16), implies uniqueness of the solution of (6.5) and, consequently, the nonexistence of branching or snapping at (C_0,λ_0). From a mechanical point of view, assumption (6.16) is equivalent to admit "infinitesimal" stability at (C_0,λ_0), which, as we have just shown, implies uniqueness of the incremental solution. This result demonstrates again the general theorem about connection between Liapounov stability and uniqueness given by Movchan[28] and by Gilbert and Knops.[29]

6.2.4 SNAPPING OR BRANCHING AT THE CRITICAL STATE

The situation changes radically when, with increasing α, the load multiplier λ reaches the critical value

$$\lambda_c = \lambda(0) \tag{6.20}$$

When $\lambda_c = \lambda(0)$ we have $D^2\mathcal{E}[\mathbf{u}(0),\lambda_c;\ \mathbf{u},\mathbf{u}] \geq 0$, according to the definition given in Section 3.4.2.3. We can obtain some information about the existence of branching or snapping at the critical state by means of the following Theorem 6.2: Let the equilibrium be critical at C_0. Thus, if C_0 is a branching point the following condition holds

$$D\mathcal{E}'[\mathbf{u}(0),\lambda_c;\ \mathbf{u}_c] = 0 \tag{6.21}$$

On the contrary, if

$$D\mathcal{E}'[\mathbf{u}(0),\lambda_c;\ \mathbf{u}_c] \neq 0 \qquad 6.22)$$

snap-through occurs at the critical state.

According to the assumptions, at the critical state, defined by the value 0 of the parameter α, from (3.75) and (3.76) we have

$$\exists\,\mathbf{u} = \mathbf{u}_c \in H_A\text{: } D^2\mathcal{E}[\mathbf{u}(0),\lambda_c;\ \mathbf{u}_c,\delta\mathbf{u}] = 0\,\forall\,\delta\mathbf{u} \in H_A \tag{6.23}$$

Equilibrium along the path at $\alpha = 0$ gives, on the other hand, the variational equation

$$D^2\mathcal{E}[\mathbf{u}(0),\lambda_c;\dot{\mathbf{u}}(0)\ \delta\mathbf{u}]+\dot{\lambda}(0)D\mathcal{E}'[\mathbf{u}(0),\lambda_c;\delta\mathbf{u}]=0 \quad \forall\ \delta\mathbf{u}\in H_A \tag{6.24}$$

Thus, taking $\delta\mathbf{u} = \mathbf{u}_c$, (6.24) yields

$$\dot{\lambda}(0)D\mathcal{E}'[\mathbf{u}(0),\lambda_c;\mathbf{u}_c]=0 \tag{6.25}$$

Condition (6.22), taking into account (6.25), is thus sufficient to imply that $\dot{\lambda}(0) = 0$ at the critical state. Moreover, we notice that, with $\dot{\lambda}(0) = 0$, the incremental solution obtained solving (6.24) becomes $\dot{\mathbf{u}}(0) = A\mathbf{u}_c$, where A is an indeterminate constant. The ratio $\dot{\mathbf{u}}(0)/|||\dot{\mathbf{u}}(0)|||$ thus takes the unique value $\mathbf{u}_c/|||\,\mathbf{u}_c|||$. Hence, with $\dot{\lambda}(0) = 0$, equilibrium at C_0 does not bifurcate. Consequently snap-through occurs at C_0.

If, on the contrary, at the critical state, snapping does not occur, $\dot{\lambda}(0) \neq 0$ and condition (6.21) holds. This condition, on the other hand, is not sufficient to entail branching because it does not imply non-uniqueness of the unit tangent vector to the equilibrium branches at the critical state. These results, proven in,[27] extend to the continuous elastic systems a similar theorem, obtained by Thompson[10] for elastic systems with a number of finite degrees of freedom. We can also immediately recognize that the statement of the Theorem 6.2 is verified in the simple examples described by Figs. 6.1, 6.2, 6.3, and 6.4.

6.2.5 STABILITY OR INSTABILITY IN THE NEIGHBORHOOD OF THE CRITICAL STATE

6.2.5.1 General Aspects of the Problem

Let us consider a generic configuration $\mathbf{u} = \mathbf{u}(\alpha)$, $\lambda = \lambda(\alpha)$ of the system T along the equilibrium path (6.1). According to the criterion examined at Section 3.4.2, equilibrium stability is governed by the sign of the minimum of the functional

$$\omega(\alpha) = \min_{H_A - \{0\}} \frac{D^2\mathcal{E}[\mathbf{u}(\alpha),\lambda(\alpha);\ \mathbf{u}^2]}{|||\mathbf{u}|||^2} \tag{6.26}$$

Existence of this minimum can be proven (see Section 3.4.2.2.). The minimizing element $\mathbf{u}_m$ is the eigenelement corresponding to the smallest eigenvalue of the variational equation

$$D^2\mathcal{E}[\mathbf{u}(\alpha),\lambda(\alpha);\mathbf{u}_m(\alpha),\delta\mathbf{u}]-\omega(\alpha)[\mathbf{u}_m(\alpha),\delta\mathbf{u}]=0 \quad \forall\alpha\in[\alpha_1,\alpha_2] \tag{6.27}$$

If the critical state is attained at the value $\alpha = 0$, we have

$$\mathbf{u}_m(0) = \mathbf{u}_c\ ;\ \omega(0) = 0 \tag{6.28}$$

The minimum of the functional (6.26), as well as the minimizing element $\mathbf{u}_m(\alpha)$, can be obtained by means of the MacLaurin expansion

$$\mathbf{u}_m(\alpha)=\mathbf{u}_c+\alpha\dot{\mathbf{u}}_m(0)+\frac{\alpha^2}{2}\ddot{\mathbf{u}}_m(0)+\ldots+o(\alpha^n) \tag{6.29}$$

$$\omega(\alpha)=\alpha\dot{\omega}(0)+\frac{\alpha^2}{2}\ddot{\omega}(0)+\ldots+o(\alpha^n) \tag{6.30}$$

where the derivatives $\dot{\mathbf{u}}_m(0)$, $\ddot{\mathbf{u}}_m(0)$, $\dot{\omega}(0)$, $\ddot{\omega}(0)$, ... will be suitably evaluated. We can state now the following Theorem 6.3: An exchange of stability takes place at the critical state along the equilibrium branch if

$$\dot{\omega}(0) \neq 0 \tag{6.31}$$

If, on the contrary, $\dot{\omega}(0) = 0$, equilibrium in the neighborhood of the critical state will be stable or unstable according to whether

$$\ddot{\omega}(0) > 0 \text{ or } \ddot{\omega}(0) < 0 \tag{6.32}$$

In fact, if $\dot{\omega}(0) \neq 0$, at the equilibrium states near to C_0, according to (6.30) we will have

$$\operatorname{sgn} \omega(\alpha) = \operatorname{sgn} \alpha\dot{\omega}(0) \tag{6.33}$$

Thus, if $\dot{\omega}(0) < 0$, the equilibrium states corresponding to $\alpha < 0$ are stable, while those corresponding to to $\alpha > 0$ are unstable. The contrary occurs if $\dot{\omega}(0) > 0$. On the other hand, if $\dot{\omega}(0) = 0$, we will have

$$\operatorname{sgn} \omega(\alpha) = \operatorname{sgn} \ddot{\omega}(0) \tag{6.34}$$

for any value, positive or negative, of α in the neighborhood of $\alpha = 0$.

This theorem, given in reference 27, repeats a classical result obtained by Poincaré[5] for systems with finite degrees of freedom on the stability exchanges occurring at the critical state. The theorem also repeats some of Koiter's results as far as equilibrium bifurcation is concerned.[8,9]

We can now give a first sketch of the behavior of the nonlinear elastic system under loading. We first suppose that the load multiplier $\lambda = \lambda(\alpha)$, zero when $\alpha = \alpha_1$, is a strictly increasing function of α. The critical state is then attained for $\alpha = 0$; thus for $\alpha_1 \leq \alpha < 0$ the equilibrium is stable in the sense specified by Theorem 6.1. At this stage of loading, to any increment of λ corresponds an unique increment $\dot{\mathbf{u}}(\alpha)$ of the deformation of the system. By increasing α, on the other hand, the critical state is attained at $\alpha = 0$ and λ reaches the critical value λ_c. An exchange of stability thus occurs along the branch if $\dot{\omega}(0) \neq 0$: the subsequent equilibrium states will be unstable. Further, a snapping occurs at the critical state if $D\mathcal{E}'\left[\mathbf{u}(0), \lambda_c; \mathbf{u}_c\right] \neq 0$. If, on the contrary, $D\mathcal{E}'\left[\mathbf{u}(0), \lambda_c; \mathbf{u}_c\right] = 0$, with $\dot{\omega}(0) \neq 0$, the critical state could represent as a branching as a snapping point. If, on the contrary, $\dot{\omega}(0) = 0$ and $\ddot{\omega}(0) > 0$, only an isolated critical state occurs at $\lambda = \lambda_c$, since equilibrium will continue to be stable along the subsequent states.

The evaluation of derivatives $\dot{\omega}(0)$ and $\ddot{\omega}(0)$ can be very useful for a detailed analysis of the various equilibrium paths. All the subsequent derivatives of identity (6.27) will be equal to zero, since (6.27) holds for every value of the loading parameter α. Thus we have

$$\begin{aligned}
&\frac{d}{d\alpha}\left\{D^2\mathcal{E}\left[\mathbf{u}(\alpha), \lambda(\alpha); \mathbf{u}_m(\alpha), \delta\mathbf{u}\right] - \omega(\alpha)\left[\mathbf{u}_m(\alpha), \delta\mathbf{u}\right]\right\} \\
&= D^2\mathcal{E}\left[\mathbf{u}(\alpha), \lambda(\alpha); \dot{\mathbf{u}}_m(\alpha), \delta\mathbf{u}\right] + D^3\mathcal{E}\left[\mathbf{u}(\alpha), \lambda(\alpha); \dot{\mathbf{u}}(\alpha), \mathbf{u}_m(\alpha), \delta\mathbf{u}\right] \\
&+ \dot{\lambda}(\alpha) D^2\mathcal{E}'\left[\mathbf{u}(\alpha), \lambda(\alpha); \mathbf{u}_m(\alpha), \delta\mathbf{u}\right] - \dot{\omega}(\alpha)\left[\mathbf{u}_m(\alpha), \delta\mathbf{u}\right] - \omega(\alpha)\left[\dot{\mathbf{u}}_m(\alpha), \delta\mathbf{u}\right] = 0
\end{aligned} \tag{6.35}$$

In particular, taking $\alpha = 0$ and $\delta\mathbf{u} = \mathbf{u}_c$ we get

$$\begin{aligned}
&\frac{d}{d\alpha}\left\{D^2\mathcal{E}\left[\mathbf{u}(\alpha), \lambda(\alpha); \mathbf{u}_m(\alpha), \delta\mathbf{u}\right] - \omega(\alpha)\left[\mathbf{u}_m(\alpha), \delta\mathbf{u}\right]\right\}_{\alpha=0} \\
&= D^3\mathcal{E}\left[\mathbf{u}(0), \lambda(0); \dot{\mathbf{u}}(0), \mathbf{u}_c^2\right] + \dot{\lambda}(0) D^2\mathcal{E}'\left[\mathbf{u}(0), \lambda(0); \mathbf{u}_c^2\right] - \dot{\omega}(0)\left[\mathbf{u}_c, \mathbf{u}_c\right] = 0
\end{aligned} \tag{6.36}$$

because

$$\mathbf{u}_m(0) = \mathbf{u}_c \quad \omega(0) = 0 \tag{6.37}$$

Using the notation

$$\mathbf{u}^* = \frac{\mathbf{u}}{|||\mathbf{u}|||} \tag{6.38}$$

from (6.36) we obtain

$$\dot{\omega}(0) = D^3\mathcal{E}\left[\mathbf{u}(0),\lambda_c;\dot{\mathbf{u}}(0),\mathbf{u}_c^{*2}\right] + \dot{\lambda}(0)D^2\mathcal{E}'\left[\mathbf{u}(0),\lambda_c;\mathbf{u}_c^{*2}\right] \tag{6.39}$$

Let us again differentiate identity (6.35). We get

$$\begin{aligned}&\frac{d}{d\alpha}\Big\{D^2\mathcal{E}[\mathbf{u}(\alpha),\lambda(\alpha);\dot{\mathbf{u}}_m(\alpha),\delta\mathbf{u}] + D^3\mathcal{E}[\mathbf{u}(\alpha),\lambda(\alpha);\dot{\mathbf{u}}(\alpha),\mathbf{u}_m(\alpha),\delta\mathbf{u}]\\ &+ \dot{\lambda}(\alpha)D^2\mathcal{E}'\left[\mathbf{u}(\alpha),\lambda(\alpha);\mathbf{u}_m(\alpha),\delta\mathbf{u}\right] - \dot{\omega}(\alpha)[\mathbf{u}_m(\alpha),\delta\mathbf{u}] - \omega(\alpha)[\dot{\mathbf{u}}_m(\alpha),\delta\mathbf{u}]\Big\}_{\alpha=0} = 0\end{aligned} \tag{6.40}$$

Thus, still with $\alpha = 0$ and $\delta\mathbf{u} = \mathbf{u}_c^*$ taking into account (6.38) and (6.39), we have

$$\begin{aligned}&2D^3\mathcal{E}\left[\mathbf{u}(0),\lambda_c;\dot{\mathbf{u}}(0),\dot{\mathbf{u}}_m(0),\mathbf{u}_c^*\right] + 2\dot{\lambda}(0)D^2\mathcal{E}'\left[\mathbf{u}(0),\lambda_c;\dot{\mathbf{u}}_m(0),\mathbf{u}_c^*\right]\\ &+ D^3\mathcal{E}\left[\mathbf{u}(0),\lambda_c;\ddot{\mathbf{u}}(0),\mathbf{u}_c^{*2}\right] + D^4\mathcal{E}\left[\mathbf{u}(0),\lambda_c;\dot{\mathbf{u}}^2(0),\mathbf{u}_c^{*2}\right]\\ &+ 2\dot{\lambda}(0)D^3\mathcal{E}'\left[\mathbf{u}(0),\lambda_c;\dot{\mathbf{u}}(0),\mathbf{u}_c^{*2}\right] + \ddot{\lambda}(0)D^2\mathcal{E}'\left[\mathbf{u}(0),\lambda_c;\mathbf{u}_c^{*2}\right]\\ &+ \dot{\lambda}^2(0)D^2\mathcal{E}''\left[\mathbf{u}(0),\lambda_c;\mathbf{u}_c^{*2}\right] - \ddot{\omega}(0)\left[\mathbf{u}_m(0),\mathbf{u}_c^*\right] - 2\dot{\omega}(0)\left[\dot{\mathbf{u}}_m(0),\mathbf{u}_c^*\right] = 0\end{aligned} \tag{6.41}$$

On the other hand, since $\mathbf{u}_m(\alpha)$ has an arbitrary norm, we can assume $|||\mathbf{u}_m(\alpha)||| = 1$ and

$$\frac{d}{d\alpha}|||u_m(\alpha)||| = 0 \quad \forall\ \alpha \in [\alpha_1, \alpha_2]$$

Thus we have

$$\frac{d}{d\alpha}|||\mathbf{u}_m(\alpha)||| = 2\{[\dot{\mathbf{u}}_m(\alpha),\mathbf{u}_m(\alpha)]\}_{\alpha=0} = 2[\dot{\mathbf{u}}_m(0),\mathbf{u}_c] = 0$$

and

$$\begin{aligned}&\left[\mathbf{u}_m(0),\mathbf{u}_c^*\right]\ddot{\omega}(0) = \dot{\lambda}^2(0)D^2\mathcal{E}''\left[\mathbf{u}(0),\lambda_c;\mathbf{u}_c^{*2}\right] + 2D^3\mathcal{E}\left[\mathbf{u}(0),\lambda_c;\dot{\mathbf{u}}(0),\dot{\mathbf{u}}_m(0),\mathbf{u}_c^*\right]\\ &+ 2\dot{\lambda}(0)D^2\mathcal{E}'\left[\mathbf{u}(0),\lambda_c;\dot{\mathbf{u}}_m(0),\mathbf{u}_c^*\right] + D^3\mathcal{E}\left[\mathbf{u}(0),\lambda_c;\ddot{\mathbf{u}}(0),\mathbf{u}_c^{*2}\right]\\ &+ D^4\mathcal{E}\left[\mathbf{u}(0),\lambda_c;\dot{\mathbf{u}}^2(0),\mathbf{u}_c^{*2}\right] + 2\dot{\lambda}(0)D^3\mathcal{E}'\left[\mathbf{u}(0),\lambda_c;\dot{\mathbf{u}}(0),\mathbf{u}_c^{*2}\right] + \ddot{\lambda}(0)D^2\mathcal{E}'\left[\mathbf{u}(0),\lambda_c;\mathbf{u}_c^{*2}\right]\end{aligned} \tag{6.42}$$

We will examine next the two cases of the asymmetrical and symmetrical critical state. In both cases snapping or branching can occur.

6.2.5.2 The Load vs. Shortening Curve; the Budiansky Theorem

Looking at a load vs. deflection curve, we sometimes encounter the opinion that there is stability if the load is rising, instability otherwise. An unknown branching on a rising load-deflection curve, of course, invalidates this statement. On the other hand, if the deflection is the "shortening" $\Delta(\mathbf{u})$ of the structure, i.e., that particular displacement along which the acting load $\lambda\mathbf{p}$ works, the affirmation contains some truth. According to Budiansky[31] we can affirm, in fact, that a negative slope on a load-shortening curve can only be associated with an unstable equilibrium state; i.e., $d\Delta/d\lambda$ (or $d\lambda/d\Delta$) negative implies instability. We will prove here this statement by using the sufficient condition of instability given in Section 3.4.2.4.

In most problems the potential energy can be expressed as

$$\mathcal{E}(\mathbf{u}, \lambda) = \mathcal{V}(\mathbf{u}) + \mathcal{U}(\mathbf{u}, \lambda)$$

where $\mathcal{V}(\mathbf{u})$ is the strain energy of the system and $\mathcal{U}(\mathbf{u},\lambda)$ is the potential energy associated with the prescribed loading. As a rule, for most problems $\mathcal{U}(\mathbf{u},\lambda)$ is linear in λ and can be written in the form

$$\mathcal{U}(\mathbf{u}, \lambda) = -\lambda\, \Delta(\mathbf{u}) \tag{6.43}$$

where $\Delta(\mathbf{0}) = 0$; the functional $\Delta(\mathbf{u})$ can be regarded as a generalized "shortening". The strain energy depends on λ only implicitly via the displacement field $\mathbf{u}$. It follows that

$$\mathcal{E}'(\mathbf{u}, \lambda) = \frac{\partial \mathcal{E}(\mathbf{u}, \lambda)}{\partial \lambda} = -\Delta(\mathbf{u})$$

We now examine equilibrium along the path $\mathbf{u}(\lambda)$. Condition (6.5) holds and

$$D\mathcal{E}[\mathbf{u}(\lambda), \lambda; \mathbf{u}'(\lambda)] = 0\ , \quad \mathbf{u}'(\lambda) = \frac{d\mathbf{u}}{d\lambda}$$

We now evaluate the derivative

$$\frac{d\Delta(\mathbf{u}(\lambda))}{d\lambda}$$

We have

$$\frac{d\Delta(\mathbf{u}(\lambda))}{d\lambda} = D\Delta[\mathbf{u}(\lambda); \mathbf{u}'(\lambda)] = -\, D\mathcal{E}'[\mathbf{u}(\lambda), \lambda; \mathbf{u}'(\lambda)]$$

But along the path $\mathbf{u} = \mathbf{u}(\lambda)$, according to (6.10), we have

$$\frac{d}{d\lambda} D\mathcal{E}[\mathbf{u}(\lambda), \lambda; \mathbf{u}'(\lambda)] = D^2\mathcal{E}[\mathbf{u}(\lambda), \lambda; \mathbf{u}'(\lambda), \mathbf{u}'(\lambda)] + D\mathcal{E}'[\mathbf{u}(\lambda), \lambda; \mathbf{u}'(\lambda)] = 0$$

and

$$\frac{d\Delta(\mathbf{u}(\lambda))}{d\lambda} = D^2\mathcal{E}[\mathbf{u}(\lambda), \lambda; \mathbf{u}'(\lambda), \mathbf{u}'(\lambda)] \tag{6.44}$$

This shows that if $d\Delta/d\lambda$ (or $d\lambda/d\Delta$) is negative, $D^2\mathcal{E}[\mathbf{u}(\lambda), \lambda; \delta\mathbf{u}, \delta\mathbf{u}]$ will also be negative for $\delta\mathbf{u} = \mathbf{u}'(\lambda)$. This last result, according to the theorem given in Section 3.4.2.4, proves instability. We observe that, on the contrary, $d\Delta/d\lambda$ (or $d\lambda/d\Delta$) positive does not prove stability.

Fig. 1.6 shows the load versus shortening curve for the cylindrical shell under axial compression. Condition (6.44) thus points out that instability occurs along the branch of Fig. 1.6 defined by decreasing λ.

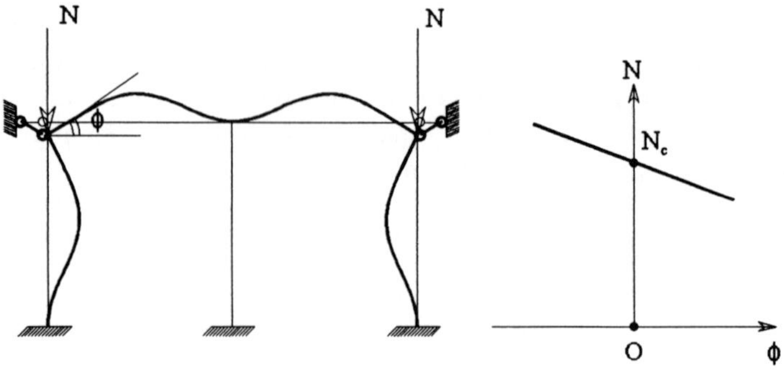

FIGURE 6.8

6.2.6 ASYMMETRIC CRITICAL STATE

The critical state is defined as asymmetric if

$$D^3\mathcal{E}\left[\mathbf{u}(0),\lambda(0);\mathbf{u}_c^{*3}\right] \neq 0 \tag{6.45}$$

More generally, the critical state is asymmetric if

$$D^{(2n+1)}\mathcal{E}\left[\mathbf{u}(0),\lambda(0);\ \mathbf{u}_c^{*(2n+1)}\right] \neq 0 \tag{6.45'}$$

where 2n + 1 is the lowest order of the nonzero differential of $\mathcal{E}(\mathbf{u},\lambda)$ evaluated along the critical mode $\mathbf{u}_c^*$. Of course, the nonsymmetric critical state is unstable.

As a rule, the critical state is asymmetric. Asymmetric critical state can occur also for symmetric structures that exhibit symmetrical critical displacements. (Fig. 6.8). In fact, let Ω_0 be the region of the space occupied by the symmetric structure T at C_0. According to the assumption, $\forall$ P $\in \Omega_0$ but not belonging to the symmetry axis, $\exists P' \neq P \in \Omega_0 : \mathbf{u}_c^*(P) = \mathbf{u}_c^*(P')$. Thus, condition (6.45), as a rule, is satisfied because of the analytical form of the third differential $D^3\mathcal{E}[\mathbf{u}(0),\lambda_c;\ \mathbf{u}_c^3]$.

6.2.6.1 Asymmetric Snapping

According to Theorem 6.2, which controls the critical state of the general nonlinear elastic system, snapping occurs at the critical state if $D\mathcal{E}'\left[\mathbf{u}(0),\ \lambda_c;\ \mathbf{u}_c\right] \neq 0$. Thus, in general, nonlinear structures snap-through at the critical state. According to (6.39) and (6.24), since $\dot{\lambda}(0) = 0$ and $\dot{\mathbf{u}}(0) = \mathbf{u}_c^*$, we have

$$\dot{\omega}(0) = D^3\mathcal{E}\left[\mathbf{u}(0),\lambda_c;\dot{\mathbf{u}}(0),\mathbf{u}_c^{*2}\right] \tag{6.46}$$

A stability exchange thus occurs at the critical state. From (6.14), taking into account that $\dot{\lambda}(0) = 0$, $\dot{\mathbf{u}}(0) = \mathbf{u}_c^*$ and assuming $\delta\mathbf{u} = \mathbf{u}_c^*$ we have

$$\ddot{\lambda}(0) = -\frac{D^3\mathcal{E}\left[\mathbf{u}(0),\lambda_c;\mathbf{u}_c^{*3}\right]}{D\mathcal{E}'\left[\mathbf{u}(0),\lambda_c;\mathbf{u}_c^{*}\right]} \tag{6.47}$$

where the direction of $\mathbf{u}_c^*$ is chosen so that $D\mathcal{E}'[\mathbf{u}(0),\lambda_c;\ \mathbf{u}_c^*] < 0$. Thus the equations

$$\mathbf{u}(\alpha) = \mathbf{u}(0) + \alpha\mathbf{u}_c^* + o(\alpha) \tag{6.48}$$

$$\lambda(\alpha) = \lambda_c + \frac{\alpha^2}{2}\ddot{\lambda}(0) + o(\alpha^2) \tag{6.49}$$

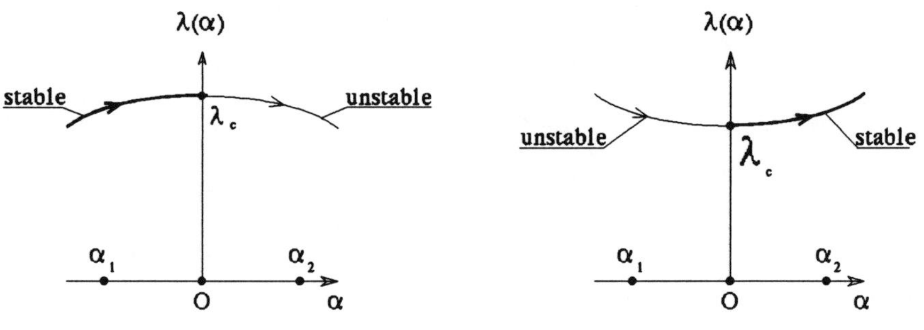

FIGURE 6.9

define the branch of the equilibrium states. Fig. 6.9 sketches the two possible cases of snapping that can occur, according to whether $\ddot{\lambda}(0) < 0$ or $\ddot{\lambda}(0) > 0$. We say that when $\ddot{\lambda}(0) < 0$ the snapping load is the failure load of the structural system.

6.2.6.2 Asymmetric Branching

If, in place of (6.22), condition (6.21) holds, i.e., $D\mathcal{E}'[\mathbf{u}(0),\lambda_c;\mathbf{u}_c^*] = 0$, branching can occur at the critical state. The incremental solution $\dot{\mathbf{u}}(0)$ of (6.24), now is

$$\dot{\mathbf{u}}(0) = \mathbf{u}_c^* + \dot{\lambda}(0)\mathbf{v}_1 \tag{6.50}$$

where $\mathbf{u}_c^*$, i.e., the critical displacement, is the integral of the homogeneous equation associated to (6.24). The displacement field $\mathbf{v}_1$, on the contrary, orthogonal in energy product to $\mathbf{u}_c^*$, i.e., satisfying the normality condition

$$[\mathbf{u}_c^*,\mathbf{v}_1] = 0 \tag{6.51}$$

is a particular integral of (6.24) and is the unique element $\mathbf{u} \in H_A^{(1)}$ that resolves the variational equation

$$D^2\mathcal{E}\left[\mathbf{u}(0),\lambda_c;\dot{\mathbf{u}}(0),\delta\mathbf{u}\right] + D\mathcal{E}'\left[\mathbf{u}(0),\lambda_c;\delta\mathbf{u}\right] = 0 \quad \forall\ \delta\mathbf{u} \in H_A \tag{6.52}$$

where

$$H_A^{(1)} = \{\mathbf{u} \in H_A : [\mathbf{u}, \mathbf{u}_c^*\ ] = 0\} \tag{6.53}$$

Condition (6.52), with (6.53), corresponds in fact to the research of the minimum of a definite positive quadratic functional.

First of all we observe that at the critical state bifurcation really occurs. In fact, if we put in (6.14) $\delta\mathbf{u} = \mathbf{u}_c^*$ we have

$$\dot{\lambda}^2(0)\left[D^3\mathcal{E}\left[\mathbf{u}(0),\lambda_c;\mathbf{v}_1^2,\mathbf{u}_c^*\right] + 2D^2\mathcal{E}'\left[\mathbf{u}(0),\lambda_c;\mathbf{v}_1,\mathbf{u}_c^*\right] + D\mathcal{E}''\left[\mathbf{u}(0),\lambda_c;\mathbf{u}_c^*\right]\right]$$

$$+2\dot{\lambda}(0)\left[D^3\mathcal{E}\left[\mathbf{u}(0),\lambda_c;\mathbf{v}_1,\mathbf{u}_c^{*2}\right] + D^2\mathcal{E}'\left[\mathbf{u}(0),\lambda_c;\mathbf{u}_c^{*2}\right]\right] + D^3\mathcal{E}\left[\mathbf{u}(0),\lambda_c;\mathbf{u}_c^{*3}\right] = 0 \tag{6.54}$$

This last algebraic equation gives, as a rule, two distinct roots $\dot{\lambda}_I(0)$ and $\dot{\lambda}_{II}(0)$ to which correspond two different equilibrium branches intersecting each other at the critical state. Thus at (C_0,λ_c) the equilibrium bifurcates and is characterized by

$$\mathbf{u}_I(\alpha) - \mathbf{u}(0) = \left[\mathbf{u}_c^* + \dot{\lambda}_I(0)\mathbf{v}_I\right]\alpha + o(\alpha) \qquad \lambda_I(\alpha) - \lambda_c = \dot{\lambda}_I(0)\alpha + o(\alpha) \tag{6.55}$$

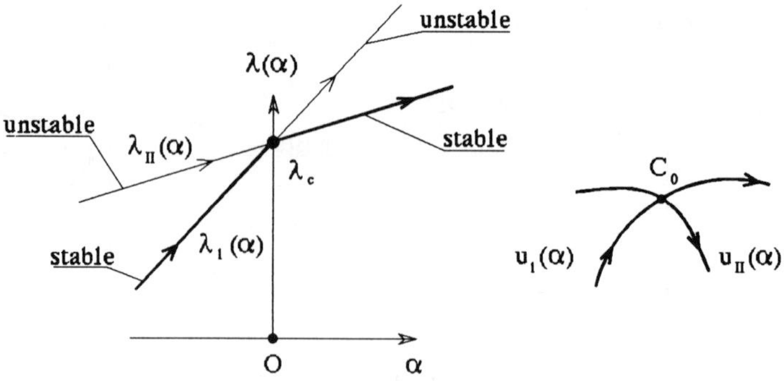

FIGURE 6.10

$$\mathbf{u}_{II}(\alpha) - u(0) = \left[\mathbf{u}_c^* + \dot{\lambda}_{II}(0)\mathbf{v}_{II}\right]\alpha + o(\alpha) \quad \lambda_{II}(\alpha) - \lambda_c = \dot{\lambda}_{II}(0)\alpha + o(\alpha) \tag{6.56)}$$

The derivative $\dot{\omega}(0)$ defines the quality of the equilibrium in the neighborhood of the bifurcation point. Substituting (6.55) into (6.39) we obtain, as a rule, two different values

$$\dot{\omega}_I(0) = \dot{\lambda}_I(0)\sqrt{\Delta}\ ,\ \dot{\omega}_{II}(0) = -\ \dot{\lambda}_{II}(0)\sqrt{\Delta} \tag{6.57}$$

where $4\Delta$is the discriminant of (6.54). Equation (6.57) shows that a stability exchange occurs between the two equilibrium branches if the roots of (6.54) have the same sign. Fig. 6.10 shows the bifurcation that occurs in this case at the critical state.

6.2.7 SYMMETRIC CRITICAL STATE

The critical state is symmetric if

$$D^3\mathcal{E}\left[\mathbf{u}(0),\lambda_c;\mathbf{u}_c^{*3}\right] = 0 \qquad D^4\mathcal{E}\left[\mathbf{u}(0),\lambda_c;\mathbf{u}_c^{*4}\right] \neq 0 \tag{6.58}$$

or, more generally, if

$$D^{(2n-1)}\mathcal{E}\left[\mathbf{u}(0),\lambda_c;\mathbf{u}_c^{*(2n-1)}\right] = 0 \qquad D^{2n}\mathcal{E}\left[\mathbf{u}(0),\lambda_c;\mathbf{u}_c^{*2n}\right] \neq 0 \tag{6.58'}$$

where 2n is the lowest order nonzero differential of $\mathcal{E}(\mathbf{u},\lambda)$ along the critical mode $\mathbf{u}_c^*$.

Conditions (6.58) show that symmetric critical states are more an exception than a rule. The following statement on the other hand, is interesting: "In symmetric structures with antisymmetric critical displacements the critical state is symmetric"(Fig. 6.11).

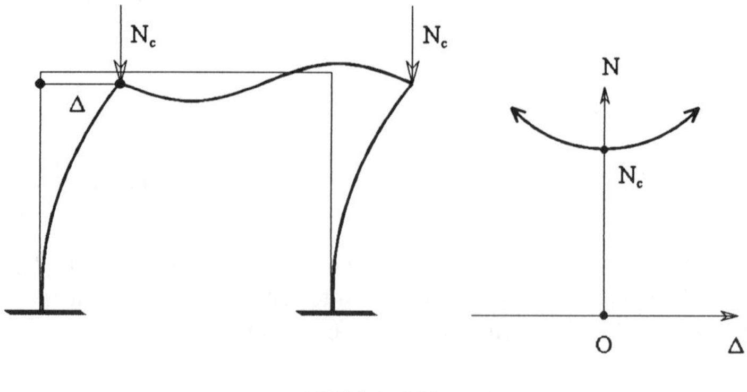

FIGURE 6.11

In fact, because of the assumption, $\forall$ P $\in \Omega_0$ not belonging to the symmetry axis of the system, there exists a point P′ $\neq$ P $\in \Omega_0$: $\mathbf{u}_c^*(P) = -\mathbf{u}_c^*(P')$. Thus, because of the analytical form of differentials of odd order of the additional potential energy, $D^3\mathcal{E}[\mathbf{u}(\alpha_c),\lambda_c;\mathbf{u}_c^{*3}] = 0$, $D^5\mathcal{E}[\mathbf{u}(\alpha_c),\lambda_c;\mathbf{u}_c^{*5}] = 0$, ..., and only differentials of even order will be different from zero.

6.2.7.1 Symmetric Snapping

Snapping occurs at the critical state C_0 whenever $D\mathcal{E}'[\mathbf{u}(0),\lambda_c;\mathbf{u}_c^*] \neq 0$. Thus we get, from (6.47), with $\dot{\lambda}(0) = 0$, $\dot{\mathbf{u}}(0) = \mathbf{u}_c^*$

$$\ddot{\lambda}(0) = 0 \tag{6.59}$$

Solution of (6.14) on the other hand gives

$$\ddot{\mathbf{u}}(0) = \mathbf{v}_2 \tag{6.60}$$

where $\mathbf{v}_2$ is the unique element of $H_A^{(1)}$, i.e., constrained by the condition

$$[\mathbf{u}_c^*, \mathbf{v}_2] = 0 \tag{6.61}$$

that resolves the variational equation

$$D^2\mathcal{E}[\mathbf{u}(0),\lambda_c;\mathbf{v}_2,\delta\mathbf{u}] + D^3\mathcal{E}[\mathbf{u}(0),\lambda_c;\mathbf{u}_c^{*2},\delta\mathbf{u}] = 0 \quad \forall\ \delta\mathbf{u} \in H_A \tag{6.62}$$

This equation is equal to (3.101), which, with (3.99), controls the Koiter stability criterion at the critical state. This equation (6.62), with $\delta\mathbf{u} = \mathbf{v}_2$, gives

$$D^3\mathcal{E}[\mathbf{u}(0),\lambda_c;\mathbf{u}_c^{*2},\mathbf{v}_2] = -\,D^2\mathcal{E}[\mathbf{u}(0),\lambda_c;\mathbf{v}_2^2] \tag{6.63}$$

On the other hand from (6.15), taking into account (6.59) and (6.60), with $\delta\mathbf{u} = \mathbf{u}_c^*$ we get

$$3D^3\mathcal{E}[\mathbf{u}(0),\lambda_c;\mathbf{u}_c^{*2},\mathbf{v}_2] + D^4\mathcal{E}[\mathbf{u}(0),\lambda_c;\mathbf{u}_c^{*4},] + \dddot{\lambda}(0)D\mathcal{E}'[\mathbf{u}(0),\lambda_c;\mathbf{u}_c^*] = 0 \tag{6.64}$$

which gives

$$\dddot{\lambda}(0) = -\frac{D^4\mathcal{E}[\mathbf{u}(0),\lambda_c;\mathbf{u}_c^{*4}] - 3D^3\mathcal{E}[\mathbf{u}(0),\lambda_c;\mathbf{u}_c^{*2},\mathbf{v}_2]}{D\mathcal{E}'[\mathbf{u}(0),\lambda_c;\mathbf{u}_c^*]} \tag{6.65}$$

The only branch of the equilibrium states passing through (C_0,λ_c) is defined, in the neighborhood of C_0, by the equations

$$\mathbf{u}(\alpha) - \mathbf{u}(0) = \alpha\mathbf{u}_c^* + \frac{\alpha^2}{2}\mathbf{v}_2 + o(\alpha^2) \tag{6.66}$$

$$\lambda(\alpha) - \lambda_c = \frac{\alpha^3}{6}\dddot{\lambda}(0) + o(\alpha^3) \tag{6.67}$$

Thus λ does not achieve a maximum or minimum at the critical point C_0. Moreover, from (6.39) we have $\dot{\omega}(0) = 0$. Hence, we have to evaluate the second derivative $\ddot{\omega}$ to analyze stability in the neighborhood of C_0. From (6.42), taking in account that $\dot{\lambda}(0) = \ddot{\lambda}(0) = 0$, $\mathbf{u}_m(0) = \mathbf{u}_c^*$ we get

$$\ddot{\omega}(0) = 2D^3\mathcal{E}[\mathbf{u}(0),\lambda_c;\dot{\mathbf{u}}_m(0),\mathbf{u}_c^{*2}] + D^3\mathcal{E}[\mathbf{u}(0),\lambda_c;\mathbf{v}_2,\mathbf{u}_c^{*2}] + D^4\mathcal{E}[\mathbf{u}(0),\lambda_c;\mathbf{u}_c^{*4}] \tag{6.68}$$

and we now have to evaluate $\dot{\mathbf{u}}_m(0)$. From (6.35), with $\alpha = 0$ and taking into account that $\omega(0) = \dot{\omega}(0) = 0$, $\dot{\lambda}(0) = 0$, $\dot{\mathbf{u}}(0) = \mathbf{u}_c^*$, we get

$$D^2\mathcal{E}\left[\mathbf{u}(0),\lambda_c;\dot{\mathbf{u}}_m(0),\delta\mathbf{u}\right]+D^3\mathcal{E}\left[\mathbf{u}(0),\lambda_c;\mathbf{u}_c^{*2},\delta\mathbf{u}\right]=0 \tag{6.69}$$

Comparing (6.69) with (6.62) thus yields

$$\dot{\mathbf{u}}_m(0) = \mathbf{v}_2 \tag{6.70}$$

and from (6.68), taking into account (6.70) and (6.63), we have

$$\ddot{\omega}(0) = D^4\mathcal{E}\left[\mathbf{u}(0),\lambda_c;\mathbf{u}_c^{*4}\right]-3D^2\mathcal{E}\left[\mathbf{u}(0),\lambda_c;\mathbf{v}_2^2\right] \tag{6.71}$$

and the equilibrium is stable near C_0 and along the branch passing through (C_0,λ_c) if

$$D^4\mathcal{E}\left[\mathbf{u}(0),\lambda_c;\mathbf{u}_c^{*4}\right]-3D^2\mathcal{E}\left[\mathbf{u}(0),\lambda_c;\mathbf{v}_2^2\right]>0 \tag{6.72}$$

We find again the Koiter condition (3.99). According to (2.98′), we have in fact

$$D^4\mathcal{E}\left[\mathbf{u}(0),\lambda_c;\mathbf{u}_c^{*4}\right]=4!\;\mathcal{E}_4\left[\mathbf{u}(0),\lambda_c;\mathbf{u}_c^{*4}\right];$$

$$3D^2\mathcal{E}\left[\mathbf{u}(0),\lambda_c;\mathbf{v}_2^2\right]=3\times 2!\;\mathcal{E}_2\left[\mathbf{u}(0),\lambda_c;\mathbf{v}_2^2\right] \tag{6.73}$$

In fact, from (6.69) and (6.70) we have $D^2\mathcal{E}[\mathbf{u}(0), \lambda_c; \mathbf{v}_2, \delta\mathbf{u}] = \mathcal{E}_{11}(\mathbf{v}_2, \delta\mathbf{u})$ and $D^3\mathcal{E}[\mathbf{u}(0), \lambda_c; \mathbf{u}_c^{*2}, \delta\mathbf{u}] = \mathcal{E}_{21}(\mathbf{u}_c^*, \delta\mathbf{u})$ according to the symbology of (2.98′). Comparing (6.62) with (3.101) we get $\mathbf{v}_2 = 2\Phi_1$, where Φ_1 is the unique element that solves (3.100) and (3.101). Then

$$\begin{aligned}\ddot{\omega}(0) &= D^4\mathcal{E}\left[\mathbf{u}(0),\lambda_c;\mathbf{u}_c^{*4}\right]-3D^2\mathcal{E}\left[\mathbf{u}(0),\lambda_c;\mathbf{v}_2^2\right]\\ &= 4!\left\{\mathcal{E}_4\left[\mathbf{u}(0),\lambda_c;\mathbf{u}_c^{*4}\right]-\mathcal{E}_2\left[\mathbf{u}(0),\lambda_c;\Phi_1^2\right]\right\}\end{aligned} \tag{6.74}$$

and we find again the condition (3.99). From (6.65) and (6.74), we have

$$\dddot{\lambda}(0) = -\frac{\ddot{\omega}(0)}{D\mathcal{E}'\left[\mathbf{u}(0),\lambda_c;\mathbf{u}_c^*\right]} \tag{6.65′}$$

From this analysis we discover that condition (3.99) implies stability not only at the critical state but also in its neighborhood. Fig. 6.12 sketches the possible equilibrium branches

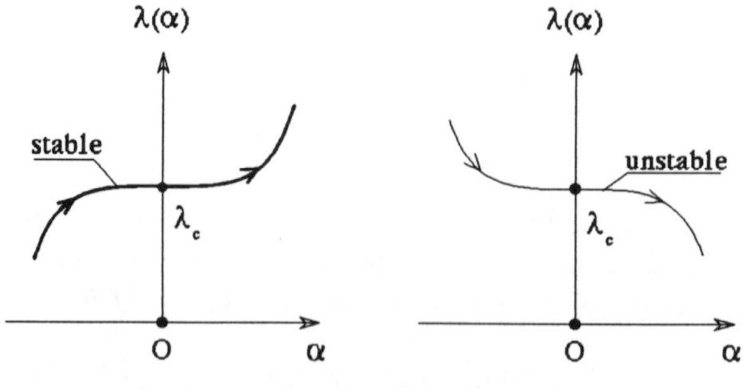

FIGURE 6.12

according to $\ddot{\omega}(0) > 0$ or $\ddot{\omega}(0) < 0$. In the figure the direction of $\mathbf{u}_c^*$ has been taken so that $D\mathcal{E}'[\mathbf{u}(0), \lambda_c; \mathbf{u}_c^*] < 0$.

6.2.7.2 Symmetric Branching

We have now

$$D^3\mathcal{E}\left[\mathbf{u}(0),\lambda_c;\mathbf{u}_c^{*3}\right]=0 \qquad D\mathcal{E}'\left[\mathbf{u}(0),\lambda_c;\mathbf{u}_c^*\right]=0 \tag{6.75}$$

The incremental solution of (6.24) is

$$\dot{\mathbf{u}}(0) = \mathbf{u}_c^* + \dot{\lambda}(0)\mathbf{v}_1 \tag{6.76}$$

where $\mathbf{u}_c^*$ is the integral of the associated homogeneous equation and $\mathbf{v}_1$, orthogonal in energy product to $\mathbf{u}_c^*$, i.e., satisfying the condition

$$[\ \mathbf{u}_c^*,\ \mathbf{v}_1\ ] = 0 \tag{6.77}$$

is the unique element $\mathbf{u}$ in $H_A^{(1)}$ that solves the variational equation

$$D^2\mathcal{E}\left[\mathbf{u}(0),\lambda_c;\mathbf{u},\delta\mathbf{u}\right]+D\mathcal{E}'\left[\mathbf{u}(0),\lambda_c;\delta\mathbf{u}\right]=0 \quad \forall\ \delta\mathbf{u}\in H_A \tag{6.78}$$

still corresponding to the search for the minimum of a definite positive quadratic functional.

The increment $\dot{\lambda}(0)$ is obtained by solving (6.14). From this equation, taking $\delta\mathbf{u} = \mathbf{u}_c^*$ we have in fact

$$\dot{\lambda}^2(0)\left\{D^3\mathcal{E}\left[\mathbf{u}(0),\lambda_c;\mathbf{u}_c^*,\mathbf{v}_1^2\right]+D\mathcal{E}''\left[\mathbf{u}(0),\lambda_c;\mathbf{u}_c^*\right]+2D^2\mathcal{E}'\left[\mathbf{u}(0),\lambda_c;\mathbf{u}_c^*,\mathbf{v}_1\right]\right\}$$

$$+\ 2\dot{\lambda}(0)\left\{D^3\mathcal{E}\left[\mathbf{u}(0),\lambda_c;\mathbf{v}_1,\mathbf{u}_c^{*2}\right]+D^2\mathcal{E}'\left[\mathbf{u}(0),\lambda_c;\mathbf{u}_c^{*2}\right]\right\}=0 \tag{6.79}$$

and we get the two solutions

$$\dot{\lambda}_I(0)=-\frac{2D^3\mathcal{E}\left[\mathbf{u}(0),\lambda_c;\mathbf{v}_1,\mathbf{u}_c^{*2}\right]+2D^2\mathcal{E}'\left[\mathbf{u}(0),\lambda_c;\mathbf{u}_c^{*2}\right]}{D^3\mathcal{E}\left[\mathbf{u}(0),\lambda_c;\mathbf{u}_c^*,\mathbf{v}_1^2\right]+D\mathcal{E}''\left[\mathbf{u}(0),\lambda_c;\mathbf{u}_c^*\right]+2D^2\mathcal{E}'\left[\mathbf{u}(0),\lambda_c;\mathbf{u}_c^*,\mathbf{v}_1\right]} \tag{6.80}$$

$$\dot{\lambda}_{II}(0) = 0 \tag{6.81}$$

The branch of the equilibrium states corresponding to $\dot{\lambda}_I(0)$ is therefore defined by

$$\mathbf{u}_I(\alpha)-\mathbf{u}(0)=\left[\mathbf{u}_c^*+\dot{\lambda}_I(0)\mathbf{v}_1\right]\alpha+o(\alpha) \tag{6.82}$$

$$\lambda_I(\alpha) - \lambda_c = \dot{\lambda}_I(0)\alpha + o(\alpha) \tag{6.82'}$$

where, according to (6.82),

$$\dot{\mathbf{u}}_I(0) = \mathbf{u}_c^* + \dot{\lambda}_I(0)\mathbf{v}_1 \tag{6.83}$$

As for as stability is concerned, from (6.39) and taking account of (6.83), we have

$$\dot{\omega}_I(0)=\dot{\lambda}_I(0)\left\{D^3\mathcal{E}\left[\mathbf{u}(0),\lambda_c;\mathbf{u}_c^{*2},\mathbf{v}_1\right]+D^2\mathcal{E}'\left[\mathbf{u}(0),\lambda_c;\mathbf{u}_c^{*2}\right]\right\} \tag{6.84}$$

A stability exchange thus occurs at the critical configuration C_0 along the branches (6.82).

It is necessary to evaluate at least the second derivative $\ddot{\lambda}_{II}(0)$ to determine the other branch, corresponding to $\dot{\lambda}_{II}(0) = 0$. With (6.76) and taking into account that $\dot{\mathbf{u}}_{II}(0) = \mathbf{u}_c^*$, because $\dot{\lambda}_{II}(0) = 0$, from (6.14) at $\alpha = 0$ we get

$$D^3\mathcal{E}\left[\mathbf{u}(0),\lambda_c;\mathbf{u}_c^{*2},\delta\mathbf{u}\right]+D^2\mathcal{E}\left[\mathbf{u}(0),\lambda_c;\ddot{\mathbf{u}}_{II}(0),\delta\mathbf{u}\right]+\ddot{\lambda}_{II}(0)D\mathcal{E}'\left[\mathbf{u}(0),\lambda_c;\delta\mathbf{u}\right]=0 \tag{6.85}$$

to be satisfied $\forall\ \delta\mathbf{u} \in H_A$. The solution of this equation is

$$\ddot{\mathbf{u}}_{II}(0) = A\mathbf{u}_c^* + \mathbf{v}_2 + \ddot{\lambda}_{II}(0)\mathbf{v}_3 \tag{6.86}$$

with

$$[\mathbf{u}_c^*\ ,\ \mathbf{v}_2] = 0 \qquad [\mathbf{u}_c^*\ ,\ \mathbf{v}_3] = 0 \tag{6.87}$$

and, where $\mathbf{v}_2$ solves (6.62) while $\mathbf{v}_3$ solves the other equation,

$$D^2\mathcal{E}\left[\mathbf{u}(0),\lambda_c;\mathbf{v}_3,\delta\mathbf{u}\right]+D\mathcal{E}'\left[\mathbf{u}(0),\lambda_c;\delta\mathbf{u}\right]=0 \quad \forall\ \delta\mathbf{u}\in H_A \tag{6.88}$$

The second derivative $\ddot{\lambda}(0)$ is thus definite by (6.15), evaluated at $\alpha = 0$ and where we assume

$$\dot{\mathbf{u}}_{II}(0) = \mathbf{u}_c^* \quad \dot{\lambda}_{II}(0) = 0 \quad \delta\mathbf{u} = \mathbf{u}_c^* \tag{6.89}$$

Thus we have the equation

$$3D^3\mathcal{E}\left[\mathbf{u}(0),\lambda_c;\ddot{\mathbf{u}}_{II}(0),\mathbf{u}_c^{*2}\right]+D^4\mathcal{E}\left[\mathbf{u}(0),\lambda_c;\mathbf{u}_c^{*4}\right]+3\ddot{\lambda}_{II}(0)D^2\mathcal{E}'\left[\mathbf{u}(0),\lambda_c;\mathbf{u}_c^{*2}\right]=0 \tag{6.90}$$

Hence, taking account of (6.86) we have

$$\begin{aligned}&3\ddot{\lambda}_{II}(0)\left\{D^2\mathcal{E}'\left[\mathbf{u}(0),\lambda_c;\mathbf{u}_c^{*2}\right]+D^3\mathcal{E}\left[\mathbf{u}(0),\lambda_c;\mathbf{v}_3,\mathbf{u}_c^{*2}\right]\right\}\\&+\left\{3D^3\mathcal{E}\left[\mathbf{u}(0),\lambda_c;\mathbf{v}_2,\mathbf{u}_c^{*2}\right]+D^4\mathcal{E}\left[\mathbf{u}(0),\lambda_c;\mathbf{u}_c^{*4}\right]\right\}=0\end{aligned} \tag{6.91}$$

On the other hand, taking into account that $\mathbf{v}_2$ resolves (6.62) and that, therefore, condition (6.63) holds, we get

$$\ddot{\lambda}_{II}(0)=-\frac{D^4\mathcal{E}\left[\mathbf{u}(0),\lambda_c;\mathbf{u}_c^{*4}\right]-3D^2\mathcal{E}\left[\mathbf{u}(0),\lambda_c;\mathbf{v}_2^2\right]}{3\left\{D^3\mathcal{E}\left[\mathbf{u}(0),\lambda_c;\mathbf{v}_3,\mathbf{u}_c^{*2}\right]+D^2\mathcal{E}'\left[\mathbf{u}(0),\lambda_c;\mathbf{u}_c^{*2}\right]\right\}} \tag{6.92}$$

The other branch of the equilibrium states passing through C_0 is therefore

$$\mathbf{u}_{II}(\alpha)-\mathbf{u}(0)=\alpha\mathbf{u}_c^*+\frac{\alpha^2}{2}\left[A\mathbf{u}_c^*+\mathbf{v}_2+\ddot{\lambda}_{II}(0)\mathbf{v}_3\right]+o(\alpha^2) \tag{6.93}$$

$$\lambda_{II}(\alpha)-\lambda_c=\frac{\alpha^2}{2}\ddot{\lambda}_{II}(0)+o(\alpha^2) \tag{6.94}$$

We can take $\ddot{\mathbf{u}}(0)$ orthogonal in energy to $\dot{\mathbf{u}}(0)$. Thus we assume $A = 0$ in (6.93). We have

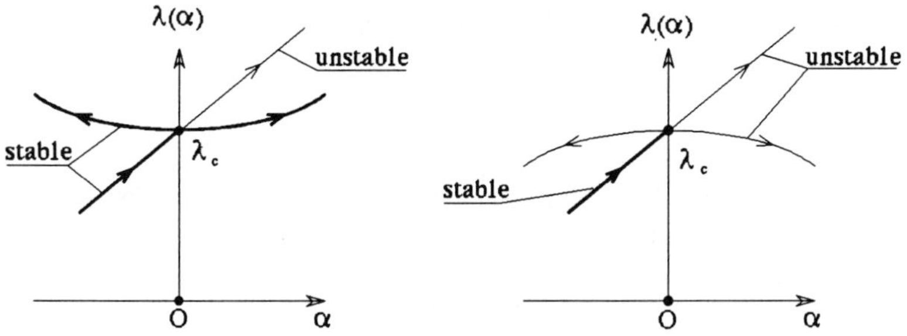

FIGURE 6.13

$$\mathbf{u}_{II}(\alpha)-\mathbf{u}(0)=\alpha\mathbf{u}_c^*+\frac{\alpha^2}{2}\left[\mathbf{v}_2+\ddot{\lambda}_{II}(0)\mathbf{v}_3\right]+o(\alpha^2) \tag{6.93'}$$

Since $\dot{\omega}_{II}(0) = 0$, it is necessary to evaluate $\ddot{\omega}_{II}(0)$ and consequently $\dot{\mathbf{u}}_m(0)$ to analyze stability. Let us consider (6.35) at $\alpha = 0$. Taking into account that $\omega_{II}(0) = 0$, $\dot{\omega}_{II}(0) = 0$, $\dot{\lambda}_{II}(0) = 0$, $\dot{\mathbf{u}}_{II}(0) = \mathbf{u}_c^*$, we get

$$D^2\mathcal{E}\left[\mathbf{u}(0),\lambda_c;\dot{\mathbf{u}}_m(0),\delta\mathbf{u}\right]+D^3\mathcal{E}\left[\mathbf{u}(0),\lambda_c;\mathbf{u}_c^{*2},\delta\mathbf{u}\right]=0 \tag{6.95}$$

Comparing (6.95) and (6.62) shows that

$$\dot{\mathbf{u}}_m(0) = \mathbf{v}_2 \tag{6.96}$$

Then, from (6.42) and (6.86) we have

$$\ddot{\omega}_{II}(0)=D^4\mathcal{E}\left[\mathbf{u}(0),\lambda_c;\mathbf{u}_c^{*4}\right]+3D^3\mathcal{E}\left[\mathbf{u}(0),\lambda_c;\mathbf{u}_c^{*2},\mathbf{v}_2\right]$$

$$+\ddot{\lambda}_{II}(0)\left\{D^3\mathcal{E}\left[\mathbf{u}(0),\lambda_c;\mathbf{u}_c^{*2},\mathbf{v}_3\right]+D^2\mathcal{E}'\left[\mathbf{u}(0),\lambda_c;\mathbf{u}_c^{*2}\right]\right\} \tag{6.97}$$

Taking into account (6.63) and (6.92) we have

$$\ddot{\omega}_{II}(0)=\frac{2}{3}\left\{D^4\mathcal{E}\left[\mathbf{u}(0),\lambda_c;\mathbf{u}_c^{*4}\right]+3D^2\mathcal{E}\left[\mathbf{u}(0),\lambda_c;\mathbf{v}_2^2\right]\right\} \tag{6.98}$$

which is equivalent to (3.99), according to the analysis performed in the previous section. The two equilibrium branches with the assumption that $D^3\mathcal{E}[\mathbf{u}(0), \lambda_c; v_3, \mathbf{u}_c^{*2}] + D^2\mathcal{E}'[\mathbf{u}(0), \lambda_c; \mathbf{u}_c^{*2}] < 0$, are sketched in Fig. 6.13 for both cases of stable or unstable critical states.

6.3 BIFURCATION OF PERFECT SYSTEMS

The analysis of perfect systems can be strongly simplified if we take into account some special features of their behavior under loading. Common structures, in fact, in a first stage as a rule oppose the loads, and all their displacements increase with the increasing loads.

Assume that along the equilibrium branch, which we call "fundamental", the function

$$\lambda = \lambda(\alpha),\ \forall\ \alpha \in [\alpha_1,\alpha_2] \tag{6.99}$$

is a strictly increasing (or decreasing) function of α. In this case the function $\lambda = \lambda(\alpha)$ is invertible and the inverse

$$\alpha = \alpha(\lambda) \tag{6.99'}$$

does exist. We can assume that this function is defined for $\lambda \geq 0$. The derivative

$$\frac{d\alpha}{d\lambda}$$

thus exists for every $\lambda \geq 0$ and snapping cannot occur along this branch. The description of the fundamental equilibrium branch can then be simplified and we can write, in place of (6.1),

$$\mathbf{u}_0 = \mathbf{u}_0[\alpha(\lambda)] = \mathbf{u}_0(\lambda) \qquad \lambda \geq 0 \tag{6.1'}$$

where the subscript 0 indicates the "fundamental" branch.

The case of a fundamental branch along which the function (6.1′) is strictly increasing is frequent in structural engineering. Moreover, in a first stage of loading, common structures behave very rigidly and the function (6.1′) is practically linear. In this case we can admit the existence of the initial linear branch

$$C_\lambda = \{ \lambda \mathbf{u}_0, \ \mathbf{u}_0 \in H_A, \ \lambda \geq 0 \} \tag{6.99'}$$

Only bifurcation can occur along this path at the critical state.

To evaluate the bifurcated path it is thus convenient to define, according to Koiter, the additional displacements that move the structure T from the configuration C_λ, belonging to the principal path, to the configuration C on the secondary branch. Let **u** define this additional displacement. The energy functional at C is

$$\mathcal{E}(\lambda \mathbf{u}_0 + \mathbf{u}, \lambda) \tag{6.100}$$

while the increment of the potential energy, from C_λ to C,

$$\mathcal{E}(\mathbf{u},\lambda) = \mathcal{E}[\mathbf{u}_0(\lambda) + \mathbf{u},\lambda] - \mathcal{E}[\mathbf{u}_0(\lambda),\lambda] \tag{6.101}$$

is defined as the additional potential energy of all the external and internal forces acting on the elastic structure T.

By definition, along C_λ, the additional potential energy is denoted by $\mathcal{E}(\mathbf{0},\lambda)$; we have $\mathcal{E}(\mathbf{0},\lambda) = 0$, identically. Equilibrium along the fundamental path is satisfied if

$$D\mathcal{E}[0,\lambda;\delta\mathbf{u}] = 0, \ \lambda \geq 0, \ \forall \ \delta\mathbf{u} \in H_A \tag{6.102}$$

Continuity of equilibrium along C_λ, gives, with the notation of (6.12)

$$D\mathcal{E}'[0,\lambda;\delta\mathbf{u}] = 0 \ ; \ D\mathcal{E}''[0,\lambda;\delta\mathbf{u}] = 0;\ldots \tag{6.103}$$

Condition (6.21), necessary to have bifurcation at the critical state, is thus implicitly satisfied.

Branching will occur along C_λ. Let λ_c be the value of λ at which the critical state is attained and let C_{λ_c} be the configuration on the fundamental branch C_λ occupied by T at $\lambda = \lambda_c$. Another equilibrium branch

$$\mathbf{u} = \mathbf{u}(\alpha) \qquad \lambda = \lambda(\alpha) \qquad \forall \ \alpha \in [\alpha_1,\alpha_2] \tag{6.104}$$

thus intersects the fundamental path C_λ at C_{λ_c}. Likewise, along these other branches, different from C_λ, we have to satisfy as equilibrium

$$D\mathcal{E}[\mathbf{u}(\alpha),\lambda(\alpha);\delta\mathbf{u}] = 0 \quad \forall \ \delta\mathbf{u} \in H_A \tag{6.105}$$

as equilibrium continuity

$$\frac{d}{d\alpha} D\mathcal{E}[\mathbf{u}(\alpha), \lambda(\alpha); \delta\mathbf{u}] = 0, \ldots \tag{6.106}$$

This last condition in a more explicit form gives

$$D^2\mathcal{E}[\mathbf{u}(\alpha), \lambda(\alpha); \dot{\mathbf{u}}(\alpha), \delta\mathbf{u}] + \dot{\lambda}(\alpha) D\mathcal{E}'[\mathbf{u}(\alpha), \lambda(\alpha); \delta\mathbf{u}] = 0 \tag{6.106$'$}$$

At the critical state, defined by the critical value λ_c of λ, the additional displacement $\mathbf{u}(\lambda_c)$ that defines the perfect system at C_{λ_c}, belonging also to the initial branch, is

$$\mathbf{u}(\lambda_c) = 0 \tag{6.107}$$

and condition (6.106′) becomes

$$D^2\mathcal{E}[0, \lambda_c; \dot{\mathbf{u}}(\lambda_c), \delta\mathbf{u}] + \dot{\lambda}(0) D\mathcal{E}'[0, \lambda_c; \delta\mathbf{u}] = 0 \quad \forall\ \delta\mathbf{u} \in H_A \tag{6.108}$$

The first of the conditions (6.103) that hold along C_λ and, in particular, at C_{λ_c}, simplifies (6.108) and gives

$$D^2\mathcal{E}[0, \lambda_c; \dot{\mathbf{u}}(\lambda_c), \delta\mathbf{u}] = 0 \quad \forall\ \delta\mathbf{u} \in H_A \tag{6.109}$$

This last condition admits the solution

$$\dot{\mathbf{u}}(\lambda_c) = A\mathbf{u}_c^* \tag{6.110}$$

where $\mathbf{u}_c^*$ is the critical displacement with unitary energy norm.

Comparison with the branching of the more general structural systems shows that, at the critical state, "perfect" systems bifurcate with additional displacements proportional to the critical mode $\mathbf{u}_c^*$. We say that perfect structures at branching buckle with critical displacements proportional to the buckling mode $\mathbf{u}_c^*$.

After these preliminary statements let us apply the general results of Sections 6.2.6.2 and 6.2.7.2 to the bifurcation analysis of "perfect" structural systems.

6.3.1 ASYMMETRIC BRANCHING

We have

$$D^3\mathcal{E}[0, \lambda_c; \mathbf{u}_c^{*3}] \neq 0 \tag{6.111}$$

Along the fundamental branch C_λ the equilibrium equations are given by (6.102) and (6.103) while along the other branch (6.104) they are given by (6.105), (6.106). Bifurcation occurs at $\lambda = \lambda_c$ and the corresponding incremental solution is given by (6.110). The second equilibrium branch, according to (6.7) and (6.8), is thus represented by

$$\mathbf{u}(\alpha) = \alpha\mathbf{u}_c^* + o(\alpha) \tag{6.112}$$

$$\lambda(\alpha) - \lambda_c = \alpha\dot{\lambda}(0) + o(\alpha) \tag{6.113}$$

where $\dot{\lambda}(0)$ has to be determined by solving (6.14). From this equation, with $\delta\mathbf{u} = \mathbf{u}_c^*$ and with (6.98), we immediately get

$$D^3\mathcal{E}[0, \lambda_c; \mathbf{u}_c^{*3}] + 2\dot{\lambda}(0) D^2\mathcal{E}'[0, \lambda_c; \mathbf{u}_c^{*2}] = 0 \tag{6.114}$$

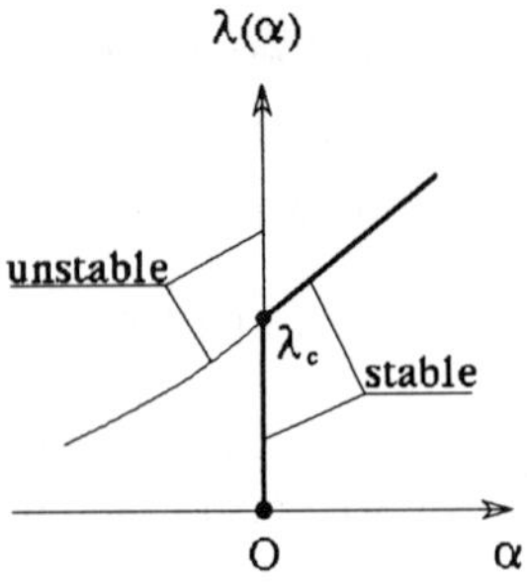

FIGURE 6.14

Thus we have, taking into account that $D^2\mathcal{E}'\left[0,\lambda_c;\mathbf{u}_c^{*2}\right]\neq 0$

$$\dot{\lambda}(0) = -\frac{D^3\mathcal{E}\left[0,\lambda_c;\mathbf{u}_c^{*3}\right]}{2D^2\mathcal{E}'\left[0,\lambda_c;\mathbf{u}_c^{*2}\right]} \tag{6.115}$$

In the applications of the postbuckling theory the expansions (2.98′) are often used. Consequently, we make use of the expressions (6.7′) and (6.8′) to describe the equilibrium branch passing through (C_0,λ_0). In this case the coefficient λ_1 is given by

$$\lambda_1 = -\frac{3\mathcal{E}_3\left(\mathbf{u}_c^*\right)}{2\mathcal{E}_2'\left(\mathbf{u}_c^*\right)} \tag{6.115′}$$

where $\mathcal{E}_2'(\mathbf{u}_c^*)$ denotes the partial derivative, with respect to λ, at the critical state of the second differential of the potential energy. As far as stability is concerned, from (6.39) and (6.115) we have

$$\dot{\omega}(0) = \frac{1}{2}D^3\mathcal{E}\left[0,\lambda_c;\mathbf{u}_c^{*3}\right] \tag{6.116}$$

and, from (6.30)

$$\omega(\alpha) = \frac{\alpha}{2}D^3\mathcal{E}\left[0,\lambda_c;\mathbf{u}_c^{*3}\right] + o(\alpha) = D^2\mathcal{E}'\left[0,\lambda_c;\mathbf{u}_c^{*2}\right]\left(\lambda_c - \lambda(\alpha)\right) + o(\alpha) \tag{6.117}$$

An exchange of stability occurs at C_{λ_c} with stable configurations on the branch (6.113) for $\lambda > \lambda_c$ (Fig. 6.14).

6.3.2 SYMMETRIC BRANCHING

We now have

$$D^3\mathcal{E}\left[0,\lambda_c;\mathbf{u}_c^{*3}\right] = 0 \tag{6.118}$$

To obtain the second branch, let us consider (6.13), which, because of the first of (6.103), gives

$$D^2\mathcal{E}\left[0,\lambda_c;\dot{\mathbf{u}}(0),\delta\mathbf{u}\right] = 0 \quad \forall\ \delta\mathbf{u}\in H_A \tag{6.119}$$

Thus we find again (6.110) and

$$\dot{\mathbf{u}}(0) = \mathbf{u}_c^* \tag{6.120}$$

As in the case of a symmetric branching, symmetric bifurcation of "perfect" structures occurs with critical displacements equal to the buckling mode $\mathbf{u}_c^*$. From (6.14), at $\alpha = 0$, with $\delta\mathbf{u} = \mathbf{u}_c^*$ and taking into account (6.118) and (6.103), we have

$$2\dot{\lambda}(0)D^2\mathcal{E}'\left[0,\lambda_c;\mathbf{u}_c^{*2}\right]=0 \tag{6.121}$$

But $D^2\mathcal{E}'[0, \lambda_c; \mathbf{u}_c^{*2}] \neq 0$. Thus at the symmetric branching we have

$$\dot{\lambda}(0) = 0 \tag{6.122}$$

Let us evaluate now $\ddot{\mathbf{u}}(0)$. From (6.14), with $\dot{\lambda}(0) = 0$, $\dot{\mathbf{u}}(0) = \mathbf{u}_c^*$, we obtain

$$D^2\mathcal{E}\left[0,\lambda_c;\ddot{\mathbf{u}}(0),\delta\mathbf{u}\right]+D^3\mathcal{E}\left[0,\lambda_c;\mathbf{u}_c^{*2},\delta\mathbf{u}\right]=0 \quad \delta\mathbf{u}\in H_A \tag{6.123}$$

whose solution is $\ddot{\mathbf{u}}(0) = \mathbf{u}_c^* + \mathbf{v}_2$ with $\mathbf{v}_2 \in H_A^{(1)}$. Thus

$$\mathbf{u}(\alpha) = \alpha\mathbf{u}_c^* + \frac{\alpha^2}{2}(\mathbf{u}_c^* + \mathbf{v}_2) + o(\alpha^2)\ ;\ \ \lambda(\alpha) - \lambda_c = \frac{\alpha^2}{2}\ddot{\lambda}(0) + o(\alpha^2) \tag{6.124}$$

and $\ddot{\lambda}(0)$ is obtained by (6.15). Hence with $\delta\mathbf{u} = \mathbf{u}_c^*$

$$3D^3\mathcal{E}\left[0,\lambda_c;\mathbf{u}_c^{*2},\mathbf{v}_2\right]+D^4\mathcal{E}\left[0,\lambda_c;\mathbf{u}_c^{*4}\right]+3\ddot{\lambda}(0)D^2\mathcal{E}'\left[0,\lambda_c;\mathbf{u}_c^{*2}\right]=0 \tag{6.125}$$

and, by using (6.63), we get

$$\ddot{\lambda}(0)=-\frac{D^4\mathcal{E}\left[0,\lambda_c;\mathbf{u}_c^{*4}\right]-3D^2\mathcal{E}\left[0,\lambda_c;\mathbf{v}_2^2\right]}{3D^2\mathcal{E}'\left[0,\lambda_c;\mathbf{u}_c^{*2}\right]} \tag{6.126}$$

If the equivalent expansions (2.98′), (6.7′), and (6.8′) are used, the expression of the coefficient λ_2 in (6.8′) with $\ddot{\lambda}(0) = 2\lambda_2$, $\mathbf{v}_2 = 2\mathbf{u}_2 = 2\mathbf{\Phi}_1$, according to (6.7) and (6.7′), and where $\mathbf{\Phi}_1$ is the solution of (3.100), (3.101).

$$\lambda_2 = -2\frac{\mathcal{E}_4\left(\mathbf{u}_c^*\right)-\mathcal{E}_2\left(\mathbf{u}_2\right)}{\mathcal{E}_2'\left(\mathbf{u}_c^*\right)} \tag{6.126′}$$

Equation (6.98) is still valid to analyze stability. The various possible branches are sketched in Fig. 6.15 under the assumption that $D^2\mathcal{E}'[0, \lambda_c; \mathbf{u}_c^{*2}] < 0$.

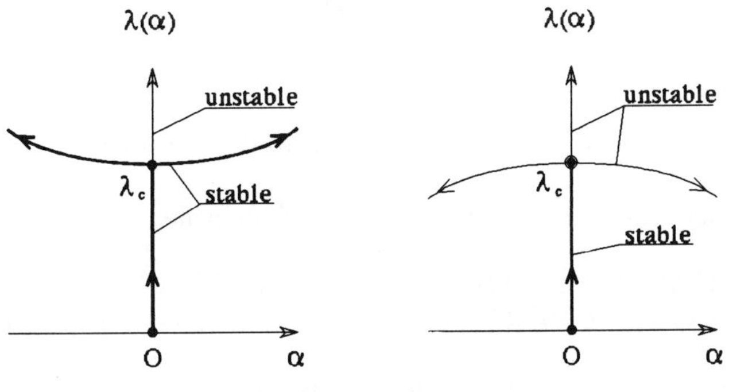

FIGURE 6.15

6.3.3 POSTBUCKLING BEHAVIOR OF PERFECT SYSTEMS IN THE CASE OF SIMULTANEOUS MULTIMODE BUCKLING

We have simultaneous multimode buckling when the equation

$$D^2\mathcal{E}[0,\lambda;\mathbf{u},\delta\mathbf{u}]=0 \quad \forall\ \delta\mathbf{u}\in H_A \tag{6.127}$$

admits n (n > 1) eigenvectors $\mathbf{u}_{ic}$ (i = 1,2,...,n) corresponding to the same first eigenvalue λ_c. In this situation, in addition to the principal branch C_λ, the structure will present many other equilibrium branches passing through C_{λ_c}. In fact, according to (6.127), the incremental solution $\dot{\mathbf{u}}(0)$, in place of (6.110), is

$$\dot{\mathbf{u}}(0)=\sum_{i=1}^{n}\beta_i\mathbf{u}^*_{ic}\ , \quad |||\mathbf{u}^*_{ic}|||=1 \tag{6.128}$$

where, in order to define the norm of $\dot{\mathbf{u}}(0)$, we can put

$$|||\dot{\mathbf{u}}(0)|||=1 \tag{6.128$'$}$$

Both the coefficients β_i of (6.128) and the derivative $\dot{\lambda}(0)$, are obtained from (6.14)

$$D^3\mathcal{E}\left[0,\lambda_c;\left(\sum_{i=1}^{n}\beta_i\mathbf{u}^*_{ic}\right)^2,\delta\mathbf{u}\right]+D^2\mathcal{E}[0,\lambda_c;\ddot{\mathbf{u}}(0),\delta\mathbf{u}]+2\dot{\lambda}(0)D^2\mathcal{E}'\left[0,\lambda_c;\left(\sum_{i=1}^{n}\beta_i\mathbf{u}^*_{ic}\right),\delta\mathbf{u}\right]=0 \tag{6.129}$$

Condition (6.129) is satisfied for any displacement $\delta\mathbf{u}$; thus with $\delta\mathbf{u}=\mathbf{u}^*_{ic,}$ we obtain the following system

$$\sum_{k=1}^{n}\sum_{j=1}^{n}\beta_j\beta_k D^3\mathcal{E}[0,\lambda_c;\mathbf{u}^*_{kc},\mathbf{u}^*_{jc},\mathbf{u}^*_{ic}]+2\dot{\lambda}(0)\sum_{j=1}^{n}D^2\mathcal{E}'[0,\lambda_c;\mathbf{u}^*_{jc},\mathbf{u}^*_{ic}]=0 \tag{6.130}$$

Equations (6.130), together with (6.128), enable us to evaluate the coefficients β_i and the corresponding $\dot{\lambda}$, i.e., the n equilibrium branches passing through C_{λ_c}.

It has been shown that the maximum number of branches passing through C_{λ_c} is $2^n - 1$. Of course, if, along the generic branch k, $\dot{\lambda}_k = 0$, then we evaluate the derivative $\ddot{\lambda}_{k,}$ as has been shown in the case of the simple symmetric branching. As far as stability is concerned, with the assumption of $\dot{\lambda} \neq 0$, on a branch defined by the value of $\dot{\mathbf{u}}(0)$, we get

$$\dot{\omega}(0)=\frac{1}{2}D^3\mathcal{E}[0,\lambda_c;\dot{\mathbf{u}}^{*3}_c] \tag{6.131}$$

and consequently

$$\omega(\alpha)=-\ D^2\mathcal{E}'[0,\lambda_c;\dot{\mathbf{u}}^{*2}_c](\lambda(\alpha)-\lambda_c)+o(\alpha) \tag{6.132}$$

A stability exchange thus occurs along all the branching paths.

6.4 GENERAL ASPECTS OF THE POSTCRITICAL BEHAVIOR OF QUASI-PERFECT STRUCTURAL SYSTEMS

A large number of "imperfect" structural systems behave almost as perfect systems. These "imperfect" systems, called "quasi-perfect", can be studied, according to Koiter,[8,9] by adding small imperfections to the corresponding perfect structure. Thus, once the behavior of the perfect system is known, perturbation analysis is capable of investigating the behavior of the "quasi-perfect" system.

Let ε be the imperfection parameter; the potential energy of the quasi-perfect system under loads $\lambda\mathbf{q}$ at the configuration C will be indicated as

$$\mathcal{E}(C,\lambda,\varepsilon) \tag{6.133}$$

Let C_λ be the equilibrium configuration of the corresponding perfect system along its principal branch under loads $\lambda\mathbf{q}$. Thus, the difference between the potential energies of the imperfect system at C and C_λ, is

$$\mathcal{E}(\mathbf{u},\lambda,\varepsilon) = \mathcal{E}(C,\lambda,\varepsilon) - \mathcal{E}(C_\lambda,\lambda) \tag{6.134}$$

where, as usual, $\mathbf{u}$ indicates the additional displacement of the structure form C_λ to C. Under the assumption of quasi-perfect system, because of the smallness of the imperfection parameter ε, we put

$$\mathcal{E}(\mathbf{u},\lambda,\varepsilon) = \mathcal{E}(\mathbf{u},\lambda,0) + \varepsilon\mathcal{I}(\mathbf{u},\lambda) + \varepsilon^2\mathcal{L}(\mathbf{u},\lambda) + o(\varepsilon^2) \tag{6.135}$$

where $\mathcal{E}(\mathbf{u},\lambda,0)$ indicates the total potential energy increment from C_λ to C of the corresponding perfect system and $\mathcal{I}(\mathbf{u},\lambda)$, $\mathcal{L}(\mathbf{u},\lambda)$ are differentiable functionals of $\mathbf{u}$ and λ.

Major and minor imperfections are respectively characterized by the linear and by the quadratic terms $\varepsilon\mathcal{I}(\mathbf{u},\lambda)$ and $\varepsilon^2\mathcal{L}(\mathbf{u},\lambda)$ of (6.135). Let us examine the more relevant case of the "major" imperfections. Limiting our analysis to the neighborhood of the critical state, the additional energy of the quasi-perfect structure can be represented by

$$\mathcal{E}(\mathbf{u},\lambda,\varepsilon) = \mathcal{E}(\mathbf{u},\lambda_c) + \mathcal{E}'(\mathbf{u},\lambda_c)(\lambda-\lambda_c) + \varepsilon\mathcal{I}(\mathbf{u},\lambda) \tag{6.136}$$

Furthermore, we can take into account only the linear part of $\mathcal{I}((\mathbf{u},\lambda)$ with respect to $\mathbf{u}$ and λ because of the smallness of ε; thus we have

$$\mathcal{E}(\mathbf{u},\lambda,\varepsilon) = \mathcal{E}(\mathbf{u},\lambda_c) + \mathcal{E}'(\mathbf{u},\lambda_c)(\lambda-\lambda_c) + \varepsilon\lambda_c\mathcal{I}_0\mathbf{u} + \varepsilon\lambda\mathcal{I}_1\mathbf{u} \tag{6.137}$$

where $\mathcal{I}_0\mathbf{u}$ and $\mathcal{I}_1\mathbf{u}$ are linear functionals. The equilibrium condition of the quasi-perfect structural system is

$$D\mathcal{E}(\mathbf{u},\lambda_c;\delta\mathbf{u}) + D\mathcal{E}'(\mathbf{u},\lambda_c;\delta\mathbf{u})(\lambda-\lambda_c) + \varepsilon\lambda_c\mathcal{I}_0\delta\mathbf{u} + \varepsilon\lambda\mathcal{I}_1\delta\mathbf{u} = 0 \quad \forall\,\delta\mathbf{u}\in H_A \tag{6.138}$$

Furthermore, we can assume that the equilibrium configurations of the "quasi-perfect" system are close to the equilibrium branch of the perfect system. The equilibrium path of the "quasi-perfect" system can thus be written as follows

$$\mathbf{u}(\alpha) = \alpha\mathbf{u}_c + \frac{1}{2}\alpha^2\mathbf{v}_2 + \cdots \tag{6.139}$$

The equilibrium condition (6.138) gives the necessary connection within the parameters λ, α, ε. By means of a MacLaurin expasion of the functionals $D\mathcal{E}[\mathbf{u},\lambda_c;\,\delta\mathbf{u}]$ and $D\mathcal{E}'[\mathbf{u},\lambda_c;\,\delta\mathbf{u}]$ and by taking into account equations (6.102) and (6.103), condition (6.138) becomes

$$\begin{aligned} &D^2\mathcal{E}[0,\lambda_c;\mathbf{u},\delta\mathbf{u}] + \frac{1}{2}D^3\mathcal{E}[0,\lambda_c;\mathbf{u}^2,\delta\mathbf{u}] + \frac{1}{6}D^4\mathcal{E}[0,\lambda_c;\mathbf{u}^3,\delta\mathbf{u}] + o(\mathbf{u}^3) \\ &+ \{D^2\mathcal{E}'[0,\lambda_c;\mathbf{u},\delta\mathbf{u}] + o(\mathbf{u})\}(\lambda-\lambda_c) + \varepsilon\lambda_c\mathcal{I}_0\delta\mathbf{u} + \varepsilon\lambda\mathcal{I}_1\delta\mathbf{u} = 0 \quad \forall\,\delta\mathbf{u}\in H_A \end{aligned} \tag{6.140}$$

Taking $\delta\mathbf{u} = \mathbf{u}_c^*$ we have

$$\frac{1}{2}D^3\mathcal{E}\left[0,\lambda_c;\mathbf{u}^2,\mathbf{u}_c^*\right]+\frac{1}{6}D^4\mathcal{E}\left[0,\lambda_c;\mathbf{u}^3,\mathbf{u}_c^*\right]+o\left(\mathbf{u}^3\right)$$

$$+\left[D^2\mathcal{E}'\left[0,\lambda_c;\mathbf{u},\mathbf{u}_c^*\right]+o(\mathbf{u})\right](\lambda-\lambda_c)+\varepsilon\lambda_c I_o\mathbf{u}_c^*+\varepsilon\lambda I_1\mathbf{u}_c^*=0 \tag{6.141}$$

Substituting (6.139) into (6.141) we obtain

$$\alpha^2\frac{1}{2}D^3\mathcal{E}\left[0,\lambda_c;\mathbf{u}_c^{*3}\right]+\alpha^3\left\{\frac{1}{2}D^3\mathcal{E}\left[0,\lambda_c;\mathbf{u}_c^{*2},\mathbf{v}_2\right]+\frac{1}{6}D^4\mathcal{E}\left[0,\lambda_c;\mathbf{u}_c^{*4}\right]\right\}+o\left(\alpha^3\right)$$

$$+\left\{\alpha D^2\mathcal{E}'\left[0,\lambda_c;\mathbf{u}_c^{*2}\right]+o(\alpha)\right\}(\lambda-\lambda_c)+\varepsilon\lambda_c I_o\mathbf{u}_c^*+\varepsilon\lambda I_1\mathbf{u}_c^*=0 \tag{6.142}$$

By dividing by $D^2\mathcal{E}'[0, \lambda_c;\ \mathbf{u}_c^2]$ we have, finally, the following equation in α, λ, and ε

$$-\alpha^2\lambda_1-\frac{\alpha^3}{2}\lambda_2+o(\alpha^3)+(\alpha+o(\alpha))(\lambda-\lambda_c)+\varepsilon\lambda_c\beta_o+\varepsilon\lambda\beta_1=0 \tag{6.143}$$

where λ_1 and λ_2 are given by (6.115) and (6.125), and

$$\beta_o=\frac{I_o\mathbf{u}_c^*}{D^2\mathcal{E}'\left[0,\lambda_c;\mathbf{u}_c^{*2}\right]}\qquad\beta_1=\frac{I_1\mathbf{u}_c^*}{D^2\mathcal{E}'\left[0,\lambda_c;\mathbf{u}_c^{*2}\right]} \tag{6.144}$$

As far as stability of the quasi-perfect system is concerned, from (6.137) the second derivative of $\mathcal{E}(\mathbf{u},\lambda,\varepsilon)$ along direction $\mathbf{v}$ and at configuration $\mathbf{u}$, is given by

$$D^2\mathcal{E}\left[\mathbf{u},\lambda,\varepsilon;\mathbf{v}^2\right]=D^2\mathcal{E}\left[\mathbf{u},\lambda_c;\mathbf{v}^2\right]+D^2\mathcal{E}'\left[\mathbf{u},\lambda_c;\mathbf{v}^2\right](\lambda-\lambda_c) \tag{6.145}$$

With the assumption of small imperfections, we can assume that the minima $\omega(\mathbf{u},\lambda,\varepsilon)$ and $\omega(\mathbf{u},\lambda)$ of the quasi-perfect and perfect systems are the same; namely, under the same loads and at the same configuration

$$\omega(\mathbf{u},\lambda,\varepsilon)=\min_{\mathbf{v}\in S^*}D^2\mathcal{E}\left[\mathbf{u},\lambda,\varepsilon;\mathbf{v}^2\right]=\omega(\mathbf{u},\lambda)=\min_{\mathbf{v}\in S^*}D^2\mathcal{E}\left[\mathbf{u},\lambda;\mathbf{v}^2\right] \tag{6.146}$$

With a MacLaurin expansion of (6.145) and using (6.139) for $\mathbf{u}(\alpha)$, we obtain

$$D^2\mathcal{E}\left[\mathbf{u},\lambda_c;\mathbf{v}^2\right]=D^2\mathcal{E}\left[\mathbf{u},\lambda_c;\mathbf{v}^2\right]+\alpha\, D^3\mathcal{E}\left[0,\lambda_c;\mathbf{u}_c^{*2},\mathbf{v}^2\right]$$

$$+\frac{1}{2}\alpha^2\left\{D^4\mathcal{E}\left[0,\lambda_c;\mathbf{u}_c^{*2},\mathbf{v}^2\right]+D^3\mathcal{E}\left[0,\lambda_c;\mathbf{v}_2,\mathbf{v}^2\right]\right\}+o\left(\alpha^2\right)$$

$$+\left\{D^2\mathcal{E}'\left[0,\lambda_c;\mathbf{v}^2\right]+o(\alpha)\right\}(\lambda-\lambda_c) \tag{6.147}$$

It is not difficult to prove that the minimum of the functional (6.147) is attained at the element $\mathbf{v}$ given by

$$\mathbf{v}(\alpha,\lambda)=\mathbf{u}_c+\alpha\mathbf{v}_2+\mathbf{o}(\alpha)+\mathbf{o}(\lambda-\lambda_c) \tag{6.148}$$

Substituting (148) into (147) gives

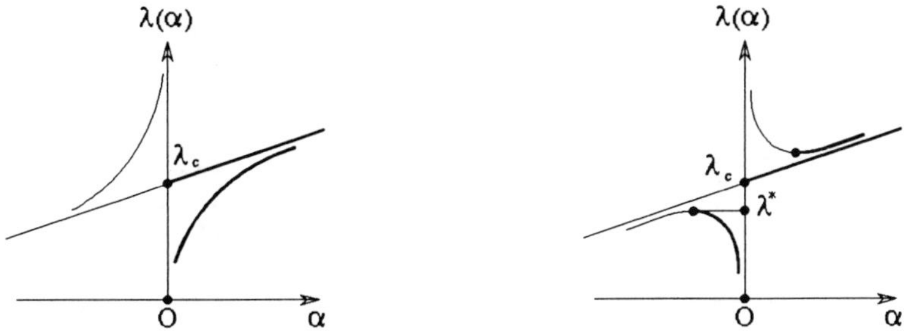

FIGURE 6.16

$$\min_{\mathbf{v}\in S^*} D^2\mathcal{E}[\mathbf{u},\lambda,\varepsilon;\mathbf{v}^2] = \alpha D^3\mathcal{E}[0,\lambda_c;\mathbf{u}_c^3]$$

$$+\frac{\alpha^2}{2}\left[D^4\mathcal{E}[0,\lambda_c;\mathbf{u}_c^4]+D^3\mathcal{E}[0,\lambda_c;\mathbf{u}_c^2,\mathbf{v}_2]+2D^2\mathcal{E}[0,\lambda_c;\mathbf{v}_2^2]+4D^3\mathcal{E}[0,\lambda_c;\mathbf{u}_c^2,\mathbf{v}_2]\right]$$

$$+o(\alpha^2)+\left[D^2\mathcal{E}'[0,\lambda_c;\mathbf{u}_c^2]+o(\alpha)\right](\lambda-\lambda_c) \tag{6.149}$$

Equation (6.149) gives the minimum of the functional $D^2\mathcal{E}[\mathbf{u},\lambda,\varepsilon;\mathbf{v}^2]$ at the configuration (6.139) and under the loads $\lambda\mathbf{q}$ of the quasi perfect system. Considering (6.149) equal to zero, we obtain the relation connecting α and λ, which separates the possible stable and unstable equilibrium states

$$-2\lambda_1\alpha-\frac{3}{2}\lambda_2\alpha^2+o(\alpha^2)+(\lambda-\lambda_c)(1+o(\alpha))=0 \tag{6.150}$$

This analysis allows to define the behavior of the "quasi-perfect" structure according to the type of branching that occurs at the critical state of the corresponding perfect structure.

1. *A perfect structure with asymmetric branching.* We have $\lambda_1 \neq 0$. Thus, from (6.143), which defines the equilibrium states of the "quasi-perfect" structure, and disregarding terms smaller than α, we have

$$-\alpha^2\lambda_1+\alpha(\lambda-\lambda_c)+\varepsilon\lambda_c\beta_0+\varepsilon\lambda\beta_1=0 \tag{6.151}$$

The possible equilibrium paths of the imperfect structure are sketched in Fig. 6.16. With the assumption that

$$\varepsilon\frac{\lambda_1}{\lambda_c}(\beta_o+\beta_1)<0 \tag{6.152}$$

the structure presents a collapse load λ^* that can be obtained solving the following equation

$$\frac{(\lambda_c-\lambda^*)^2}{4\lambda_1}-\varepsilon\beta_1(\lambda_c-\lambda^*)+\varepsilon\lambda_c(\beta_o+\beta_1)=0 \tag{6.153}$$

An approximate solution of this equation giving the maximum load λ^* is

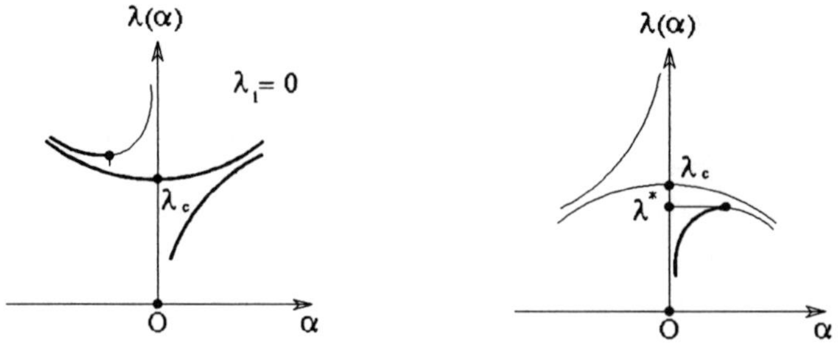

FIGURE 6.17

$$1-\frac{\lambda^*}{\lambda_c}=2\left[-\varepsilon\frac{\lambda_1}{\lambda_c}(\beta_o+\beta_1)\right]^{1/2} \tag{6.154}$$

which shows the relevant imperfection sensitivity of the structure, that is the reduction of the maximum load λ^* with respect to λ_c of the "given"-perfect structure. With respect to the buckling load of the perfect structure, it is in the order of 1/2 in the imperfection parameter ε. From (6.150) we deduce then that in the plane λ - α the straight line

$$-2\lambda_1\alpha+(\lambda-\lambda_c)=0 \tag{6.155}$$

separates the possible stable and unstable states. The collapse load λ^* marks the passage from stable to unstable equilibrium configurations (Fig. 6.16).

2. *A perfect structure with symmetric branching*. We have now $\lambda_1=0$, and the equilibrium states of the perfect structure are defined by

$$-\frac{1}{2}\alpha^3\lambda_2+(\lambda-\lambda_c)\alpha+\varepsilon\lambda_c\beta_0+\varepsilon\lambda\beta_1=0 \tag{6.156}$$

Fig. 6.17 sketches the corresponding possible equilibrium paths. If we assume that $\lambda_2<0$, the imperfect structure has a collapse load λ^* that can be obtained by solving the equation

$$\frac{(\lambda_c-\lambda^*)^{3/2}}{(-\lambda_2)^{1/2}}\left(\frac{2}{3}\right)^{3/2}+\varepsilon\beta_1(\lambda_c-\lambda^*)-\varepsilon\lambda_c(\beta_o+\beta_1)=0 \tag{6.157}$$

An approximate solution of this equation is

$$1-\frac{\lambda^*}{\lambda_c}=\frac{3}{2}\left(\frac{\lambda_2}{\lambda_c}\right)^{1/3}[\varepsilon(\beta_o+\beta_1)]^{2/3} \tag{6.158}$$

In this case, the imperfection sensitivity is of order 2/3. The possible stable and unstable states are thus parted by the parabola

$$\lambda-\lambda_c=\frac{3}{2}\lambda_c\alpha^2 \tag{6.159}$$

and we find again that the collapse load λ^* marks the passage from stable to unstable configurations (Fig. 6.17).

6.5 SOME APPLICATIONS OF THE INITIAL POSTBUCKLING ANALYSIS

6.5.1 INTRODUCTORY REMARKS

Bifurcation analysis at the critical state has been widely applied to investigate the real behavior of elastic structures in the neighborhood of the buckling load. The postbuckling behavior of the reference "perfect" model, together with the load-displacement relation in presence of small imperfections, has been analyzed for a large class of structural problems.

The most important and interesting applications are related to the unstable postbuckling behavior of bidimensional shells and to their high imperfection sensitivity. After the first results given by Koiter,[8,9] a large number of difficult structural problems have been analyzed, and a summary of the obtained results has been presented in some review papers by Hutchinson and Koiter,[19] Budiansky and Hutchinson[20] and Tvergaard.[30]

In this section we will give some examples of bifurcation analysis of "perfect" structural models. The effect of small imperfections is easily deduced according to the results given at Section 6.4. First, some postbuckling problems of beams are analyzed by means of the one-dimensional models developed in Chapter 4. Further, the postbuckling behavior of plates and shallow shells, according to the basic formulation given by Budiansky,[31] is summarized and some numerical results are presented.

6.5.2 POSTBUCKLING ANALYSIS OF BEAMS AND FRAMES

In connection with the stability analysis of elastic rods developed in Chapter 4, we analyze the postbuckling behavior of some monodimensional beam models.

6.5.2.1 The "Elastica" Model

As a first example, let us study the postbuckling behavior of the simply supported compressed rod. According to (4.59) and (4.62), the critical state, characterized by the buckling load $N_c = \pi^2 EI / L^2$ and the buckling mode $v_c = A \sin \pi z/L$, is stable. The third differential of the potential energy is zero along the buckling mode. According to (6.122), the derivative $\dot{\lambda}(0)$ vanishes and the bifurcation is symmetric and stable. We can write, for the bifurcated equilibrium configurations corresponding to the load λN,

$$v(\alpha) = \alpha v_c + o(\alpha) \quad ; \quad \frac{\lambda(\alpha)}{\lambda_c} = 1 + \frac{\ddot{\lambda}(0)}{\lambda_c}\frac{\alpha^2}{2} + o(\alpha^2) \tag{6.160}$$

where

$$\ddot{\lambda}(0) = -\frac{D^4\mathcal{E}[0,\lambda_c;v_c^4]}{3D^2\mathcal{E}'[0,\lambda_c;v_c^2]} \tag{6.161}$$

Equations (4.45) and (4.62) give

$$\begin{aligned} D^4\mathcal{E}[0,\lambda_c;v_c^4] &= 4!\,\mathcal{E}_4(v_c) = 3A^4\frac{\pi^6}{8}\frac{EI}{L^5} \\ 3D^2\mathcal{E}'[0,\lambda_c;v_c^2] &= -3N_c\int_0^L v_c'^2 dz = -3N_c A^2\frac{\pi^2}{L^2}\frac{L}{2} = 3\frac{\pi^4}{2L^4}EIL \end{aligned} \tag{6.162}$$

$$\ddot{\lambda}(0) = A^2\frac{\pi^2}{4L^2}EIL \tag{6.163}$$

and the relation between the load N and the displacement parameter α becomes:

$$\frac{\lambda}{\lambda_c} = 1 + \frac{A^2\pi^2}{8}\frac{\alpha^2}{L^2} + o(\alpha^2) \qquad (6.164)$$

Assume for the coefficient A in the buckling mode the value A = L. Thus we get

$$\frac{\lambda}{\lambda_c} = \frac{N}{N_c} = 1 + \frac{\lambda_2}{\lambda_c}\alpha^2 \qquad (6.164')$$

where

$$\frac{\lambda_2}{\lambda_c} = \frac{\pi^2}{8}; \qquad \alpha = \frac{v(L/2)}{L} \qquad (6.164'')$$

We notice that α represents the nondimensional displacement of the middle section of the beam. According to (6.164′), large displacements occur along the bifurcated path under small increments of the axial load.

6.5.2.2 Flexural-Torsional Postbuckling Behavior of Deep Beams

In this section the postbuckling of simply supported beams is studied in detail according to the analysis of Ascione and Grimaldi.[32] With reference to Fig. 6.18, the parameter α denotes the lateral displacement of the centroid of the section at z = 0. Bifurcation is symmetric if the loads, represented by bending moments and axial loads applied at the end sections, act in the vertical plane (y,z). Also in this case the displacement-load relations can be approximated as

$$\frac{\lambda}{\lambda_c} = 1 + \frac{\ddot{\lambda}}{2\lambda_c}\alpha^2 = 1 + \frac{\lambda_2}{\lambda_c}\alpha^2 \qquad (6.165)$$

$$\mathbf{u} = \alpha\mathbf{u}_c + \alpha^2\mathbf{u}_2 \qquad (6.166)$$

where $\mathbf{u}_c$ is the buckling mode and $\mathbf{u}_2$ the second-order displacement in the series expansion of the bifurcated equilibrium path. We now analyze the buckling state under the loading condition of Fig. 6.18. The fundamental equilibrium states, i.e., equilibrium states before buckling, can be expressed as

$$v(\zeta) = \frac{ML}{B_2}\frac{1}{2}\left(\frac{1}{4} - \zeta^2\right) = \lambda\sqrt{\chi_1\chi_2}\, g(\zeta) \qquad (6.167)$$

where

$$\frac{ML}{B_2} = \lambda\sqrt{\chi_1\chi_2}; \qquad g(\zeta) = \frac{1}{2}\left(\frac{1}{4} - \zeta^2\right) \qquad (6.168)$$

according to (4.154) and (4.155). The critical state of the beam is described by the following differential equations

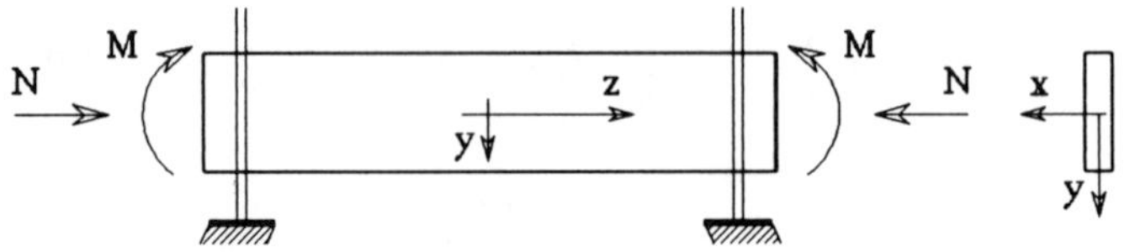

FIGURE 6.18

$$\chi_1 U_c^{IV} + \lambda_c \sqrt{\chi_1\chi_2}\, \alpha_c'' + \lambda_c\, \eta \sqrt{\chi_1\chi_2}\, U_c'' = 0$$

$$- \chi_2\, \alpha_c'' + \lambda_c \sqrt{\chi_1\chi_2}\; U_c'' = 0 \tag{6.169}$$

and boundary conditions

$$U_c = U_c'' = \alpha_c = 0 \quad \text{for } \zeta = \pm\frac{1}{2} \tag{6.170}$$

that can be obtained from the general ones given in Chapter 4. In (6.169) the parameter η defines the nondimensional eccentricity of the axial force:

$$\eta = \frac{NL}{M} \tag{6.171}$$

The solution of the eigenvalue problem (6.169), (6.170) gives the buckling mode components:

$$U_c = \cos \pi\, \zeta\, ; \quad \alpha_c = \frac{\lambda_c}{K} \cos \pi\, \zeta \tag{6.172}$$

where

$$K = \sqrt{\chi_2 / \chi_1} = \sqrt{C / B_3} \tag{6.173}$$

The critical value λ_c of the load multiplier, as a function of η and K, is defined by the relation

$$\lambda_c^2 + K\, \eta\, \lambda_c - \pi^2 = 0 \tag{6.174}$$

We observe that expression (6.172) of U_c satisfies the condition $U_c(0) = 1$. The postbuckling behavior of the deep beam can be easily described and we can obtain a simple formula for the coefficient λ_2, depending on the stiffness properties of the beam section and on the loading condition, this last represented by the parameter η. Inspection of the variational condition (6.123), taking into account that the vertical component V_c of the critical displacement is zero, shows that the second-order vector displacement $\mathbf{v}_2$ is given by $\mathbf{v}_2 = (0, V, 0)$ and corresponds to a vertical flexure. The variational equation (6.123) in this case takes the form:

$$\int_{-1/2}^{1/2}\Big[0V''\delta V'' - U_c''\alpha_c\delta V'' + \chi_1 U_c''\alpha_c\delta V'' + \chi_2 U_c''\alpha_c'\delta V' - (1/2)\lambda_c\,\eta\sqrt{\chi_1\chi_2}\,V'\,\delta V'$$

$$\lambda_c\sqrt{\chi_1\chi_2}\; g''\alpha_c^2\delta V''\Big]\, d\zeta + \left|(1/2)\lambda_c\sqrt{\chi_1\chi_2}\; g''\left(U_c'^{\,2}\delta V' + \alpha_c^2\,\delta V'\right)\right|_{-1/2}^{1/2} = 0 \tag{6.175}$$

and gives the following differential equation

$$0\,V^{IV} - \left(U_c''\alpha_c\right)'' + \chi_1\left(U_c''\alpha_c\right)'' - \chi_2\left(U_c''\alpha_c'\right)' - \lambda_c\sqrt{\chi_1\chi_2}\left(g''\alpha_c^2\right)'' + 0\lambda_c\eta\sqrt{\chi_1\chi_2}\,V'' = 0 \tag{6.176}$$

and boundary conditions

$$V'' + \lambda_c\sqrt{\chi_1\chi_2}\; g'' U_c'^{\,2} = 0\, ; \qquad V = 0 \qquad \text{at } t = \pm\,\frac{1}{2} \tag{6.177}$$

from which we can evaluate the second-order displacement V. We recall that quantities χ_1, and χ_2 respectively represent the ratios of the lateral bending and torsional stiffness on the vertical bending stiffness of the beam section. Then, taking into account that χ_1 and χ_2 are assumed very small, the solution of (6.176) can be approximated in the form

$$V = V_o + \chi_1 V_1 + \chi_2 V_2 + \sqrt{\chi_1 \chi_2}\, V_3 \qquad (6.178)$$

This approximation makes possible to obtain a simple expression of the second derivative V'' of the solution V. Substitution of (6.178) in (6.176) and (6.177) and neglecting the nonlinear terms in χ_1 and χ_2, yields, in fact, four uncoupled equations in the unknown functions V_0, V_1, V_2, and V_3. These equations and their solutions, for simply supported beams, are now given

1. Evaluation of V_o''. The differential equation and the boundary conditions are $(1/2)V_o^{IV} - (U_c''\alpha_c)'' = 0$; $V_o'' = 0$ at $\zeta = \pm 1/2$. The solution is

$$V_o'' = 2U_c''\alpha_c \qquad (6.179)$$

2. Evaluation of V_1''. The differential equation and the boundary conditions are $(1/2)V_1^{IV} - (U_c''\alpha_c)'' = 0$ and $V_1'' = 0$ at $\zeta = \pm 1/2$. The solution is

$$V_1'' = -\,2U_c''\alpha_c \qquad (6.180)$$

3. Evaluation of V_2''. The differential equation and the boundary conditions are $(1/2)V_2^{IV} - (U_c''\alpha_c')' = 0$ and $V_2'' = 0$ at $\zeta = \pm 1/2$. The solution is

$$V_2'' = 2\int_{-1/2}^{\zeta} U_c''\alpha_c'\, d\zeta \qquad (6.181)$$

4. Evaluation of V_3''. The differential equation and the boundary conditions are $(1/2)V_3^{IV} - \lambda_c(g''\alpha_c^2)'' + (1/2)\,\lambda_c\,\eta\, V_o'' = 0$ and $0\, V_3'' + (1/2)\,\lambda_c\, g''\, U_c'^2 = 0$ at $\zeta = \pm 1/2$. The solution is

$$V_3'' = 2\lambda_c g''\alpha_c^2 - 2\lambda_c\eta\int_{-1/2}^{\zeta} d\zeta\int_{o}^{\zeta} U_c''\alpha_c d\zeta - \lambda_c g''\left(-\tfrac{1}{2}\right) U_c'^2\left(-\tfrac{1}{2}\right) \qquad (6.182)$$

In the expression of the coefficient $\ddot{\lambda}$ not only the second derivative of V appears, but also the term $\sqrt{\chi_1\chi_2}\, V'^2$. Taking into account that, according to the assumption, χ_1 and $\chi_2 << 1$, we can write, up to terms linear in χ_1 and χ_2:

$$\sqrt{\chi_1\chi_2}\, V'^2 = \sqrt{\chi_1\chi_2}\left(\int_o^{\zeta} V''\, d\zeta\right)^2 \cong \sqrt{\chi_1\chi_2}\left(\int_o^{\zeta} V_o''\, d\zeta\right)^2 = 4\sqrt{\chi_1\chi_2}\left(\int_o^{\zeta} U_c''\,\alpha_c\, d\zeta\right)^2 \qquad (6.183)$$

It is useful to write:

$$F_1(\zeta) = \int_{-1/2}^{\zeta} U_c''\alpha_c\, d\zeta;\; F_2(\zeta) = \int_{-1/2}^{3} d\zeta\int_0^{3} U_c''\alpha_c d\zeta;\; F_3(\zeta) = \int_{-1/2}^{\zeta} d\zeta\int_0^{\zeta} U_c''\alpha_c d\zeta;$$

$$F_4(\zeta) = \int_0^{\zeta} U_c''\alpha_c d\zeta;\; F_5 = g''\left(-\frac{1}{2}\right)U_c'^2\left(-\frac{1}{2}\right) \qquad (6.184)$$

Consequently, the first and second derivatives of the solution V(t) of (6.176) and (6.177) are given by the expressions:

$$V'(\zeta) = F_4(\zeta) \tag{6.185}$$

$$V''(\zeta) = 2\left[U_c''\alpha_c - \chi_1 U_c'' \alpha_c + \chi_2 F_1(\zeta) + \sqrt{\chi_1\chi_2}\left(\lambda_c g''\alpha_c^2 - \lambda_c\eta F_3(\zeta) - \frac{1}{2}\lambda_c F_5\right)\right] \tag{6.186}$$

We are now able to give the final expression of the coefficient $\lambda_2 = (1/2)\ddot{\lambda}$ in the case of a simply supported beam loaded by a vertical load and an axial force. Substitution of (6.185) and (6.186) in the expression of λ_2 gives, in fact, the following relation

$$\lambda_c = -4\left[\int_{-1/2}^{1/2}\left(-2g''U_c''\alpha_c - \eta U_c'^2\right)d\zeta\right]^{-1}\left\{\int_{-1/2}^{1/2}\left[\frac{1}{K}\left(U_c'^2U_c''^2 + U_c''^2\alpha_c^2\right) - 2KU_c''\alpha_c F_1\right.\right.$$

$$\left.\left. -2\lambda_c\left(\frac{1}{3}g''U_c''\alpha_c^3 + \frac{1}{2}g''\,U_c'^2U_c''\alpha_c - \frac{1}{2}F_5U_c''\alpha_c\right) - \lambda_c\eta\left(\frac{1}{4}U_c'^4 - 2U_c''\alpha_c F_3 - F_4^2\right)\right]d\zeta\right\} \tag{6.187}$$

Substituting the expression of the buckling displacements U_c, α_c in (6.187) and evaluating the integrals gives

$$I_1 = \frac{K}{\lambda_c}\int_{-1/2}^{1/2} U_c'^2\,d\zeta = 4.93480;\quad I_2 = \int_{-1/2}^{1/2} U_c'^2\,U_c''^2\,d\zeta = 120.17331;$$

$$I_3 = \frac{K^2}{\lambda_c^2}\int_{-1/2}^{1/2} U_c''^2\alpha_c^2 d\zeta = 36.52840;$$

$$I_4\frac{K^3}{\lambda_c^2}\int_{-1/2}^{1/2} U_c''\alpha_c^3 d\zeta = 3.701102;\quad I_5 = \frac{K^2}{\lambda_c^2}\int_{-1/2}^{1/2} F_4^2 d\zeta = 3.571483 \tag{6.188}$$

and

$$\lambda_2 = \frac{2}{\lambda_c}\left[(2\lambda_c + \eta)\,I_1\right]^{-1}\left[\frac{1}{K}I_2 - \frac{I_3}{4}\eta\lambda_c + \left(-\frac{4}{3}\frac{I_3}{K} + \frac{\pi^2}{K}I_1 + \frac{I_3}{K^3}\right)\lambda_c^2 - \frac{I_5}{K^2}\eta\lambda_c^3 + \frac{2}{3}\frac{\lambda_c^4}{K^3}I_4\right] \tag{6.189}$$

The values of λ_2 as function of λ_c are plotted in Fig. 6.19. The values of λ_c and η are connected by means of (6.174). Consequently, in the plots of Fig. 6.19, λ_2 is given only as function of λ_c and K. To $\lambda_c = 0$ corresponds M = 0, i.e., the axially loaded beam.

6.5.2.3 Postbuckling Behavior of Framed Structures

The postbuckling behavior of framed structures will be now analyzed in the framework of the results obtained by Roorda,[12] Grimaldi,[33] Godley and Chilver,[34] Di Carlo, Pignataro and Rizzi,[36] and Britvec.[24] In the following, we will face the problem according to the formulation given in reference 33.

Consider a plane frame loaded only by nodal forces increasing proportionally to a multiplier λ. Suppose that the nodal loads can be equilibrated only by axial forces. Thus, for increasing λ, the frame will show a fundamental path C_λ of equilibrium states characterized by the presence in the bars of pure extensional stresses, linearly increasing with λ. In this stage of loading, in fact, the flexural stresses can be neglected.

If we consider the case of frames with a number b of bars $b \leq 2n$, where n is the number of joints, the axial forces in the bars can be easily evaluated solving the nodal equilibrium equations. We identify the scheme of the perfect frame (Fig. 6.20a) with this assumption. As matter of fact, every real frame behaves always as an imperfect structure, owing to the

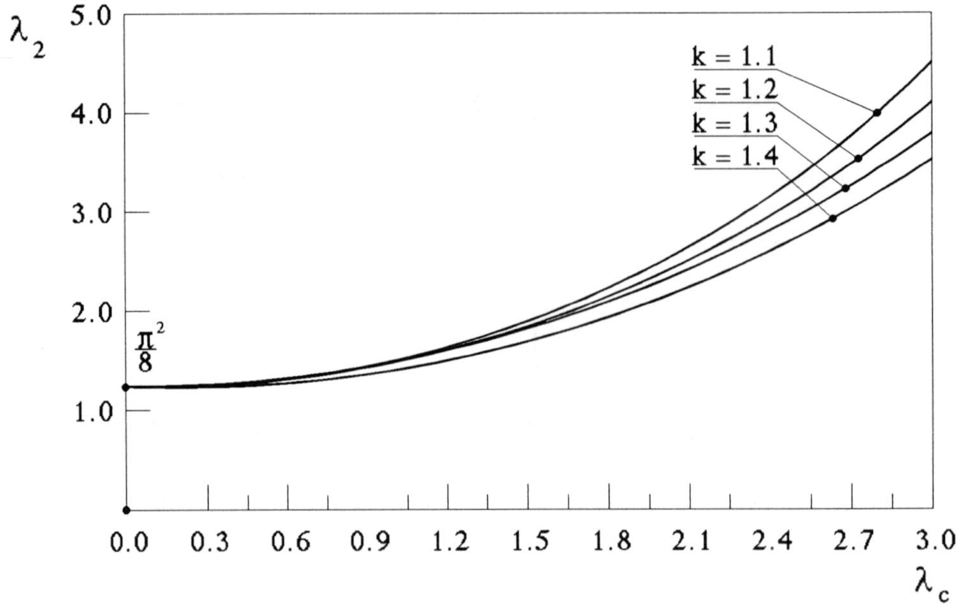

FIGURE 6.19

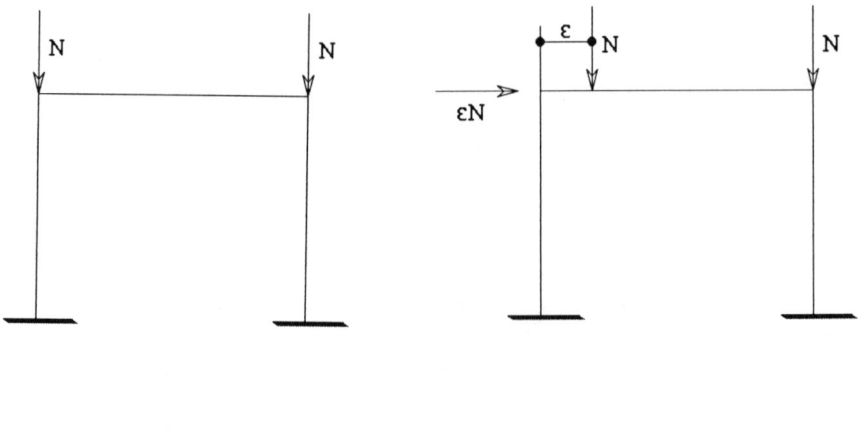

FIGURE 6.20a FIGURE 6.20b

presence of distributed loads, eccentricities of nodal forces, horizontal forces, or geometrical imperfections (Fig. 6.20b). On the other hand, for small values of the imperfections, the behavior of the imperfect structure can easily be deduced from the perfect one.

Consider the behavior of a perfect frame. Increasing the load multiplier λ, the equilibrium configurations of the frame are stable up to the critical state, defined by the value λ_c of the load multiplier λ. At $\lambda = \lambda_c$ the fundamental equilibrium path C_λ bifurcates and new flexural

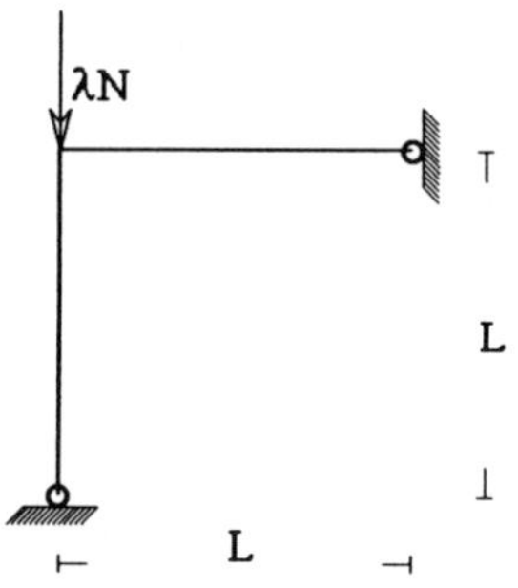

FIGURE 6.21

equilibrium configurations of the frame begin to occur. The buckling and postbuckling behavior of the frame is analyzed assuming that the displacement increments and the load parameter λ, according to (6.7) and (6.8), can be expressed as analytic functions of a some parameter α, which can represent a generalized displacement of the buckled frame (Fig. 6.21):

$$\mathbf{u} = \alpha\mathbf{u}_1 + \alpha^2\mathbf{u}_2 + o(\alpha^2); \quad \lambda = \lambda_c + \alpha\lambda_1 + \alpha^2\lambda_2 + o(\alpha^2) \tag{6.190}$$

To study asymmetric bifurcation, we have to evaluate the second and third differentials of the strain energy of the frame, and in the case of symmetric branching, we also need the fourth differential. Let us suitably express the strain energy of the frame supposing small deformations but finite displacements. By a Taylor expansion of the finite strain energy, we will get the successive variations $\mathcal{E}_2(\mathbf{u})$, $\mathcal{E}_3(\mathbf{u})$, and $\mathcal{E}_4(\mathbf{u})$, which define the postbuckling behavior of the frame. The first step is to define the strain energy for plane extension and flexure of every beam of the frame. The extension of the axis of the beams can be generally neglected. On the other hand, if we assume as variables the displacements of the frame, it is easier to meet the joint compatibility conditions taking into account the beam axis extension. With this assumption the strain energy of the structure is not smooth at the origin of H_A, as we have shown at Section 4.2.3. On the other hand, it will be shown that, for unbraced frames, the bifurcation parameters λ_1 and λ_2 are the same as in the inextensible structure. To study the plane extension and flexure of the bars of the frame we will use the model of extensible beam analyzed in Chapter 4. Equation (4.106) gives the expression of the strain energy of the beam. To analyze the postbuckling behavior of the frames we have to evaluate, on the other hand, the increment of the strain energy corrresponding to the imposed additional deformation that takes the beam from the primary configuration C_0 to a close configuration C. Only an axial load N is acting on the beam at the primary configuration C_0: this configuration is thus defined by the displacements

$$w_0 = -\frac{N}{EA} z \tag{6.191}$$

If we indicate with v(z) and w(z) the transversal and axial components of the additional displacement from C_0 to the configuration C, we have

$$(v)_c = v(z) \qquad (w)_c = w_0(z) + w(z) \tag{6.191'}$$

We can evaluate the strain energy increment by means of a power series expansion of the displacement increments v(z) and w(z). Let us consider, for instance, the additional extensional strain energy

$$\mathcal{V}_{ext}(C) - \mathcal{V}_{ext}(C_o) = \frac{EA}{2}\left[\int_0^L \left(\frac{ds}{dz} - 1\right)^2 dz\right](C) - \frac{EA}{2}\left[\int_0^L \left(\frac{ds}{dz} - 1\right)^2 dz\right](C_o) \tag{6.192}$$

For the axial extension, on the other hand, we have the following expansion

$$\left(\frac{ds}{dz}-1\right)^2 = w'^2 + v'^2 w' + \left[(1/4)v'^4 - v'^2 w'^2\right] + \left[(-3/4)v'^4 w' + w'^3 v'^2\right] + \ldots \quad (6.191'')$$

Consequently, we get

$$\left(\frac{ds}{dz}-1\right)^2 (C) - \left(\frac{ds}{dz}-1\right)^2 (C_o) = \left(w_o' + w'\right)^2 + v'^2\left(w_o' + w'\right)$$

$$+\left[(1/4)v'^4 + v'^2\left(w_o' + w'\right)^2\right] + \left[(-3/4)v'^4 + \left(w_o' + w'\right) + \left(w_o' + w'\right)^3 v'^2\right] + \ldots - w_o'^2 \quad (6.191''')$$

According to the assumption of very stiff beams, the configuration C_0 will be very close to the initial unstressed state C_i. Hence, we can retain only the linear terms in $w_0(z)$ in (6.191″). Thus we obtain, up to the fourth order,

$$\left(\frac{ds}{dz}-1\right)^2 (C) - \left(\frac{ds}{dz}-1\right)^2 (C_o) = 2w_o' w' + \left(w'^2 + w_o' v'^2\right) + \left(v'^2 w' - 2w_o' v'^2 w'\right)$$

$$+\left[(1/4)v'^4 - v'^2 w'^2 - (3/4)w_o' v'^4 + 3w_o' w'^2 v'^2\right] + \ldots \quad (6.191^{IV})$$

Thus, from (6.192), by integration on the bar and taking into account (6.191^{IV}), we obtain the following expressions of the extensional strain energy variations

$$\mathcal{V}_{1_{ext}} = -N\int_0^L w'\,dz; \quad \mathcal{V}_{2_{ext}} = \frac{1}{2}EA\int_0^L w'^2\,dz - \frac{N}{2}\int_0^L v'^2\,dz$$

$$\mathcal{V}_{3_{ext}} = \frac{1}{2}EA\left(1+\frac{2N}{EA}\right)\int_0^L w' v'^2\,dz$$

$$\mathcal{V}_{4_{ext}} = \frac{1}{2}EA\left(1+\frac{2N}{EA}\right)\int_0^L \frac{1}{4}v'^4\,dz - \frac{1}{2}EA\left(1+\frac{3N}{EA}\right)\int_0^L w'^2 v'^2\,dz \quad (6.193)$$

A similar procedure gives for the flexural stain energies the expansions

$$\mathcal{V}_{1_{fl}} = 0; \quad \mathcal{V}_{2_{fl}} = \frac{1}{2}EI\left(1+\frac{2N}{EA}\right)\int_0^L v''^2\,dz; \quad \mathcal{V}_{3_{fl}} = -EI\left(1+\frac{3N}{EA}\right)\int_0^L v''(v'\,w')'\,dz;$$

$$\mathcal{V}_{4_{fl}} = \frac{1}{2}EI\left(1+\frac{4N}{EA}\right)\int_0^L \left(3v''^2 w'^2 + v'^2 w''^2 - 2v''^2 + 6v''w''v'w''\right)dz \quad (6.193')$$

The strain energy increment of a framed structure is therefore given by the summation, extended to all the bars of the frame, of the flexural and extensional strain energies, respectively given by the (6.193) and (6.193′). It is now useful to define the following nondimensional quantities:

$$t = \frac{z}{L}; \; V = \frac{v}{L}; \; W = \frac{w}{L}; \; \frac{d(\,)}{dt} = (\,)'; \; \theta = \frac{I}{AL^2} << 1; \; N = \lambda\bar{N}; \; \frac{N}{EA} = \beta\theta = \lambda\bar{\beta}\theta \quad (6.194)$$

where θ is a small parameter defining the ratio between the flexural and the extensional beam stiffnesses. We will retain only terms linear in θ. With this assumption we get the following expression of the expansions of the additional strain energy:

$$\mathcal{V}_1 = -\sum \mathrm{EAL}\beta\theta \int_0^1 \mathrm{W'}\ \mathrm{dt};$$

$$\mathcal{V}_2 = \frac{1}{2}\sum \mathrm{EAL}\left(-\beta\theta \int_0^1 \mathrm{V'}^2\ \mathrm{dt} + \int_0^1 \mathrm{W'}^2\ \mathrm{dt} + \theta \int_0^1 \mathrm{V''}^2\ \mathrm{dt}\right)$$

$$\mathcal{V}_3 = \frac{1}{2}\sum \mathrm{EAL}\left[(1+2\beta\theta)\int_0^1 \mathrm{V'}^2\mathrm{W'}\ \mathrm{dt} - 2\theta \int_0^1 \mathrm{V''}(\mathrm{V'}\,\mathrm{W'})'\ \mathrm{dt}\right]$$

$$\mathcal{V}_4 = \frac{1}{2}\mathrm{EAL}\left[(1+3\beta\theta)\int_0^1 \left(\frac{1}{4}\mathrm{V'}^4 - \mathrm{W'}^2\mathrm{V'}^2\right)\mathrm{dt}\right.$$

$$\left. + \theta \int_0^1 \left(3\mathrm{V''}^2\mathrm{W'}^2 + \mathrm{V'}^2\mathrm{W''}^2 - 2\mathrm{V''}^2\mathrm{V'}^2 + 6\mathrm{V''}\mathrm{W''}\mathrm{V'}\,\mathrm{W'}\right)\mathrm{dt}\right] \qquad (6.195)$$

Thus, the additional potential energy, evaluated as from the fundamental configurations C_λ, has the series expansion

$$\mathcal{E}(\mathbf{u}) = \mathcal{E}_1(\mathbf{u}) + \mathcal{E}_2(\mathbf{u}) + \mathcal{E}_3(\mathbf{u}) + \ldots \qquad (6.196)$$

Equilibrium at C_λ on the other hand implies

$$\mathcal{E}_1(\mathbf{u}) = 0$$

and, for the common case of dead loads

$$\mathcal{E}_2(\mathbf{u}) = \mathcal{V}_2(\mathbf{u});\ \mathcal{E}_3(\mathbf{u}) = \mathcal{V}_3(\mathbf{u});\ \mathcal{E}_4(\mathbf{u}) = \mathcal{V}_4(\mathbf{u});\ldots \qquad (6.196')$$

Buckling

The critical state of the frame is reached when, increasing λ, the variational condition (6.23) is satisfied. Equation (6.23) defines both the buckling value λ_c and the buckling mode $\mathbf{u}_c$ of components V_c, W_c. From (6.196′) and (6.195) we get:

$$\mathcal{E}_{11}(\mathbf{u}_c,\delta\mathbf{u}) = \sum \mathrm{EAL}\left(-\beta\theta\int_0^1 \mathrm{V}_c'\delta\mathrm{V}_c'\ \mathrm{dt} + \int_0^1 \mathrm{W}_c'\delta\mathrm{W}_c'\ \mathrm{dt} + \theta\int_0^1 \mathrm{V}_c''\delta\mathrm{V}_c''\ \mathrm{dt}\right) = 0 \qquad (6.197)$$

and we obtain, for every bar of the frame, the following equations:

$$\mathrm{EV}_c'''' + \beta\mathrm{V}_c'' = 0; \qquad -\mathrm{W}_c'' = 0 \qquad (6.198)$$

together with the corresponding joint equilibrium conditions, obtained from the boundary terms of (6.197) and the joint compatibility equations:

$$\mathrm{EV}_c''''\sum\left| -\left(\mathrm{V}_c''' + \beta\mathrm{V}_c'\right)\theta\delta\mathrm{V}_c + \mathrm{W}_c'\delta\mathrm{W}_c\right|_0^1 = 0;\ \sum\left|\theta\mathrm{V}_c''\delta\mathrm{V}_c'\right|_0^1 = 0 \qquad (6.199)$$

Owing to the bars extension, both the transversal and axial components, V_c and W_c, of the buckling displacements must be evaluated. However, (6.197) shows that

$$V_c = V_{0_c} + \theta V_{0_c} \qquad W_c = \theta \overline{W}_c \tag{6.200}$$

In fact, because of the small value of the ratio θ between the flexural and extensional stiffnesses of the bars, the axial displacements are always very small compared to the transversal ones. Nevertheless, as we will see later, we only need to know the transversal displacements V_{0c} and the axial strain W_c' for the analysis of the postcritical behavior of the frame. It is therefore possible to consider the bars of the frame as axially rigid and first to evaluate the transversal displacements V_{0c}; in the case of unbraced frames, solving the nodal equilibrium equations, we can then easily obtain the axial forces increments $\Delta N = EAW_c'$. The buckling analysis is thus reduced to the classic one, that is, to the evaluation of the critical mulplier λ_c involving only the transversal critical displacements V_c.

Asymmetric Postbuckling Behavior

Asymmetric postbuckling behavior occurs in the structures if the quantity λ_1, that is $\mathcal{E}_3(\mathbf{u}_c)$, is not zero. In the case of framed structures. the third variation $\mathcal{E}_3(\mathbf{u}_c)$ of the total potential energy along the buckling mode can be expressed as

$$\mathcal{E}_3(\mathbf{u}_c) = \frac{1}{2} \sum EAL \left[(1 + 2\beta\theta) \int_0^1 V_c'^2 W_c' \, dt - 2\theta \int_0^1 W_c''(V_c' W_c')' \, dt \right]$$

Then, neglecting the nonlinear terms in θ and if the term

$$\sum \int_0^1 V_c'^2 W_c' \, dt$$

does not vanish, we get

$$\mathcal{E}_3(\mathbf{u}_c) = \frac{1}{2} \sum EAL\theta \int_0^1 V_c'^2 \overline{W}_c' \, dt = \frac{1}{2} \sum \Delta NL \int_0^1 V_c'^2 \, dt \tag{6.201}$$

Thus, to evaluate $\mathcal{E}_3(\mathbf{u}_c)$,we only need to know the transversal displacements V_c and the axial force increments ΔN. If we remember that the quantity:

$$\Delta^{(2)} L = \frac{1}{2} L \int_0^1 V_c'^2 \, dt$$

represents the second-order extension of the axis of the bars, we can write, more clearly;

$$\mathcal{E}_3(\mathbf{u}_c) = \frac{1}{2} \sum \Delta N \cdot \Delta^{(2)} L \tag{6.202}$$

The third variation $\mathcal{E}_3(\mathbf{u}_c)$ of the potential energy of the frame can thus be evaluated as the work done by the first-order increment of the axial forces for the second-order axial elongations of the beams. Similarly, we easily obtain for the quantity $\mathcal{E}_2'(\mathbf{u}_c)$:

$$\mathcal{E}_2'(\mathbf{u}_c) = \frac{1}{2} \sum \left(-EAL\overline{\beta}\theta \int_0^1 V_c'^2 \, dt \right) = \sum \overline{N} \Delta^{(2)} L \tag{6.203}$$

Therefore the coefficient λ_1 which controls the initial asymmetric postbuckling behavior of the frames, is given by

$$\lambda_1 = -\frac{3}{2}\frac{\mathcal{E}_3(\mathbf{u}_c)}{\mathcal{E}_2'(\mathbf{u}_c)} = -\frac{3}{2}\frac{\sum \Delta N \cdot \Delta^{(2)}L}{\sum \overline{N} \cdot \Delta^{(2)}L} \tag{6.204}$$

For unbraced frames, the first-order axial forces increments ΔN can be directly obtained in terms of the transversal critical displacements. Thus the parameter λ_1 depends only on the transversal critical displacement of the bars by means of the second-order elongations $\Delta^{(2)}L$. It means that the value of λ_1 is the same as for the inextensible frames.

The simple two-bar frame shown in Fig. 6.23 exhibits an asymmetric postbuckling behavior. An analytic and experimental study of this frame was first developed by Roorda.[12] We denote with V_{1c}, W_{1c} and V_{2c}, W_{2c} the critical displacements of the vertical and of the horizontal bar. The explicit form of the buckling equations (6.198) and (6.199) is

Equilibrium Equations

$$V_{1c}^{IV} + \beta V_{1c}'' = 0; \; W_{1c}'' = 0; \; V_{2c}^{IV} = 0; \; W_{2c}'' = 0 \tag{6.205}$$

Joint Kinematical Conditions

$$V_{1c}(0) = W_{1c}(0) = V_{2c}(0) = W_{2c}(0) = 0$$

$$V_{1c}(L) = W_{2c}(L); \; W_{1c}(L) = V_{2c}(L); \; V_{1c}'(L) = V_{2c}'(L) \tag{6.206}$$

Joint Equilibrium Condition

$$\theta\left(V_{1c}''''(L) + \beta V_{1c}'(L)\right) - W_{2c}'(L) = 0$$

$$\theta V_{2c}''''(L) - W_{1c}'(L) = 0; \; V_{1c}''(L) + V_{2c}''(L) = 0; \; V_{1c}''(0) = V_{2c}''(0) = 0 \tag{6.207}$$

The solution of the buckling problem is:

$$V_{1c} = \left[\frac{\sin\sqrt{\beta_c}\,t}{\left(\sin\sqrt{\beta_c} - t\right)}\right]; W_{1c} = -3\theta t; \; V_{2c} = \frac{1}{2}\left(t^3 - t\right); \; W_{2c} = -3\theta t \tag{6.208}$$

with $\beta_c = 1.1861^2\pi^2$. The postbuckling behavior of the frame can be approximately described as

$$\frac{\lambda}{\lambda_c} \cong 1 + \frac{\lambda_1}{\lambda_c}\alpha \tag{6.209}$$

From (6.204) we obtain

$$\frac{\lambda_1}{\lambda_c} = 0.3805 \tag{6.210}$$

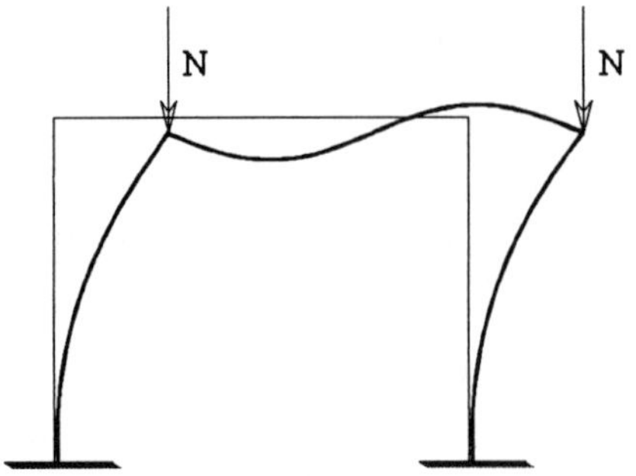

FIGURE 6.22

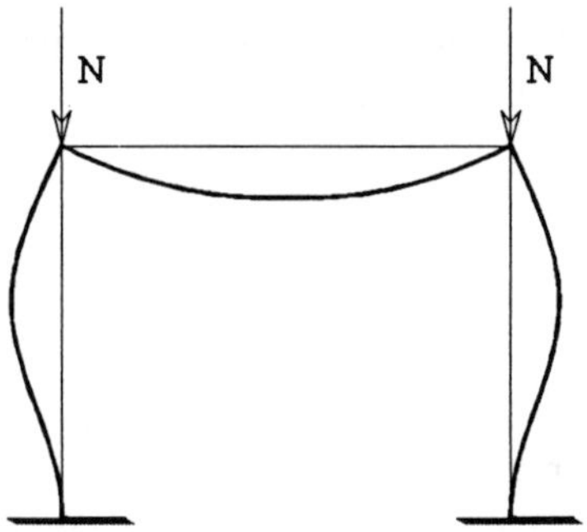

FIGURE 6.23

In (6.209) the parameter α can be identified with the rotation angle of the joint connecting the two bars.

Symmetric Postbuckling

Frames of generic geometry, as a rule, exhibit asymmetric bifurcation. On the contrary, in the fairly common case of symmetric and symmetrically loaded frames, the initial postbuckling behavior is very often symmetric. For these frames the buckling mode is antisymmetric with sideways displacements of the joints (Fig. 6.22). Thus, according to the result pointed out in Section 6.2.7, symmetric bars give opposite contributions and $\mathcal{E}_3(\mathbf{u}_c)$ vanishes. In this case the postbuckling behavior of the frame and its imperfection sensitivity are determined by the sign of the coefficient λ_2 of (6.190). In the less frequent case of symmetric buckling, (Fig. 6.23) we easily deduce that $\mathcal{E}_3(\mathbf{u}_c) \neq 0$ and the bifurcation is nonsymmetric.

Examine the symmetric postbuckling behavior and let us evaluate the quantity $\mathcal{E}_4(\mathbf{u}_c)$ - $\mathcal{E}_2(\mathbf{u}_2)$. The second-order displacements $\mathbf{v}_2$ of the expansion of the displacement $\mathbf{u}$ corresponding to the buckled equilibrium configuration, are determined by the variational condition (6.123). From (6.196′) and (6.194) we get, for the quantities $\mathcal{E}_{11}$ and $\mathcal{E}_{21}$;

$$\mathcal{E}_{11}(\mathbf{u}_2,\delta\mathbf{u}) = \sum EAL\left(-\beta\theta\int_0^1 V'\,\delta V'\,dt + \int_0^1 W'\,\delta W'\,dt + \theta\int_0^1 V''\,\delta V''\,dt\right)$$

$$\mathcal{E}_{21}(\mathbf{u}_c,\delta\mathbf{u}) = \sum EAL\left[(1+2\beta_c\theta)\frac{1}{2}\int_0^1 \left(V_c'^{\,2}\delta W' + 2V_c'W_c'\delta V'\right)dt\right. \tag{6.211}$$

$$\left. -\,\theta\int_0^1 \left(V_c'W_c'\right)'\,\delta V''\,dt - \theta\int_0^1 V_c''\left(V_c'\delta W' + W_c'\delta V'\right)'\,dt\right] \tag{6.212}$$

where V and W, respectively, are the transversal and axial nondimensional components of the displacement $\mathbf{u}_2$. To deduce, from the variational condition (6.211), the differential equations for the displacements V and W, together with the corresponding nodal equilibrium conditions, it is useful to analyze more clearly the nodal compatibility conditions for the displacements V, W and δV, δW. In fact, in the case of a rigidly connected joint, the rotations of the sections connecting in the joint must be equal. Thus we have to impose from (4.95) the nonlinear compatibility condition involving the rotation angle of the cross sections

$$\tan\varphi = \frac{V'}{1 + W'} = \text{constant} \tag{6.213}$$

for all the end sections of the bars rigidly connected in a joint. Therefore, if we consider the compatibility of the displacements $\mathbf{u} = \alpha\mathbf{u}_c + \alpha^2\mathbf{u}_2 + \ldots$, expanding (6.213), we get

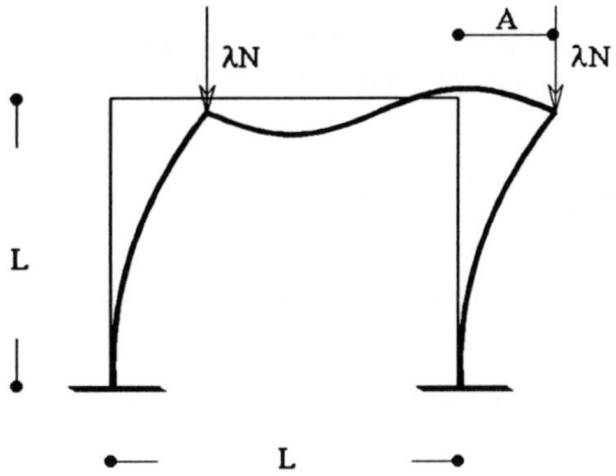

FIGURE 6.24

$$\tan\varphi = \frac{\alpha V_c' + \alpha^2 V' + \ldots}{1 + \alpha W_c' + \alpha^2 W' + \ldots} = \alpha V_c' + \alpha^2(V' - V_c' W_c') + o(\alpha^2) \tag{6.214}$$

from which we deduce that the critical displacements V_c must satisfy the classical linearized nodal compatibility condition

$$V_c' = \text{constant} \tag{6.215}$$

while, for the second-order transversal displacements V, we obtain:

$$V' - V_c' W_c' = \text{constant} \tag{6.216}$$

Similarly, imposing the compatibility of the joint rotation for the displacements

$$\mathbf{u} = \alpha\mathbf{u}_c + \alpha^2\mathbf{u}_2 + \ldots + \delta\mathbf{u} \tag{6.217}$$

with the positions

$$\delta V = \varepsilon V_o \qquad \delta W = \varepsilon W_o \tag{6.218}$$

we have

$$\tan\varphi = \alpha V_c' + \alpha^2\left(V' - V_c' W_c'\right) + \varepsilon V_c' - \alpha\varepsilon\left(W_c' V_o' + V_c' W_o'\right)$$

$$+ o\left(\varepsilon\alpha^2\right) + o\left(\varepsilon^2\right) + o\left(\alpha^3\right) = \text{constant} \tag{6.219}$$

for every small value of α and ε. The coefficients of ε and $\varepsilon\alpha$ must be costant in a joint and

$$\delta V' = \text{constant} \tag{6.220}$$

$$W_c' \delta V' + V_c' \delta W' = \text{constant} \tag{6.221}$$

Taking into account (6.221), the variational condition (6.211) can be now written in the form

$$\mathcal{E}_{11}(\mathbf{u}_2, \delta\mathbf{u}) + \mathrm{E}\,\mathcal{E}_{21}(\mathbf{u}_c, \delta\mathbf{u}) = \sum \mathrm{EAL}\int_0^1 [W'\delta W' + \frac{1}{2} V_c'^2\, \delta W'$$

$$+ \theta(-\beta_c V'\,\delta V' + V''\,\delta V'' + \beta_c V_c'^2\,\delta W' + \overline{W}_c V_c'\,\delta V' - V_c''' V_c'\,\delta W')]dt \tag{6.222}$$

For each bar of the frame, (6.222) gives the following differential equations for the transversal displacements V and the axial displacements W:

$$W'' + \left(\frac{1}{2} + \theta\beta_c\right)\left(V_c'^2\right)' + \theta\left(V_c''' V_c'\right)' = 0;\quad V^{IV} + \beta_c V'' - \overline{W}_c' V_c'' = 0 \tag{6.223}$$

together with the corresponding nodal equilibrium conditions:

$$\sum \left| \left[W' + \left(\frac{1}{2} + \theta\beta_c\right) V_c'^2 + \theta\left(V_c''' V_c'\right)\right] \delta W + \theta\left(- V''' - \beta_c V' + \overline{W}_c' V_c'\right) \delta V \right|_0^1 = 0 \tag{6.224}$$

$$\sum \left| \theta V'' \, \delta V' \right|_0^1 = 0 \tag{6.225}$$

In case of symmetric bifurcation, (6.223), (6.224), and (6.225), together with the compatibility relations of the joint displacements and rotations, determine the second-order displacements and the initial postbuckling behavior of the frame. The general solution of (6.223) is easily found:

$$W = C_1 + C_2 t - \left(\frac{1}{2} + \theta\beta_c\right)\int_0^t V_c'^2 \, dt - \theta \int_0^t V_c''' V_c' \, dt \tag{6.226}$$

$$V = C_3 + C_4 t + C_5 \sin\sqrt{\beta_c}\, t + C_6 \cos\sqrt{\beta_c}\, t - \frac{\overline{W}_c'}{2\beta_c} V_c' t \qquad \text{if } \beta_c > 0$$

$$V = C_3 + C_4 t + C_5 \sinh\sqrt{-\beta_c}\, t + C_6 \cosh\sqrt{-\beta_c}\, t - \frac{\overline{W}_c'}{2\beta_c} V_c' t \qquad \text{if } \beta_c < 0$$

$$V = C_3 + C_4 t + C_5 t + C_6 t^3 + \overline{W}_c' \int_0^t dt \int_0^t V_c \, dt \qquad \text{if } \beta_c = 0 \tag{6.227}$$

The six constants that appear in the expressions of the displacements W and V are clearly determined by the joint equilibrium and compatibility conditions. To evaluate the quantity $\mathcal{E}_4 - \mathcal{E}_2$ up to terms linear in the small parameter θ, it is sufficient to calculate the transversal displacements V, neglecting the terms in θ, and the axial displacements W up to the terms linear in θ. Further, taking into account the general expression (6.226) and (6.227) of the displacements V and W, we observe that the joint equilibrium equations (6.224) can be written in the simpler form

$$\sum \left| C_2 \, \delta W + \theta C_7 \, \delta V \right|_0^1 = 0 \tag{6.228}$$

where

$$C_7 = -\beta_c C_4 + \frac{3\overline{W}_c'}{2\beta_c}\left(V_c''' + \beta_c V_c'\right) \qquad \text{if } \beta_c \geq 0$$

$$C_7 = C_6 \qquad \text{if } \beta_c = 0 \tag{6.229}$$

Assume $C_2 = C_2' + \theta C_2''$. In the case of unbraced frames ($b \leq 2n$), (6.228) shows that we have $C_2' = 0$. Thus the general expression of the axial displacements W takes the form

$$W = C_1 + \theta C_2 t - \left(\frac{1}{2} + \theta\beta_c\right)\int_0^1 V_c'^2\, dt - \theta\int_0^1 V_c''' V_c'\, dt \tag{6.230}$$

while the equilibrium equations (6.224) and (6.225) of the joints can be written as

$$\Sigma \left| \theta(C_2\ \delta W + C_7\ \delta V)\right|_0^1 = 0; \quad \Sigma \left| \theta V''\ \delta V' \right|_0^1 = 0 \tag{6.231}$$

To evaluate the constants $C_1,...,C_6$ we have to use (6.225) and (6.231), together with the joint compatibility condition. Once that this first problem has been solved, we can evaluate the coefficient λ_2. The fourth differential $\mathcal{E}_4(\mathbf{u}_c)$ evaluated along the buckling mode is easily obtained from (6.195); thus we have, up to the terms linear in θ,

$$\mathcal{E}_4(\mathbf{u}_c) = \frac{1}{2}\Sigma\ EAL\int_0^1 [(1 + 3\beta_c\theta)\,\frac{1}{4}\,V_c'^4 - 2\theta V_c''^2 V_c'^2]\, dt \tag{6.232}$$

Similarly for the second variation $\mathcal{E}_2(\mathbf{v}_2)$ the following relation holds:

$$\mathcal{E}_2(\mathbf{u}_2) = \frac{1}{2}\sum EAL\int_0^1 \left[\frac{V_c'^4}{4} + \theta\left(-C_2 V_c'^2 + \beta_c V_c'^4 + V_c'^3 V_c''' - \beta_c V'^2 + V''^2\right)\right] dt \tag{6.233}$$

and

$$\mathcal{E}_4(\mathbf{u}_c) - \mathcal{E}_2(\mathbf{u}_2) = \frac{1}{2}\sum EAL\theta\int_0^1\left(-\frac{\beta_c}{4}V_c'^4 + V_c''^2 V_c'^2 + C_2 V_c'^2 + \beta_c V'^2 - V''^2\right) dt \tag{6.234}$$

The coefficient λ_2 therefore takes the expression

$$\lambda_2 = -2\,\frac{\mathcal{E}_4(\mathbf{u}_c) - \mathcal{E}_2(\mathbf{u}_2)}{\Sigma\,\overline{N}\,\Delta^{(2)} L} \tag{6.235}$$

In this expression only the critical transversal displacements are included. The bifurcation parameter λ_2 of the inextensible structure will therefore be given by the same (6.235). It is also worthwile to notice that $\mathcal{E}_4(\mathbf{u}_c)$ is always positive owing to the effect of the term

$$\frac{1}{2}\Sigma\, EAL\int_0^1 \frac{1}{4}\, V_c'^4\, dt$$

which represents the strain energy of the frame due to the second order extension of the bars axes. An equal term is present in the expression of $\mathcal{E}_2(\mathbf{v}_2)$, so that their sum vanishes and only linear terms in θ appear in the expression of $\mathcal{E}_4(\mathbf{u}_c)$ - $\mathcal{E}_2(\mathbf{v}_2)$. The quantity λ_2 can be therefore positive or negative, so we cannot decide in advance if symmetric frames exhibit or not imperfection sensitivity. The previous formulation was applied by Lanni[35] to postbuckling analysis of the square portal frame of Fig. 6.24, characterized by the ratio $k = EI_c/EI_b$ of the columns/beam flexural stiffnesses. In Fig. 6.25 the relation between the load $\lambda = NL^2/4EI_c$ and the displacement parameter $\alpha = A/L$ is given. In this case the bifurcation is stable with the occurrence of large displacements under small load increments.

6.5.3 THE VIRTUAL WORK APPROACH TO POSTBUCKLING ANALYSIS

The approach to postbuckling analysis of perfect and quasi-perfect elastic structures, given in previous sections, is essentially based on Koiter's original work, where the energy formulation

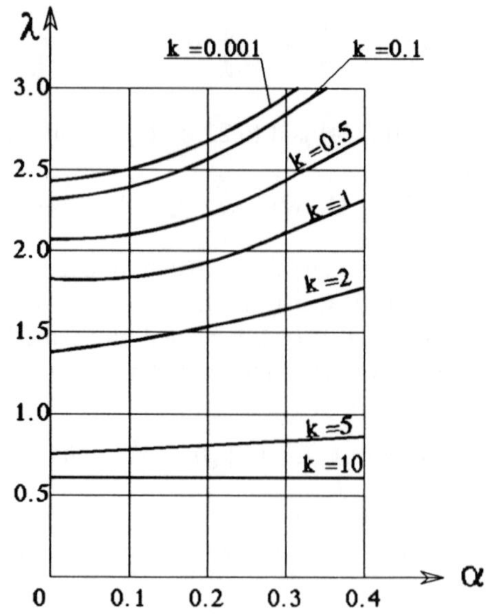

FIGURE 6.25

was extensively used. In the period 1960–1970 Budiansky and Hutchinson[37,38] developed an alternative approach based on the use of the virtual work equation. This formulation was successfully applied to many difficult problems, particularly concerning the postbuckling analysis of thin plates and shells.

We give here a presentation of this virtual work approach, according to the formulation developed by Budiansky.[31,38] Generalized loads, stresses, strains, and displacements will be denoted by $\mathbf{q}$, σ, ε, and $\mathbf{u}$, respectively. The potential energy of the structure is written in the form

$$\mathcal{E}(\mathbf{u},\lambda) = \mathcal{F}(\boldsymbol{\varepsilon}) - \lambda \langle \mathbf{q},\mathbf{u} \rangle \tag{6.236}$$

where:

$\mathbf{u}$ is the displacement field evaluated from the initial unstressed configuration C_i. $\langle \lambda\mathbf{q},\mathbf{u} \rangle$ is the work of the applied load $\lambda\mathbf{q}$ increasing with the load parameter λ. $\mathcal{F}(\varepsilon)$ is the strain energy of the structure, considered as a functional of the generalized strains ε according to the relation

$$\mathcal{F}(\varepsilon) = \int_\Omega \phi(\varepsilon)\, d\Omega \tag{6.237}$$

if $\phi(\varepsilon)$ is the strain energy density.

The strains ε are then expressed by a strain displacement function $\varepsilon(\mathbf{u})$. The variational equation of equilibrium along the equilibrium path of the structure is

$$D\mathcal{E}[\mathbf{u},\lambda;\delta\mathbf{u}] = D\mathcal{F}[\varepsilon;\delta\varepsilon] - \lambda < \mathbf{q},\delta\mathbf{u} > = 0 \tag{6.238}$$

to be satisfied for all the admissible generalized displacements $\delta\mathbf{u}$ and where

$$\delta\,\varepsilon = \nabla_u \varepsilon(\mathbf{u})\delta\mathbf{u} \tag{6.239}$$

Thus, from (6.237)

$$D\mathcal{F}[\varepsilon; \delta\varepsilon] = \int_{\Omega} \nabla_{\varepsilon}\phi(\varepsilon)\delta\varepsilon \, d\Omega = < \nabla_{\varepsilon}\phi(\varepsilon), \delta\varepsilon > \tag{6.240}$$

where $\nabla_{\varepsilon}\phi(\varepsilon)$ is the gradient of $\phi(\varepsilon)$ at ε. If we introduce the notation

$$\sigma = \nabla_{\varepsilon}\phi(\varepsilon) \tag{6.241}$$

the equilibrium equation becomes

$$< \sigma, \delta\varepsilon > - \lambda < p, \delta\mathbf{u} > = 0 \tag{6.238$'$}$$

With a more compact notation, according to Budiansky and Hutchinson[37] the equilibrium equation can be expressed as

$$\sigma \cdot \delta\varepsilon - \lambda p \cdot \delta\mathbf{u} = 0 \tag{6.238$''$}$$

This has the form of a principle of virtual work, stating that, in an equilibrium state, the change of the potential energy of the loads, associated with the "virtual" displacements $\delta\mathbf{u}$, equals the internal work of the generalized stresses σ for the strains $\delta\varepsilon$, compatible with the displacements $\delta\mathbf{u}$. To simplify the analysis we assume a quadratic relation between ε and $\mathbf{u}$:

$$\varepsilon(\mathbf{u}) = L_1(\mathbf{u}) + \frac{1}{2} L_2(\mathbf{u}) \tag{6.242}$$

where L_1 and L_2, respectively, are linear and quadratic operators. We introduce the bilinear operator L_{11} $(\mathbf{u},\mathbf{v})$ defined by the identity

$$L_2(\mathbf{u} + \mathbf{v}) = L_2(\mathbf{u}) + 2L_{11}(\mathbf{u}, \mathbf{v}) + L_2(\mathbf{v}) \tag{6.243}$$

Then the strain variation $\delta\varepsilon$, compatible with $\delta\mathbf{u}$, may be written as

$$\delta\varepsilon = L_1(\delta\mathbf{u}) + L_{11}(\mathbf{u}, \delta\mathbf{u}) \tag{6.244}$$

Note that, according to definition (6.243),

$$L_{11}(\mathbf{u},\mathbf{v}) = L_{11}(\mathbf{v},\mathbf{u}) \tag{6.245}$$

and

$$L_{11}(\mathbf{u},\mathbf{u}) = L_2(\mathbf{u}) \tag{6.245$'$}$$

It also useful the notation

$$\mathbf{e}(\mathbf{u}) = L_1(\mathbf{u}) \tag{6.246}$$

so we can write

$$\delta\varepsilon = \delta e + L_{11}(\mathbf{u}, \delta\mathbf{u}) \tag{6.244$'$}$$

Moreover, according to the assumption of large displacements and relatively small strains, we assume a linear relation between generalized stresses σ and strains ε:

$$\sigma = \mathbf{C}\varepsilon \tag{6.247}$$

where $\mathbf{C}$ is the stiffness matrix. The reciprocal relation

$$C\varepsilon_1 \cdot \varepsilon_2 = C\varepsilon_2 \cdot \varepsilon_1 \tag{6.248}$$

is assumed valid for all ε_1 and ε_2.

We will limit our analysis to the case of perfect elastic structures that, according to the definitions given in Section 6.3.2, in the first stage of loading possess a fundamental path of equilibrium configurations increasing linearly with λ:

$$\mathbf{u}_0\,(\lambda) = \lambda\,\mathbf{u}_0 \tag{6.249}$$

The field equations governing the problem of finding σ, ε, and $\mathbf{u}$ may now be collected as

$$\sigma \cdot \delta\varepsilon - \lambda\mathbf{q} \cdot \delta\mathbf{u} = 0 \qquad \text{(equilibrium)}$$

$$\sigma = C\varepsilon \qquad \text{(elastic stress-strain relation)}$$

$$\varepsilon(\mathbf{u}) = L_1(\mathbf{u}) + \frac{1}{2}L_2(\mathbf{u}) \qquad \text{(strain-displacement relation)} \tag{6.250}$$

The Fundamental Path

Before buckling, the response of the structure is represented by

$$\mathbf{u} = \lambda\mathbf{u}_0; \quad \varepsilon = \lambda\varepsilon_0; \quad \sigma = \lambda\sigma_0 \tag{6.251}$$

Thus a linear theory is assumed valid before buckling with $\varepsilon_0 = \mathbf{e}_0 = L_1(\mathbf{u}_0)$, $\sigma_0 = C\varepsilon_0$ and

$$\sigma_0 \cdot \delta\mathbf{e} = \mathbf{q}_0 \cdot \delta\mathbf{u} \tag{6.252}$$

At the buckling, bifurcation occurs with displacements generally expressed as

$$\mathbf{u} = \lambda\mathbf{u}_0 + \mathbf{v} \tag{6.253}$$

We also assume that near the critical equilibrium configuration $\lambda_c\mathbf{u}_0$, the additional displacements $\mathbf{v}$ in (6.253) are such that

$$L_{11}(\mathbf{u}_0,\mathbf{v}) = 0 \tag{6.254}$$

i.e., that the bifurcation displacements $\mathbf{v}$ are orthogonal to the prebuckling displacements $\mathbf{u}_0$ in the sense of (6.254). Condition (6.254) means that $L_2(\mathbf{u})$, i.e., the quadratic part of the strain, does not involve those displacement components that characterize the fundamental prebuckling equilibrium path. For instance, for a beam submitted to axial and transversal displacements $\mathbf{u}$ of components w(z) and v(z) the strain of the axis is

$$\varepsilon = L_1(\mathbf{u}) + \frac{1}{2}L_2(\mathbf{u}) = \frac{dw}{dz} + \frac{1}{2}\left(\frac{dv}{dz}\right)^2$$

Therefore, if the fundamental state is represented by the rectilinear configuration of the bar, i.e., expressed by

$$v_0 = 0\;;\; w_0 = w_0(z)$$

the orthogonality condition (6.254) is satisfied for any additional displacement v(z).

Buckling and Postbuckling

To analyze the buckling and postbuckling behavior of the structure, let us consider the bifurcated equilibrium configurations represented by the expansions:

$$\mathbf{u}(\alpha) = \lambda \mathbf{u}_o + \alpha \mathbf{u}_1 + \alpha^2 \mathbf{u}_2 + \ldots \quad (6.255)$$

$$\varepsilon(\alpha) = \lambda \varepsilon_o + \alpha \varepsilon_1 + \alpha^2 \varepsilon_2 + \ldots \quad (6.255')$$

$$\sigma(\alpha) = \lambda \sigma_o + \alpha \sigma_1 + \alpha^2 \sigma_2 + \ldots \quad (6.255'')$$

$$\lambda(\alpha) = \lambda_o + \alpha \lambda_1 + \alpha^2 \lambda_2 + \ldots \quad (6.255''')$$

where $\mathbf{u}_2$, $\mathbf{u}_3$, ... are all orthogonalized to the buckling mode $\mathbf{u}_1$ in the sense that

$$\sigma_0 \cdot L_{11} (\mathbf{u}_n, \mathbf{u}_1) = 0 \ (n = 2, 3, ...) \quad (6.256)$$

Substitution of the expansions (6.255) in the field equations (6.250) and collection of the terms linear, quadratic and cubic in α, gives the following conditions:

$$\lambda_c \sigma_o \cdot L_{11}(\mathbf{u}_1, \delta\mathbf{u}) + \sigma_1 \cdot \delta e = 0 \quad (6.257')$$

$$\lambda_c \sigma_o \cdot L_{11}(\mathbf{u}_2, \delta\mathbf{u}) + \lambda_1 \sigma_o \cdot L_{11}(\mathbf{u}_1, \delta\mathbf{u}) + \sigma_1 \cdot L_{11}(\mathbf{u}_1, \delta\mathbf{u}) + \sigma_2 \cdot \delta e = 0 \quad (6.257'')$$

$$\lambda_c \sigma_o \cdot L_{11}(\mathbf{u}_3, \delta\mathbf{u}) + \lambda_2 \sigma_o \cdot L_{11}(\mathbf{u}_1, \delta\mathbf{u}) + \lambda_1 \sigma_o \cdot L_{11}(\mathbf{u}_2, \delta\mathbf{u})$$
$$+ \sigma_1 \cdot L_{11}(\mathbf{u}_2, \delta\mathbf{u}) + \sigma_2 \cdot L_{11}(\mathbf{u}_1, \delta\mathbf{u}) + \sigma_3 \cdot \delta e = 0 \quad (6.257''')$$

The first equation defines the buckling load λ_c and the buckling mode u_1. Now we can analyze the equilibrium bifurcation of the structure.

Asymmetric Branching ($\lambda_1 \neq 0$)

With $\delta\mathbf{u} = \mathbf{u}_1$, and using the notation $\mathbf{e}(\mathbf{u}_n) = \mathbf{e}_n$ (6.257′) we obtain

$$\lambda_1 \sigma_o \cdot L_2(\mathbf{u}_1) + \sigma_1 \cdot L_2(\mathbf{u}_1) + \sigma_2 \cdot e_1 = 0 \quad (6.258)$$

Note that substitution of the expansion (6.255) in the strain-displacement relation (6.250) yields

$$\varepsilon(\lambda \mathbf{u}_o + \alpha \mathbf{u}_1 + \alpha^2 \mathbf{u}_2 + \ldots)$$
$$= \lambda \mathbf{e}(\mathbf{u}_o) + \alpha \mathbf{e}(\mathbf{u}_1) + \alpha^2 \mathbf{e}(\mathbf{u}_2) + \alpha^3 \mathbf{e}(\mathbf{u}_3) + \frac{1}{2} L_2(\alpha \mathbf{u}_1 + \alpha^2 \mathbf{u}_2 + \ldots)$$
$$= \lambda \mathbf{e}(\mathbf{u}_o) + \alpha \mathbf{e}(\mathbf{u}_1) + \alpha^2 \left[\mathbf{e}(\mathbf{u}_2) + \frac{1}{2} L_2(\mathbf{u}_1)\right] + \alpha^3 \left[\mathbf{e}(\mathbf{u}_3) + L_{11}(\mathbf{u}_1, \mathbf{u}_2)\right] + o(\alpha^3) \quad (6.259)$$

and therefore

$$\varepsilon_o = \mathbf{e}; \quad \varepsilon_1 = \mathbf{e}_1; \quad \varepsilon_2 = \mathbf{e}_2 + \frac{1}{2} L_2(\mathbf{u}_1); \quad \varepsilon_3 = \mathbf{e}_3 + L_{11}(\mathbf{u}_1, \mathbf{u}_2) \quad (6.260)$$

We can now evaluate the term $\sigma_2 \cdot e_1$ in (6.258)

$$\sigma_2 \cdot \mathbf{e}_1 = \mathbf{C}\varepsilon_2 \cdot \varepsilon_1 = \varepsilon_2 \cdot \mathbf{C}\varepsilon_1 = \sigma_1 \cdot \mathbf{e}_2 + \frac{1}{2}\sigma_1 \cdot L_2(\mathbf{u}_1) \tag{6.261}$$

The buckling equation (6.257′) shows that the orthogonalty condition (6.256) is equivalent to

$$\sigma_2 \cdot \mathbf{e}_n = 0 \qquad (n = 2, 3, \ldots) \tag{6.262}$$

and finally we obtain

$$\sigma_2 \cdot \mathbf{e}_1 = \frac{1}{2}\sigma_1 \cdot L_2(\mathbf{u}_1) \tag{6.263}$$

The postbuckling coefficient λ_1 is given by

$$\lambda_1 = -\frac{3}{2}\frac{\sigma_1 \cdot L_2(\mathbf{u}_1)}{\sigma_o \cdot L_2(\mathbf{u}_1)} \tag{6.264}$$

and the buckling condition (6.257′) for $\delta\mathbf{u} = \mathbf{u}_1$ gives

$$\sigma_o \cdot L_2(\mathbf{u}_1) = -\frac{1}{\lambda_c}\sigma_1 \cdot \varepsilon_1 \tag{6.265}$$

We can also write

$$\frac{\lambda_1}{\lambda_c} = \frac{3}{2}\frac{\sigma_1 \cdot L_2(\mathbf{u}_1)}{\sigma_1 \cdot \varepsilon_1} \tag{6.266}$$

Symmetric Branching ($\lambda_1 = 0$ $\lambda_2 \neq 0$)

In this case from (6.257″) we can evaluate the second-order term $\mathbf{u}_2$ of the displacements expansion. Using (6.257‴) with $\delta\mathbf{e} = \mathbf{e}_1$ we get

$$\lambda_c\sigma_o \cdot L_2(\mathbf{u}_1) + \sigma_1 \cdot L_{11}(\mathbf{u}_2, \mathbf{u}_1) + \sigma_2 \cdot L_2(\mathbf{u}_1) + \sigma_3 \cdot \mathbf{e}_1 = 0 \tag{6.267}$$

The expression of ε_3 in (6.260) is used to write

$$\sigma_3 \cdot \mathbf{e}_1 = \mathbf{C}\varepsilon_3 \cdot \varepsilon_1 = \sigma_1 \cdot L_{11}(\mathbf{u}_1, \mathbf{u}_2) \tag{6.268}$$

and, finally, the expression of the coefficient λ_2 is

$$\lambda_2 = -\frac{2\sigma_1 \cdot L_{11}(\mathbf{u}_1, \mathbf{u}_2) + \sigma_2 \cdot L_2(\mathbf{u}_1)}{\sigma_o \cdot L_2(\mathbf{u}_1)} \tag{6.269}$$

or

$$\frac{\lambda_2}{\lambda_c} = \frac{2\sigma_1 \cdot L_{11}(\mathbf{u}_1, \mathbf{u}_2) + \sigma_2 \cdot L_2(\mathbf{u}_1)}{\sigma_1 \cdot \varepsilon_1} \tag{6.269′}$$

It is easy to show that, using (6.236) and assuming for the potential energy of the structure the expression $\mathcal{F}(\varepsilon) = 1/2\, \mathbf{C}\varepsilon \cdot \varepsilon$, the equations (6.264) and (6.269) for λ_1 and λ_2 are equivalent to the equations (6.115) and (6.126).

6.5.4 POSTBUCKLING OF PLATES AND SHALLOW SHELLS

Many remarkable results have been obtained by application of the previously exposed postbuckling theory, particularly solving the difficult problem of the imperfection sensitivity evaluation of shell structures. In the next sections we will apply this theory to analyze, by means of the nonlinear Von Kármán-Donnell equations, the postbuckling behavior of flat plates axially loaded and of circular cylindrical shells subjected to lateral pressure; this last analysis will be performed according to the formulation given by Budiansky and Amazigo.[40]

In the framework of the shallow shell approximation, discussed in detail in Chapter 5, the strain-displacement relationship can be represented by

$$\varepsilon_x = \frac{\partial u}{\partial x} + \frac{1}{2}\left(\frac{\partial w}{\partial x}\right)^2 \quad \varepsilon_y = \frac{\partial v}{\partial y} + \frac{w}{R} + \frac{1}{2}\left(\frac{\partial w}{\partial y}\right)^2 \quad \gamma_{xy} = \frac{\partial u}{\partial y} + \frac{\partial v}{\partial x} + \frac{\partial w}{\partial x}\frac{\partial w}{\partial y}$$

$$K_x = -\frac{\partial^2 w}{\partial x^2} \qquad K_y = -\frac{\partial^2 w}{\partial y^2} \qquad K_{xy} = -\frac{\partial^2 w}{\partial x \partial y} \tag{6.270}$$

where R is the cylinder radius. For $R \to \infty$ (6.270) degenerate into the plate description. In (6.270) u, v, and w represent the inplane and transverse displacement components, widely discussed in Chapter 5. The stress-strain equation on the other hand is

$$\begin{bmatrix} N_x \\ N_y \\ N_{xy} \\ M_x \\ M_y \\ M_{xy} \end{bmatrix} = \begin{bmatrix} D_c & \nu D_c & & & & \\ \nu D_c & D_c & & & & \\ & & [(1-\nu)/2]D_c & & & \\ & & & D_f & \nu D_f & \\ & & & \nu D_f & D_f & \\ & & & & & [(1-\nu)]D_f \end{bmatrix} \begin{bmatrix} \varepsilon_x \\ \varepsilon_y \\ \varepsilon_{xy} \\ K_x \\ K_y \\ K_{xy} \end{bmatrix} \tag{6.271}$$

where

$$D_c = \frac{Et}{(1-\nu^2)}, \quad D_f = \frac{Et^3}{12(1-\nu)^2} \tag{6.272}$$

respectively are the extensional and flexural stiffnesses of the shell, with E being Young's modulus, ν the Poisson's ratio, and t the shell thickness. For both cases of flat plates and shallow shells, it is useful to express the inplane stress resultants by means of an Airy stress function as

$$N_x = \frac{\partial^2 F}{\partial y^2}, \quad N_y = \frac{\partial^2 F}{\partial x^2}, \quad N_{xy} = -\frac{\partial^2 F}{\partial x \partial y} \tag{6.273}$$

Consequently the virtual equilibrium equation (6.238″) becomes

$$t\iint \left[\frac{\partial^2 F}{\partial y^2}\delta\varepsilon_x + \frac{\partial^2 F}{\partial x^2}\delta\varepsilon_y - \frac{\partial^2 F}{\partial x \partial y}\delta\gamma_{xy} + M_x\delta K_x + M_y\delta K_y + 2M_{xy}\delta K_{xy}\right] dxdy$$

$$= [\text{external virtual work}] \tag{6.274}$$

from which we can easily obtain the Von Kármán-Donnell equations of cylindrical shells

$$D\nabla^4 w + R^{-1}F_{,xx} = S(F, w) - p \tag{6.275}$$

$$\nabla^4 F - EtR^{-1}w_{,xx} = -EtS(w, w) \tag{6.276}$$

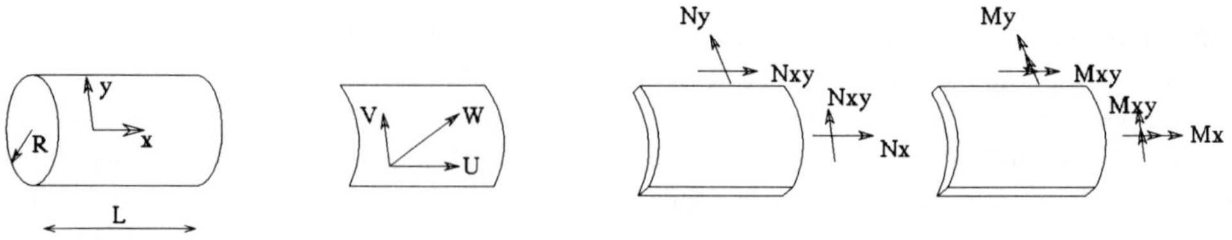

FIGURE 6.26

with

$$S(P, Q) = P_{,xx}Q_{,yy} + P_{,yy}Q_{,xx} - 2P_{,xy}Q_{,xy} \tag{6.277}$$

Of course, for $R \to \infty$, (6.275) and (6.276) degenerate into the Von Kármán plate equation. Perturbation expansions for w and F are assumed in the form

$$\left\{ \begin{matrix} w \\ F \end{matrix} \right\} = p \left\{ \begin{matrix} w_o \\ F_o \end{matrix} \right\} + \alpha \left\{ \begin{matrix} w_1 \\ F_1 \end{matrix} \right\} + \alpha^2 \left\{ \begin{matrix} w_2 \\ F_2 \end{matrix} \right\} + \ldots \tag{6.278}$$

where w_0 and F_0 are associated with the axisymmetric prebuckling state of the shell, w_1 and F_1 describe a normalized buckling mode, and the remaining terms are orthogonal to the buckling mode. The prebuckling state is assumed to be described by a constant membrane stress field with components N_x, N_y both for the flat plate and the cylindrical shell. The classical buckling equations, analyzed in detail in Chapter 5, can now be written as

$$D\nabla^4 w_1 + R^{-1}F_{1,xx} - p_c S[F_o, w_1] = 0$$

$$D\nabla^4 F_1 - EtR^{-1}\, w_{1,xx} = 0 \tag{6.279}$$

The second-order terms w_2 and F_2 of the expansion (6.278) are obtained by solving the equations

$$D\nabla^4 w_2 + R^{-1}F_{2,xx} - p_c S(F_o, w_2) = S(F_1, w_1)$$

$$D\nabla^4 F_2 - EtR^{-1}w_{2,xx} = -\frac{1}{2} EtS(w_1, w_1) \tag{6.280}$$

In both considered examples $\lambda_1 = 0$, while λ_2 will be evaluated according to (6.269).

Flat Plates

For the example of the simply supported square plate, loaded by uniform compression along the edges, (Fig. 6.27), the boundary conditions of simple support are

$$w = \frac{\partial^2 w}{\partial x^2} = 0 \qquad \text{at } x = 0, a$$

$$w = \frac{\partial^2 w}{\partial y^2} = 0 \qquad \text{at } y = 0, a \tag{6.281}$$

and the conditions of zero shear and uniform normal displacement along the edges are

$$\frac{\partial^2 F}{\partial x \partial y} = \frac{\partial^3 F}{\partial x^3} = 0 \qquad \text{at } x = 0, a$$

$$\frac{\partial^2 F}{\partial x \partial y} = \frac{\partial^3 F}{\partial y^3} = 0 \qquad \text{at } y = 0, a \tag{6.282}$$

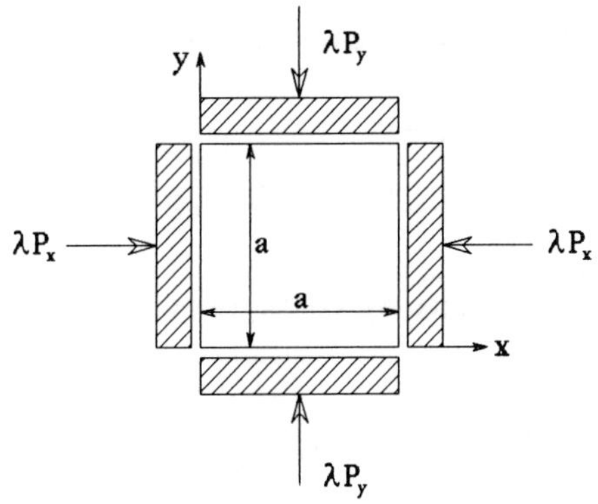

FIGURE 6.27

The prescribed force conditions, on the other hand, require that

$$\frac{\partial F}{\partial y}(0, a) - \frac{\partial F}{\partial y}(0, 0) = -\frac{\lambda P_x}{a}$$

$$\frac{\partial F}{\partial x}(a, 0) - \frac{\partial F}{\partial x}(0, 0) = -\frac{\lambda P_y}{a} \tag{6.283}$$

The prebuckling solution is

$$w_o = 0$$

$$F_o = -\frac{\lambda}{2a}\left[P_x y^2 + P_y x^2\right] \tag{6.284}$$

Assuming that P_x and P_y are fixed in magnitude, we look for the critical value of the scalar multiplier λ. The buckling mode is

$$w_1 = t \sin\frac{\pi x}{a} \sin\frac{\pi y}{a}$$

$$F_1 = 0$$

$$\lambda_c = \frac{4\pi^2 D}{a\left(P_x + P_y\right)} \tag{6.285}$$

To calculate λ_2, we will need to evaluate w_2 and F_2. By collecting terms of order α^2 in (6.278) we get

$$\nabla^4 w_2 + \frac{\lambda_c}{a}\left[P_x \frac{\partial^2 w_2}{\partial x^2} + P_y \frac{\partial^2 w_2}{\partial y^2}\right] = 0$$

$$\nabla^4 F_2 + Et\left[\frac{\partial^2 w_1}{\partial x^2}\frac{\partial^2 w_1}{\partial y^2} - \left(\frac{\partial^2 w_1}{\partial x \partial y}\right)^2\right] = 0 \tag{6.286}$$

Using w_1 as given by (6.285) yields

$$\nabla^4 F_2 = \frac{\pi^4 E t^3}{2a^4}\left[\cos\frac{2\pi x}{a} + \cos\frac{2\pi y}{a}\right] \tag{6.287}$$

and the solution of (6.286), satisfying the orthogonality condition $< w_2, w_1 > = 0$, is

$$w_2 = 0$$

$$F_2 = \frac{Et^3}{32}\left[\cos\frac{2\pi x}{a} + \frac{2\pi y}{a}\right] \tag{6.288}$$

To calculate λ_2, we use (6.269). Since $w_2 = 0$, $L_{11}(u_2, u_1) = 0$ and, therefore, since $\frac{\partial^2 F_2}{\partial x \partial y} = 0$, we have

$$\sigma_2 \cdot L_2(\mathbf{u}_1) = t\int\int_\Omega \left[\left(\frac{\partial^2 F_2}{\partial y^2}\right)\left(\frac{\partial^2 w_1}{\partial x^2}\right)^2 + \left(\frac{\partial^2 F_2}{\partial x^2}\right)\left(\frac{\partial^2 w_1}{\partial y^2}\right)^2\right] dA = \frac{\pi^4 Et^5}{32a^2} \tag{6.289}$$

and

$$\sigma_0 \cdot L_2(\mathbf{u}_1) = -\frac{\lambda_c}{a}\int\int_\Omega \left[P_x\left(\frac{\partial w_1}{\partial x}\right)^2 + P_y\left(\frac{\partial w_1}{\partial y}\right)^2\right] dA = -\frac{\pi^4 Et^5}{12(1-\nu^2)a^2} \tag{6.290}$$

The result for $\frac{\lambda_2}{\lambda_c}$ is

$$\frac{\lambda_2}{\lambda_c} = \frac{3}{8}(1 - \nu^2) \tag{6.291}$$

independent of the ratio between the loads P_x and P_y. Note that $w_{max} \cong (\alpha t)$ for small α, so that we have the result

$$\frac{\lambda_2}{\lambda_c} \cong 1 + \frac{3}{8}(1 - \nu^2)\left(\frac{w_{max}}{t}\right)^2 \tag{6.292}$$

The postbuckling of the plate is therefore stable and to small lateral deflections correspond relevant loading increments.

Cylindrical Shells

For the cylindrical shell, if the pressure contributes to axial stress through the end plates, we can assume $F_o = -\frac{1}{2}\left(x^2 + \frac{1}{2}y^2\right)R$ to describe the inplane stress resultants, whereas, if the pressure acts only laterally, we have $F_o = -\frac{1}{2}x^2R$. Calculations are made for

$$F_o = -\frac{1}{2}R\left[x^2 + \frac{1}{2}\beta y^2\right] \tag{6.293}$$

so that, by appropriate choice of β, results will are available for both cases. The cylindrical shell is assumed to be simply supported; the end plates are assumed infinitely flexible in bending but infinitely stiff in stretching. Thus we have the boundary conditions

$$N_x = \beta pR \qquad v = w = w_{,zz} = 0 \tag{6.294}$$

at each end. This leads to the boundary conditions

$$F_j = F_{j,xx} = w_j = w_{j,xx} = 0 \qquad \text{at } x = 0, L \tag{6.295}$$

for the terms corresponding to j = 1, 2, ... in expansion (6.293). The solution for the buckling problem as given by Batdorf[41] and in agreement with (5.245), is

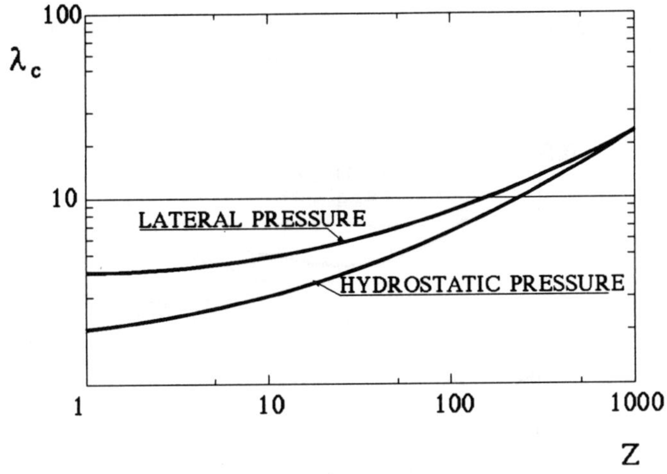

FIGURE 6.28

$$w_1 = t \sin\frac{\pi x}{L} \sin\frac{ny}{R}$$

$$F_1 = -\left(\frac{Et^2L^2}{\pi^2 R}\right)\frac{1}{(1+\zeta)^2}\sin\frac{\pi x}{L}\sin\frac{ny}{R} \tag{6.296}$$

and

$$p_c = \left(\frac{\pi^2 D}{L^2 R}\right)\left(\frac{\beta}{2}+\zeta\right)^{-1}\left[(1+\zeta)^2 + A^2(1+\zeta)^{-2}\right] \tag{6.297}$$

where

$$\zeta = \frac{n^2L^2}{\pi^2R^2}, \quad A = \frac{L^2}{\pi^2 tR}\sqrt{12(1-\nu^2)} \tag{6.298}$$

and n is the integer that minimizes p_c. Minimization of p_c on the basis of the simplifying assumption that ζ may vary continuously gives

$$\lambda_c \equiv \frac{p_c L^2 R}{\pi^2 D} = \frac{4(1+\zeta)^2}{3\zeta + 1 + \beta} \qquad A^2 \equiv \frac{(1-\zeta)^4(\zeta - 1 + \beta)}{3\zeta + 1 + \beta} \tag{6.299}$$

The consequent result for λ_c vs $Z = \pi^2 \frac{A}{\sqrt{12}}$, as plotted by Batdorf,[41] is shown in Fig. 6.28 for $\beta = 0$ (lateral pressure) and $\beta = 1$ (hydrostatic pressure). The analogous result has already been presented in Fig. 5.28 with reference to the case of the lateral pressure ($\beta = 0$). The differential equations for w_2 and F_2 are then obtained in the form

$$D\nabla^4 w_2 + R^{-1}F_{2,xx} - p_c S(F_o, w_2) = S(F_1, w_1)$$

$$\nabla^4 F_2 - EtR^{-1}w_{2,xx} = -\frac{1}{2}EtS(w_1, w_1) \tag{6.300}$$

The initial postbuckling behavior of the shell is symmetric ($\lambda_1 = 0$), so we have

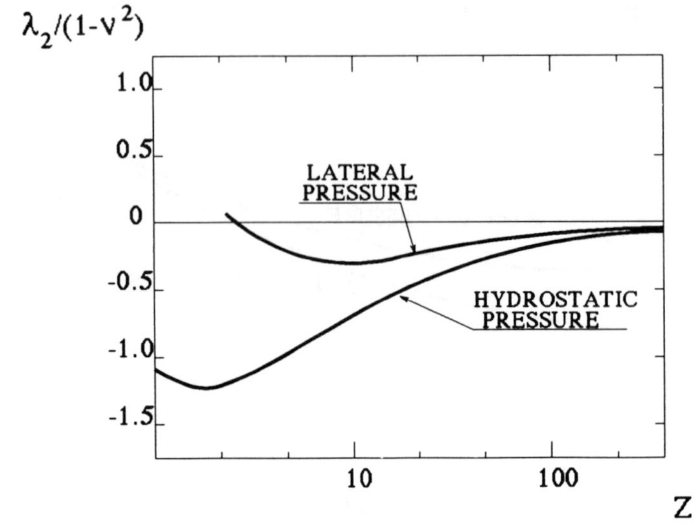

FIGURE 6.29

$$\frac{\lambda}{\lambda_c} \cong 1 + \frac{\lambda_2}{\lambda_c}\left(\frac{w_{max}}{t}\right)^2 \tag{6.301}$$

To evaluate λ_2, we have according to (6.269).

$$\sigma_1 \cdot L_{11}(\mathbf{u}_1, \mathbf{u}_2) = t\iint_\Omega [F_{1,yy}w_{1,x}w_{2,x} + F_{1,xx}w_{1,y}w_{2,y} - F_{1,xy}(w_{1,x}w_{2,y} + w_{1,y}w_{2,x}]\, dA \tag{6.302}$$

$$\sigma_2 \cdot L_2(\mathbf{u}_1) = t\iint_\Omega [F_{2,yy}(w_{1,x})^2 + F_{2,xx}(w_{1,y})^2 - 2F_{2,xy}w_{1,x}w_{1,y}]\, dA \tag{6.303}$$

$$\sigma_1 \cdot \varepsilon_1 = -\lambda_c\sigma_o \cdot L_2(\mathbf{u}_1) = -p_c\iint_\Omega [F_{o,yy}(w_{1,x})^2 + F_{o,xx}(w_{1,y})^2]\, dA - p_c\iint_\Omega [\tfrac{1}{2}\alpha(w_{1,x})^2 + (w_{1,y})^2]\, dA \tag{6.304}$$

Solutions of (6.300) have the form

$$w_2 = t\,\zeta\sqrt{12(1-\nu^2)}\left[w_{20}(\zeta) + w_{22}(\zeta)\cos\frac{2y}{R}\right]$$

$$F_2 = \frac{Et^2L^2}{\pi^2R}\zeta\sqrt{12(1-\nu^2)}\left[f_{20}(\zeta) + f_{22}(\zeta)\cos\frac{2y}{R}\right] \tag{6.305}$$

where $w_{20}(\zeta)$, $w_{22}(\zeta)$, and $f_{20}(\zeta)$, $f_{22}(\zeta)$ are suitable functions of $\zeta = \pi x/L$. The λ_2 vs. Z diagram, obtained by Budiansky and Amazigo,[40] is plotted in Figure 6.29.

The greatest imperfection sensitivity occurs when the cylinder under hydrostatic pressure; in the other case the imperfection sensitivity occurs in a more restrictive range of values of the parameter Z.

REFERENCES

1. Vainberg, M. M. and Trenogin, V. A., Theory of Branching of Solutions of Nonlinear Equations. Noordhoff Int. Publ. Leyden, 1974.
2. Vainberg, M. M., Variational Methods for the Study of Nonlinear Operators. Holden Day, San Francisco, 1964.
3. Keller, J. B. and Antman, S., Bifurcation Theory and Non-linear Eigenvalue Problems, W. A. Benjamin, New York, 1969.
4. Ambrosetti, Prodi, Nonlinear Analysis, Scuola Normale, Pisa, 1981.
5. Poincaré, H., Sur l'equilibre d'une masse fluide animeé d'un mouvement de rotation, Acta Mathem. 7, 1885.
6. Potier-Ferry, M., Perturbed bifurcation theory, J. Diff. Eq., 33, 1979.
7. Potier-Ferry, M., Multiple bifurcation, symmetry and secondary bifurcation, Research Notes in Mathematics, 46, London, 1981.
8. Koiter, W. T., Over de stabiliteit van het elastisch Evenwicht, Thesis, Delft, H. J. Paris, Amsterdam, 1945, English Transl. NASA YY-F-10, 833, 1967.
9. Koiter, W. T., Elastic stability and postbuckling behavior, in Nonlinear Problems, edited by R. E. Langer, University of Wisconsin Press, 1963.
10. Thompson, J. M. T, Discrete branching points in the general theory of elastic stability, J. Mech. Phys. Solids 13, 1965.
11. Thompson, J. M. T. and Walker, A. C., The nonlinear perturbations analysis of discrete structural systems, Int. J. Solids Structures 4, 1968.
12. Roorda, J., Stability of structures with small imperfections, J. Eng. Mech. Div. ASCE 91, 1965.
13. Sewell, M. J., A general theory of equilibrium paths through critical points, I and II, Proc. R. Soc. A, London, 1968.
14. Huseyn, K., Postbuckling behavior of structures under combined loading, ZAMM, 51, 1971.
15. Thom, R. Stabilité structurelle et morphogénèse, W.A. Benjamin, New York, 1969; *Topology,* 8, 1969.
16. Koiter, W. T., General Equations of Elastic Stability for Thin Shells. Donnell Anniversary Volume, Houston, 1966.
17. Budiansky, B. and Hutchinson J. W., Buckling of circular cylindrical shells under axial compression, in Contributions to the Theory of Aircraft Structures, Van Der Neut Anniversary Volume, Delft University Press, 1972.
18. Hutchinson, J. W., Imperfection sensitivity of externally pressurized spherical shells, J. Appl. Mech. 3, 1967.
19. Hutchinson, J. W. and Koiter, W. T., Postbuckling theory, Appl. Mech. Rev., 23, 1970.
20. Budiansky, B. and Hutchinson, J. W., A Survey of some buckling problems, AIAA J. 4, 1966.
21. Thompson, J. M. T. and Hunt, G. W., A General Theory of Elastic Stability, John Wiley & Sons, New York, 1973.
22. Thompson, J. M. T., Instabilities and Catastrophes in Science and Engineering, John Wiley & Sons, New York, 1982.
23. Huseyn, K., Multiple Parameters Stability Theory and Its Applications. Clarendon Press, Oxford, 1986.
24. Britvic, S., The Stability of Elastic Systems. Pergamon Press, New York, 1963.
25. El Naschie, M. S., Stress, Stability and Chaos in Structural Engineering: An Energy Approach, McGraw-Hill, New York, 1990.
26. Pignataro, M. Rizzi, N. and Luongo, A., Stability bifurcations and post critical behavior of structures, Elsevier, Amsterdam, 1991.
27. Como, M. and Grimaldi, A., Stability, Buckling and Postbuckling of Elastic Structures, Par III: Equilibrium configurations in the neighborhood of the critical state. Dpt. of Structures, University of Calabria, Report no. 3, 1974.
28. Movchan, A. A., The direct methods of Liapounov in stability problems of elastic systems, J. Appl. Math. Mech., 23, 1960.
29. Gilbert, J. E. and Knops, R. J., Stability of general systems, Arch. Ration. Mech. Anal. Appl., 17, 1967.
30. Tvergaard, V., Buckling and Plates and Shell Structures, Proc. of XIV IUTAM Congrss, W. T. Koiter Ed., North Holland, 1976.
31. Budiansky, B., Buckling behavior of elastic structures, Adv. Appl. Mech. 14, 1974.
32. Ascione, L. and Grimaldi, A., On the Stability and Postbuckling Behavior of Elastic Beams, Thin Walled Structures, Elsevier, Essex, No.1, 1983.
33. Grimaldi, A., Postbuckling Behaviour and Imperfection Sensitivity of Framed Structures, Aerotec. Missile Spazio 4, 1974.

34. Godley, M. H. R. and Chilver, A. H., The elastic postbuckling behavior of unbraced plane frames, Int. J. Mech. Sci. 9, Pergamon Press, 1967.
35. Lanni, G., Postbuckling Behaviour of a simple portal frame, Dept. of Structures, University of Calabria, Report no.7, 1974.
36. Di Carlo, A., Pignataro, M., and Rizzi, N., A modified potential energy approach to post-buckling analysis of frames, Int. J. Non-Linear Mech. 16, 1981.
37. Budiansky, B. and Hutchinson, J. W., Dynamic Buckling of Imperfection Sensitive Structures, Proc. XI Int. Congr. Appl. Mech., Munich, 1964, (Springer-Verlag, Berlin, 1964).
38. Budiansky, B., Dynamic Buckling of Elastic Structures: Criteria and Estimates, Proc. Int. Conf. Northwestern University, Evanston, Illinois, (Pergamon Press, 1965).
39. Donnell, L. H., A new theory for the buckling of thin cylinders under axial compression and bending, Trans. ASME 56, 1934.
40. Budiansky, B. and Amazigo, J. C., Initial postbuckling behavior of cylindrical shells under external pressure, J. Math. Phys., 47,3,1968.
41. Batdorf, S. B., A simplified method of elastic stability analysis for thin cylindrical shells, Donnell Equation, N.A.C.A., T.N. 1341, 1947.

Index

H

I

K

L

M

N

O

P

T

U

V

W